Engineer Practices
For
PIC Microcontrollers

New Paperback Edition

Prof. Sal R. Riggio Jr., PhD, PE
Pennsylvania State University
Duquesne University

Contact: Janny Li, jli@cognella.com, (800) 200-3908, Ext-539, for more information

Bassiln Hamadeh, Publisher
Michael Simpson, Vice President of Acquisitions
Christopher Foster, *VJ.Ce* President of Marketing Jessica
Knott, Managing Editor
Stephen Milano. Creative Director
Kevin Fahey, Cognella Marketing Program Manager
Melissa Barcomb, Acquisitions Editor
Sarah Wheeler, Project Editor Luiz
Ferreira, Licensing Associate

Copyright O 2012 by University Readers, Inc. All rights reserved. No part of this publication may be reprinted. reproduced. transmitted, or utilized in any form or by any electronic. Mechanical or other means, now known or hereafter invented, including photocopying, microfilming, and recording, or in any information retrieval system without the written permission of University Readers. Inc.

First published in the United States of America in 2012 by University Readers, Inc.

Trademark Notice: Product or corporate names may be trademarks or registered trademarks, and are used only for identification and explanation without intent to infringe.

15 14 13 12 11   1 2 3 4 5

Printed in the United States of America

ISBN: 978-1-60927-628-7

I dedicated this book to my family.

My Wife: Jan
My Daughter: Gina
My Son: Mark

Table of Contents

Preface	vi
Introduction – The Contents of the PIC Microcontroller Experiment Board	1
Chapter – 1 Learning Assembly Code	3
Chapter – 2 Flash Memory Programming with the Microchip PICKit-2 Programmer	21
Chapter – 3 Delay Loop Timing, Computer-to-PIC Serial-Communications, The Keypad, The Light-Emitting-Diode, The Liquid-Crystal-Display, Relay Control, Sound Generation	25
Chapter – 4 The Digital Voltage Regulator and The Brushless-DC Fan Speed Regulator	55
Chapter – 5 The Digital 4-Phase Stepper-Motor-Controller	83
Chapter – 6 The Digital Electronic Thermometer	95
Chapter – 7 The I2C Real-Time Programmable Clock/Calendar	111
Chapter – 8 The Digital Frequency Counter, The Voltmeter, The Ohmmeter, The Digital Waveform Generator, The One-Bit Digital-to-Analog Converter	127
Chapter – 9 The Variable Bandwidth Digital Low-Pass-Filter	151
Chapter – 10 Frequency, Phase and Amplitude Modulation	163
Chapter – 11 Oscillators and Important Microcontroller Interface Circuits	173
Appendix – A Experiment Board Schematics, Bill-of-Materials, PC-Card Layout	197
Appendix – B Lab Board Schematics, Bill-of-Materials, PC-Card Layout	213
Appendix – C The Z-Transform as it Relates to Digital Filters	221
Bibliography	227

Preface

Today's Electrical Engineer must make every effort to acquire a strong practical working knowledge of Microcontrollers. Just about every product, new or old, contains one or more of these devices. This is not only true for a product that is largely electrical, but it is also true for products that are largely mechanical in nature, as well. These devices automatically perform supervisory control and monitoring functions for any given product. They can also provide for a graphical command center, which allows the user to conveniently alter the state or current operation of a product. These devices can also provide an effective safety alert mechanism for the product. This is an important feature, since safety is always an issue with any product. They are many every day examples of products that make use of these devices. The following products all use Microcontrollers.

a microwave oven, a clothes washer, a clothes dryer, a refrigerator, a freezer, a furnace, an air conditioner, a heater, a thermostat, a hot water heater, a solar-cell based power generator, a stereo receiver, a television, any device that connects to a computer, an automobile, a boat, an indoor-outdoor lighting system, a burglar alarm, mechanical and electrical test equipment for such things as the automobile, power tools, telephone equipment, an alarm clock, remote control devices, toys, games, hospital beds medical and biomedical equipment, heath monitoring devices, body building equipment, the stationary bicycle and other exercise equipment, professional and amateur radio and television transmitting and receiving equipment, Radio and Television satellite transmitting and receiving equipment and just about any hand held device.

The goal of this book is to educate the reader regarding the process one follows in making practical use of Microcontrollers. This book also demonstrates specific techniques, which have been widely acceptable by the Microcontroller community, for creating an electrical hardware interface between discrete and integrated analog circuits and the Microcontroller.

The reader will soon discover that one must develop a strong understanding of **Assembly Code**, which is sometimes referred to a **Firmware, Electrical Circuits** and **Computer Software** to acquire a useful practical working knowledge of Microcontrollers. Assembly Code can be written directly into the compiler or it can be produced by using a higher-level computer language such as **PICBasic-Pro** or **CCS C-code**. Practical working electrical circuits are described in detail in this book. These circuits are vital and fundamental to the operation and usage of any microcontroller.

This book places a strong emphasis on the use of an exciting and powerful programming language for Microchip PIC microcontrollers known as PICBasic-Pro. This language has been specifically designed for use with Microchip's PIC12F thru PIC18F families of microcontrollers.
The PICBasic-Pro Language and Compiler are maintained and owned by MicroEngineering Labs (www.melabs.com).

The applications covered in this book also include working C-code specifically designed for use with Microchip's PIC12F thru PIC18F families of microcontrollers. The CCS C-code Language and Compiler are maintained and owned by Custom Computer Services, Inc. (www.ccsinfo.com).

The intent of this book is to assist and instruct the student, as well as the practicing electrical engineer, in the practical usage of the microcontroller. This book coves a number of important practical applications that are designed for the Microchip PIC18F4685 microcontroller. The hardware analog and digital interface circuits and assembly code (firmware), along with the PICBasic-Pro and CCS C-code programming languages, are used together to invoke the desired operation.

Introduction

There are **Five Key Areas** of **Advancements** in the **Architecture** of the **Microcontroller,** which have occurred over the last four decades. Before these advancements, microcontrollers were little more than Programmable-Logic-Devices, which had yet to make a large impact on the industrial development of commercial and industrial electromechanical products. Since the developers of embedded microcontrollers have incorporated the following five key areas of advancements within the microcontroller, the growth in the use of embedded microcontrollers in electromechanical products has been exponential.

1. **The Addition** of an **Analog-to-Digital Converter** with a **Multiple Input Analog Multiplexer**

This advancement allowed one to feed analog signals, usually between 0 and five volts, directly into the microcontroller for digital manipulation and processing.

2. **The Common Tri-State Input-Output Pin**

This advancement allows the designer to double the effective Input-Output pin count. This is because the code writer can make a given pin and output pin during one line of code and an input pin two lines of code, or more, later in the program code. Since input-output pins are always a major factor in the cost versus function and performance trade-off criteria, this advancement has caused a great increase in the use of microcontrollers.

3. **The addition** of the **Phase-Lock-Loop, Frequency Dividers,** and a **Digital Multiplexer to the basic Pierce-Colpitts Crystal Oscillator**

This advancement opened the door to operating the microcontroller at many different CPU (Central-Processing-Unit) clock frequencies. This allows the designer to choose a CPU clock frequency that best fits the application. This advancement also caused an increase in the use of microcontrollers.

4. The addition of the **Capture-Compare Pulse-Width-Modulation (PWM) Circuits** to the **Basic Timer Circuits** with **Interrupt Capability**

This allows one to use the microcontroller in a wide variety of control loop and communication applications, which again, caused an increase in the use of microcontrollers.

5. *In-Circuit Flash Programmable memory*

The invention of In-Circuit Flash Programmable Memory afforded the designer the ability to program, erase, and program again, many thousands of times. This greatly reduced the product development-cycle time, which reduced the cost of developing a product.

This experiment board **is not** an evaluation board. Instead, it contains a very state-of-the-art collection of circuits that have been designed to introduce a second-year or third-year engineering student to the modern world of Programmable Microcontrollers. Additionally, it puts forth a very strong set of analog and digital interface circuits, which are necessary in making practical use of the microcontroller. Each of the circuits on this experiment board have been designed for the purpose of teaching the student about the way hardware and firmware (microcode) work together to complete a particular task. This is accomplished by presenting the student with a number of functional real-world experiments.

State-of-the-Art Microcontrollers include the following:

1. Both Crystal and RC Oscillators;
2. A Multiplexed Frequency Synthesized Phase-Lock-Loop;
3. A multiplexed Dual-Slope Analog-to-Digital Converter with External Voltage Reference Capability;
4. One to Four Timer Circuits with Pulse-Width-Modulation Capability;
5. A Pulse-Width-Modulated Digital-to-Analog Converter;
6. Tri-State Common Input-Output Pins, which may be changed from an Input Pin To an Output Pin as the program moves from one Line of Code to another Line of Code;
7. Dual Software Controlled Comparators, which can be Hardware-Configured in Threshold or Hysteresis form, with a Software-Controlled Volt.1.ge Reference;
8. Flash Memory Programming;
9. Both Random-Access-Memory (RMI) and Electrically Erasable Programmable Read-Only Memory (EEPROM);
10. Synchronous Peripheral Interface (SPI) and the I2C Communication Interfaces;
11. The ability to generate any Basic or Small Combinatorial or Synchronous Logic Functions.

This experiment board includes the following:

- 1 Removable PIC18F4685 in a 40-Pin DIP package with 20 Megahertz Crystal
- 1 On-Board USB PICkit2 Programmer for the PIC family of microcontrollers
- 1 Liquid-Crystal-Display (16-Character, 2-Line, 5x8 Dots, reflective)
- 2-Red, 1-Yellow, and 2-Green Light-Emitting-Diode (LED)
- 1 Telephone style Keypad
- 1 Master Reset Momentary Switch
- 1 Four-Position Dip Switches
- 1 DB9 RS232 Serial-Communications Port
- 1 Battery-Backed RAM I2C Real-Time Clock with crystal on a separate I2C Connector
- 1 Lithium Coin Battery (20mm, 48mA-Hour)
- 1 One-Amp, Digital Switching Voltage Regulator with separate I/0 Connectors
- 1 Four-Phase, 1 Ampere per Phase, Stepper-Motor-Driver on a separate Output Connector
- 1 Temperature-Sensing Negative Temperature Coefficient Thermistor
- 1 Relay (l0A, 125VAC Contacts) on a separate Output Connector
- 1 Pulse-Width-Modulator Output on a separate Output Connector
- 1 Pulsed Sound Output on a separate Output Connector
- 1 Open-Drain MOSFET Driver on a separate Output Connector
- 2 Pulse/Frequency Counting Input Circuits on separate Input Connectors
- 1 Variable Low-Frequency Oscillator (40 to 1000 Hertz)
- 1 Variable High-Frequency Oscillator (40 to 1000 Hertz)
- 1 Serial EEPROM 512K (64K Bytes)
- 2 High Impedance Analog Input Circuits with selectable AC or DC Input Coupling On separate Input Connectors
- 1 Pulse-Width-Modulated Digital-to-Analog Converter on a separate Output Connector
- 1 Eight-Bit R-2R Ladder Digital-to-Analog Converter with a 4-Value Selectable Reconstruction Filter, with AC or DC Output Coupling, on a separate Output Connector

The PIC16F877, PIC18F452, 458, 452x, 455x, 458x, 462x, 465x and the 468x family of microcontrollers may also be used with this experiment board.

This experiment board allows for the implementation of the following experiments.

1. **Laboratory Experiment #1: (Chapter 3)**
   Delay loop Timing, Computer-to-PIC Serial Communications,
   The Keypad, The light-Emitting Diode, The liquid-Crystal Display,
   Relay Control, Sound Generation

2. **Laboratory Experiment #2: (Chapter 4)**
   The Digital Voltage Regulator,
   The Brushless-DC Fan Speed Regulator

3. **Laboratory Experiment #3: (Chapter 5)**
   The Digital 4-Phase Unipolar Stepper-Motor-Controller

4. **Laboratory Experiment #4: (Chapter 6)**
   The Digital Electronic Thermometer

5. **Laboratory Experiment #5: (Chapter 7)**
   The I2C Real-Time Programmable Clock/Calendar

6. **Laboratory Experiment #6: (Chapter 8)**
   The Frequency Counter, The Digital Voltmeter, The Digital Ohmmeter,
   The Digital Waveform Generator, The One-Pin Digital-to-Analog Converter

7. **Laboratory Experiment #7: (Chapter 9)**
   The Variable Bandwidth Digital Low-Pass-Filter

8. **Laboratory Experiment #8: (Chapter 10)**
   Frequency, Phase and Amplitude Modulation

Chapter-1 Instructs the student in the art of assembly code, which is compiled into machine language, ones and zeros, The ones and zeros are fundamental symbols of operation of the microcontroller.

Chapter- 2 depicts the hardware of the **PICKit-2** Flash Memory microcontroller programmer.

Chapter-11 depicts the analysis and operation of two different oscillator circuits. The **Pierce-Colpitts Crystal Oscillator** and the **Single-Time-Constant RC-Oscillator**. This chapter also shows the operation and analysis of some specific and very important microcontroller interface circuits.

Engineering Practices for the PIC Microcontroller

By: Prof. Sal R. Riggio Jr., PhD, PE

Chapter - 1

Learning Assembly Code

What is Assembly Code?

Assembly Code is a set of abbreviated English based commands that are used to instruct the microcontroller to perform a specific operation. These English based commands have been standardized and accepted for general use since approximately the year 1970. The manufacturer of a given microcontroller chip will list these commands in the device datasheet.

A microcontroller is an electronic device that will respond to slowly changing or rapidly changing on and off electrical signals, only. The digit "0" represents each off-state electrical signal and the digit "1" represents each on-state electrical signal. Each abbreviated English based Assembly Code Command is represented by a series of 14 digits of a specific combination of ones and zeros. Therefore, a computer based software package, which is known as a *Compiler,* is used to convert each abbreviated English based Assembly Code Command to the 14 digit numeric code that can be understood by the microcontroller.

Assembly Code Commands concern themselves with three basic types of numeric codes. They are known as the numeric code for the Instruction, the numeric code for an Address and the numeric code that represents a Raw Number. The Instruction is the actual abbreviated English based Assembly Code Command. The Address points to a register or a location in memory that may contain a Raw Number, another Instruction or another Address. The numeric codes for all of these possibilities are stored in a fundamental mechanism know as a register. There are many registers located throughout any given microcontroller. Most of these registers are known as 8-bit registers. This means that these registers may contain any combination of eight digits where each digit can only be a one or a zero.

Most microcontrollers contain three different types of memory. They are known as Volatile Random-Access-Memory, Non-Volatile Electrically-Erasable and Programmable Read Only memory (EEPROM) and Non-Volatile Flash Memory. Non-Volatile means that the information previously stored in the Memory will remain unchanged after the power is removed from the microcontroller. Obviously, Volatile means that the information previously stored in the memory is lost after the power is removed from the microcontroller. Flash Memory, or what is sometimes called Program Memory, is used to store the compiled numeric based assembly code. Volatile Random-Access-Memory is used to store Random Non-Volatile Program Variables in General Purpose Registers, which can be identified with English based names. It is also used to perform internal mathematical computations. A Special Purpose Register, such as TRISA and PortA, can only be used for purposed identified by the architecture of the microcontroller.

The reader must be familiar with Binary, Decimal and Hexadecimal numbering systems and be able to convert any number from one system to the other, in order to properly understand Assembly Code.

The following material will demonstrate how one uses the Working Register, as well as General Purpose and Special Purpose registers, when writing Assembly Code to perform a certain task.

The Working, TRIS & Port Registers

A register is a place inside the microcontroller to which the programmer can write data. The programmer can also read data from the register. The data stored in the register is not permanent. Because, if the power to the microcontroller is lost for any length of time, the register loses its ability to retain the data as it was originally stored. This type of register is said to have volatile memory.

The special and general-purpose register file map can be found in the microcontroller datasheet. This map shows the programmer where the different registers are located inside the microcontroller and will help explain some of the commands. The **Special Purpose** Registers are pre-defined by the architecture of the microcontroller and cannot be used for any other purpose. Registers such as the W, TRIS & Port are Special Purpose Registers. The **General Purpose** Registers may be used and reused at anytime for any data storage or data transfer purpose. The addresses of all registers are listed in the device datasheet. The following material uses the **PIC18F4685** as its reference microcontroller. For the PIC18F44685, a General Propose Registers may use any of the addresses between **0x0FF & 0x7FF**.

The **TRISA** Register allows one to select which pins on **Port-A** are **outputs** and which pins are **inputs**. *tristate*

The **TRISB** Register allows one to select which pins on **Port-B** are **outputs** and which pins are **inputs**.

The TRTSA & TRISB Registers

TRIS refers to the fact a given register can take on **three** different **Logical States**, as follows.

1.) State-1 **(High Voltage Level)** (+5 volts) (Logical "1") Positive Logic
2.) State-2 **(Low Voltage Level)** (0 volts) (Logical "0") Positive Logic
3.) State-3 **(High-Z)** (High Impedance) (No Electrical Loading Effects)

The **TRISA** Register is located at Hexadecimal Address **0x0F92**.
The **TRISB** Register is located at Hexadecimal Address **0x0F93**.

If a Logical "1" (+5v) (High) is written to a particular bit of the TRISA Register, then this bit (pin) becomes an Input.

If a Logical "0" (0v) (Low) is written to a particular bit of the TRISA Register, then this bit (pin) becomes an Output.   *Make Bit TRIS = 0*

The PORTA and PORTB Registers

The **PortA** Register is located at Hexadecimal Address **0x0F80**.
The **PortB** Register is located at Hexadecimal Address **0x0F81**.

PortA Example:
The Data within Port-A = 110110b (0x36). Note that bit zero is on the right, as shown.

Bit Number	5	4	3	2	1	0
PortA Register Pin	RA5	RA4	RA3	RA2	RA1	RA0
Binary Value	1	1	0	1	1	0

Port-B Example:
The Data within Port-B = 01011010b (0x5A). Note that bit zero is on the right, as shown.

Bit Number	7	6	5	4	3	2	1	0
Port-B Register Pin	RB7	RB6	RB5	RB4	RB3	RB2	RB1	RB0
Binary Value	0	1	0	1	1	0	1	0

The (W) Working-Register

[handwritten annotation: between → Arithmatic unit / Logic unit]

The W-Register is called the **Working Register**. The Programmer can place any data value into this register. Once a value has been assigned to the W-Register, it can be moved to another register. If you assign another value to the W-Register, its contents are overwritten.

The following assembly code instruction will set up PortA with the value from the above example, in **Binary Coded Decimal (BCD)** form.

      movlw 110110b   ;This Instruction places the Hexadecimal value (0x36)
                                   ;(00110110b) into the W-Register (assumes two leading
                                   ;zeros.

[handwritten annotation: move statement / l = literal]

The following assembly code instruction will set up PortA with the value from the above example, in **Hexadecimal (Hex)** form.

      movlw 0x36   ;This Instruction places the Hexadecimal value (0x36)
                                   ;(00110110b) into the W-Register (assumes 2 leading
                                   ;zeros.

Either the Binary Coded Decimal or the Hexadecimal form of movlw will work. The assembly code term movlw means 'Move the Literal Value Into the W-Register'. Which, in plain English, means put the given value directly into the W-Register.

At this point, it is necessary to place this value into the TRISA register to set up each of the PortA pins, individually, as either an input pin or output pin: ;

      movwf 0x0F92   ;This Instruction moves the Contents of the Working
                                    ;Register, which in this case is the Hexadecimal value
                                    ;(0x36) (00110110b), into TRISA. The TRISA Register is
                                    ;located at Hex address 0x0F92. This makes each of the
                                    ;PortA pins, which are represented by the number **"0"** in the
                                    ;TRISA Register, an **Output Pin** and each of the PortA pins,
                                    ;which are represented by the number **"1"** in the TRISA
                                    ;Register, an **Input Pin**.

This Assembly Code Instruction movwf means, "Move the Contents of the W-Register into the file register address that follows". In this case, the file register address **0x0F92** is that of the **TRISA** file register.

The **TRISA** register contains the BCD value (110110b), which is Hex value (0x36).

Bit Number	5	4	3	2	1	0
PortA Register Pin	RA5	RA4	RA3	RA2	RA1	RA0
Binary Value	1	1	0	1	1	0

Writing Data to the I/O Ports

The following example defines all of the PortB Pins (Bits) as Output Pins.

movlw 0x00   ;This Instruction places the Hexadecimal value (0x00) (00000000b)
                    ;into the W-Register.
movwf 0x0F93   ;This Instruction moves the Hexadecimal value (0x00) (00000000b)
                    ;into TRISB, which is located at address 0x0F93. This makes each
                    ;of the eight Pins of PortB an Output Pin.

The only difference between this & the last example is that all of the PortB are defined as Output Pins.

At this point, we want to Turn-On the LED. We do this by making the pin connected to the LED, High. In other words, we send a Logical "1" (+5v) (High) to this pin. The Anode of the LED is connected to PortB, Bit-0, and the Cathode is connected thru a current limiting resistor to ground.

The Assembly Code is as follows:

movlw 0x01   ;This Instruction places the Hexadecimal value (0x01)
                    ;(00000001b) into the W-Register.

movwf 0x0F81   ;This Instruction moves the Hexadecimal value (0x01)
                    ;(00000001b) into PortB, "which is located at address 0xoF80.
                    ;This puts a Logical "1" (+5v) (High) on PortB, Bit-0.

This Logical Level value can be reversed and the LED can still be Turned-On, by using a 0v Low Level, if the LED's Anode is connected to +5 volts and the LED's Cathode is connected to PortB, Bit-0, thru a current limiting resistor.

At this point, we want to Turn-Off the LED. We do this by making the pin connected to the LED, Low. In other words, we send a Logical "0" (0v) (Low) to this pin. The Anode of the LED is connected to PortB, Bit-0, and the Cathode is connected thru a current limiting resistor to ground.

The Assembly Code is as follows:

movlw 0x00   ;This Instruction places the Hexadecimal value (0x00) (00000000b)
                    ;into the W-Register.

movwf 0x0F81   ;This Instruction moves the Hexadecimal value (0x00) (00000000b)
                    ; into PortB, which is located at address 0x06. This puts a Logical "0"
                    ; (0v) (Low) on PortB, Bit-0.

At this point, the LED has been Turned-On then Turned-Off, one time. In order to Flash the LED, we must write the assembly code, such that, it continuously loops back to the beginning of the program. This is done thru the use of a label and a goto code statement. In this case, a label named Start is used.

The Actual Assembly Code without comments is as follows:

```
        movlw 0x00
        movwf 0x0F93

Start   movlw 0x01
        movwf 0x0F81
        movlw 0x00
        movwf 0xoF81
        goto Start
```

This code will flash the LED, but at a speed, which is much too fast to be seen by the human eye. This problem will be addressed during the discussion of Delay Loops.

The equ Command

The equ instruction is used to replace an Address Number with a Name.

TRISB equ 0x0F93   ;This Instruction equates the Hexadecimal address (0x0F93)
                                  ;with its Name, which is TRISB. This is a Special Purpose
                                  ;Register.

PORTB equ 0x0F81   ;This Instruction equates the Hexadecimal address (0x0F81)
with its Name,
                                  ;which is PortB. This is a Special Purpose Register.

The New Assembly Code is as follows:

```
TRISB equ 0x0F93
PORTB equ 0x0F81

        movlw 0x00
        movwf TRISB

Start   movlw 0x01
        movwf PortB
        movlw 0x00
        movwf PortB
        goto Start
```

Delay Loops

There is one drawback to this Flashing LED Program. A **One-Cycle** Instruction requires **4 CPU Oscillator Periods** to be completed. If the **CPU Oscillator Frequency = 4MHz**, then each instruction will take **(4)*(1/4MHz)**, or **1us** to be completed. We are using **four** instructions that require a **One-Cycle** Instruction time, each, and **one** instruction that requires a **Two-Cycle** Instruction time (goto Start). This means that the LED will Turn-On and then Turn-Off within **6uS**. This is far too fast for the human eye to see, and it will appear that the LED is permanently off. Therefore, we must cause a longer delay to occur between turning the LED on and turning the LED off.

The Principle of Delay is based upon the use of a counter. This counter is pre-set to a value which is reduced by one each time the processor loops thru the code. Once the count is reduced to zero, a decision is made to skip the next instruction, and then, the processor continues with the main program.

First, a variable name must be assigned to the counter, such as, COUNT. Next, we must determine the number value contained within variable COUNT. The largest number, which can be held by an 8-Bit Register, is 255 Decimal (0xFF Hex). A General Purpose Register Address must be assigned to the variable COUNT, by using the equ command.

```
COUNT equ 0x0200        ;This Instruction equates the General Purpose
                        ;Hexadecimal
                        ;address (0x0200) with the Variable Name Count.

movlw 0xFF              ;This Instruction places the Hexadecimal
                        ;value (0xFF) (11111111b) into the W-Register.

movwf COUNT             ;This Instruction moves the value of the Working
                        ;Register, which is 255 Decimal (0xFF Hex)
                        ;(11111111b), into the Variable COUNT.
```

The Contents of Count must be decreased by one each time the processor loops thru the program, until it reaches zero. The following instruction accomplishes this requirement is:

```
decfsz COUNT1,1         ;This Instruction Decrements the Variable COUNT
                        ;by 1,first, and then, places the result back into
                        ;the Variable COUNT. Then this Instruction
                        ;checks to see if the new value is Zero. If it is
                        ;zero, then the next instruction is Skipped and the
                        ;processor continues on with the main program.
                        ;The number 1 after the comma means that the
                        ;result is placed into the Variable COUNT. If the
                        ;value after the comma is a 0, then the result is
                        ;placed into the Working Register.
```

A general example of this instruction is as follows.

```
LABEL   decfsz COUNT,1
        goto LABEL
```

'Continue with Main Program'

The Complete Flashing LED Example is shown below.

```
                TRISB equ 0x0F93            ;This Instruction equates the Hexadecimal address
                                            ; (0x0F93) with its Name, which is TRISB.
                PORTB equ 0x0F81            ;This Instruction equates the Hexadecimal address
                                            ;(0x0F81) with its Name, which is PortB.
                COUNT1 equ 0x0200           ;This Instruction equates the General Purpose
                                            ;Hexadecimal address (0x0200) with the Variable
                                            ;Count1.
                COUNT2 equ 0x0201           ;This Instruction equates the General Purpose
                                            ;Hexadecimal address (0x0201) with the Variable
                                            ;Count2.

                movlw   0x00                ;This Instruction places the value 0 Decimal (0x00)
                                            ; (00000000b) into the W-Register.
                movwf   TRISB               ;This Instruction moves the Hexadecimal value
                                            ;(0x00) (00000000b) into TRISB, which is located
                                            ;at address 0x86. This makes each of the 8 Pins
                                            ;of PortB an Output Pin.
Start
                movlw   0xFF                ;This Instruction places the value 255 Decimal
                                            ;(0xFF) (11111111b) into the W-Register. This line
                                            ; of code also demonstrates the use of the
                                            ;Label named Start

                movwf   COUNT1              ;This Instruction places the value 255 Decimal
                                            ;(0xFF) (11111111b) into the Variable Register
                                            ; named COUNT1.

                movlw   0xFF                ;This Instruction places the value 255 Decimal
                                            ;(0xFF) (11111111b) into the W-Register.

                movwf   COUNT2              ;This Instruction places the value 255 Decimal
                                            ;(0xFF) (11111111b)
                                            ;into the Variable Register named COUNT2.
                movlw   0x01                ;This Instruction places the value (0x01)
                                            ;(00000001b) into the W-Register.

                movwf   PORTB               ;This Instruction places the value (0x01)
                                            ;(00000001b) into PortB. This Turns-On the LED.
Loop1
                decfsz COUNT1,1             ;This group of Instructions Creates a Delay
                goto Loop1                  ;The value of this Delay is "Approximated" as
                                            ; 256 x 256 x the Time of
                decfsz COUNT2,1             ;one One-Cycle Instruction.
                goto Loop1

                movlw   0x00                ;This Instruction places the value (0x00)
                                            ;(00000000b) into the W-Register.

                movwf   PORTB               ;This Instruction places the value (0x00)
                                            ;(00000000b) into PortB. This Turns-Off the LED.
Loop2           decfsz COUNT1,1             ;This group of Instructions Creates a Delay
                goto Loop2                  ;The value of this Delay is "Approximated"
                                            ;as; (256 x 256) Times
                                            ;(the Time of One, One-Cycle Instruction)
                                            ;Times (3);or Approximately,
                decfsz COUNT2,1             ;(256 x 256 x (4/CPU_Osc_Freq) x 3)
                goto Loop2
                goto Start
```

The Time Length of the Delay Loops will depend on the Crystal Value, the CPU Oscillator Value and the Number Values initially loaded into COUNT1 and COUNT2.

Subroutines

A Subroutine is a section of code, or program, than can be called-on when needed. Subroutines are used if you are performing the same function more than once. For example, when using a delay. Subroutines have the following Advantages.

1.) Subroutines reduce the amount of required program memory needed by the main program.
2.) Allows one to change values internal to the subroutine, which can be used globally throughout the main program.

A General Subroutine Example is as follows.

ROUTINE

Label

    movlw   0xFF   ;This Instruction places the value 255 Decimal
                                     ;(0xFF) (11111111b) into the W-Register. This line
                                     ;of code also demonstrates the use of the Label

    movwf   COUNT1   ;This Instruction places the value 255 Decimal
                                     ;(0xFF) (11111111b) into the Variable Register
                                     ;named COUNT1.

    decfsz   COUNT,1   ;This Instruction Decrements the Variable COUNT by 1, first,
                                     ;and then, places the result back into the Variable COUNT.
                                     ;Then this Instruction checks to see if the new value is Zero.
                                     ;If it is zero, then the next instruction is Skipped and the
                                     ;microprocessor continues with the main program.
    goto Label

    return   ;Returns to the Next Line of Code after the statement that
                                   ;called the subroutine.

First, we have to give the subroutine a name. In this case, I have chosen the name ROUTINE. Then type the code that is to be performed. In this case, I have chosen the Delay in the Flashing LED program. All subroutines must end with the return instruction, which returns the program to the Next Line of Code after the statement that originally called the subroutine.

Flashing LED Example with a Subroutine

```
            TRISB equ 0x0F93         ;This Instruction equates the Hexadecimal address
                                     ;(0x0F93) with its Name, which is TRISB.
            PORTB equ 0x0F81         ;This Instruction equates the Hexadecimal address
                                     ;(0x0F81) with its Name, which is PortB.
            COUNTI equ 0x0200        ;This Instruction equates the General Purpose
                                     ;Hexadecimal address (0x0200) with the Variable
                                     ;Count1.
            COUNT2 equ 0x0201        ;This Instruction equates the General Purpose
                                     ;Hexadecimal address (0x0201) with the Variable
                                     ;Count2.

            movlw    0x00            ;This Instruction places the value 0 Decimal (0x00)
                                     ; (00000000b) into the W-Register.

            movwf    TRISB           ;This Instruction moves the Hexadecimal value
                                     ;(0x00) (00000000b) into TRISB, which is located
                                     ;at address 0x86. This makes each of the 8 Pins of
                                     ;PortB an Output Pin.
Start
            movlw    0xFF            ;This Instruction places the value 255 Decimal
                                     ;(0xFF) (11111111b) into the W-Register. This line
                                     ;of code also demonstrates the use of the
                                     ;Label named Start.
            movwf    COUNT1          ;This Instruction places the value 255 Decimal
                                     ;(0xFF) (11111111b) into the Variable Register
                                     ;named COUNT1.

            movlw    0xFF            ;This Instruction places the value 255 Decimal
                                     ;(0xFF) (11111111b) into the W-Register.
            movwf    COUNT2          ;This Instruction places the value 255 Decimal
                                     ;(0xFF) (11111111b) into the Variable Register
                                     ;named COUNT2.

            movlw    0x01            ;This Instruction places the value (0x01)
                                     ;(00000001b) into the W-Register.

            movwf    PORTB           ;This Instruction places the value (0x01)
                                     ;(00000001b) into PortB. This Turns-On the LED.
            Call     Delay

            movlw    0x00            ;This Instruction places the value (0x00)
                                     ;(00000000b) into the W-Register.

            movwf    PORTB           ;This Instruction places the value (0x00)
                                     ;(00000000b) into PortB. This Turns-Off the LED.
            Call     Delay

            goto Start
Delay
            Loop

            decfsz COUNT1,1          ;This group of Instructions Creates a Delay
            goto Loop                ;The value of this Delay is "Approximated"
                                     ;as; (256 x 256) Times
                                     ;(the Time of One, One-Cycle Instruction)
                                     ;Times (3);or Approximately,
            decfsz COUNT2,1          ;(256 x 256 x (4/CPU_Osc_Freq) x 3)
            goto Loop
            return
```

By using a subroutine for the delay loop, we have reduced the size of the program. Each time we want a delay, either when the LED is Turned-On or Turned-Off, we simply call the delay subroutine. At the end of the subroutine, the program goes back to the line following our 'Call' instruction.

Reading from the I/O Ports

To this point, we have been Writing to I/O PortB so that the LED can be Turned-On and Turned-Off. Now, we need to discuss how to Read the Data value of an I/O Port.

An external Switch is connected between the +5 Volt Power Rail and Ground. Then, a current limiting resistor is connected between the center points (wiper) of the external switch to the Input Pin on the Port in use.

To set a Pin (bit) on PortB as an Input Pin, a Logical "1" (High) (+5v) is sent to the appropriate TRISB Register Pin. This is accomplished as follows.

```
movlw   0x02        ;This Instruction places the value 2 Decimal (0x02)
                    ;(00000010b);into the W-Register.

movwf   TRISB       ;This Instruction moves the Hexadecimal value
                    ;(0x02) (00000010b) into TRISB, which is located
                    ;at address 0x0F93. This makes PortB, Bit-1 an
                    ;Input Pin and all other pins on PortB, Outputs.
```

At this point, we want to check if this Input Pin is High or Low. To do this, we can use one of following two instructions.

btfsc and btfss

The **btfss** instruction means 'Perform a bit test on the specified register and bit. If the result is a Logical "1" (High) (+5v), then skip the next instruction. The **btfsc** means 'Perform a bit test on the Specified register and bit . If the result is a Logical "0" (Low) (0v), then skip the next instruction.

Example for btfss:

```
Label   btfss PortB,1   ;Test Bit-1of PortB and if it is a Logical "1" (High)
                        ;(+5v), then skip the next instruction. The number
                        ;1 after the comma means Bit-1 of PortB.
        goto Label
```

'Continue with Main Program'

This program will move onto 'Continue with Main Program' if PortB, Bit-1 is set to a Logical "1" (High) (+5v).

Example for btfsc:

```
Label   btfsc PortB,1   ;Test Bit-1of PortB and if it is a Logical "0" (Low)
                        ;(0v), then skip the next instruction. The number 1
                        ;after the comma means Bit-1 of PortB.
        goto Label
```

'Continue with Main Program'

This program will move onto 'Continue with Main Program' if PortB, Bit-1 is set to a Logical "0" (Low) (0v).

Flashing LED Example with a Subroutine and an On/Off Switch

The following is a program which will flash an LED when the switch connected to PortB, Bit-1, is set to a Logical "1" (High) (+5v). When the switch connected to PortB, Bit-1 is set to a Logical "0" (Low), the LED will no longer flash.

```
            TRISB equ 0x0F93        ;This Instruction equates the Hexadecimal address
                                    ;(0x0F93) with its Name, which is TRISB.
            PORTB equ 0x0F81        ;This Instruction equates the Hexadecimal address
                                    ;(0x0F81) with its Name, which is PortB.
            COUNT1 equ 0x0200       ;This Instruction equates the General Purpose
                                    ;Hexadecimal address (0x0200) with the Variable
                                    ;Count1.
            COUNT2 equ 0x0201       ;This Instruction equates the General Purpose
                                    ;Hexadecimal address (0x0201) with the Variable
                                    ;Count2.

            movlw   0x02            ;This Instruction places the value 0 Decimal (0x02)
                                    ;(00000010b) into the W-Register.
            movwf   TRISB           ;This Instruction moves the Hexadecimal value
                                    ;(0x02) (00000010b) into TRISB, which is located
                                    ;at address 0x0F93. This makes PortB, Bit-1 an
                                    ;Input Pin and all other PortB Pins Outputs.
Start
            btfss PortB,1           ;This Command Performs a bit test on PortB,
                                    ;Bit-1. If the result is a Logical "1" (High) (+5v),
                                    ;then skip the next instruction.
            goto Start

            movlw   0xFF            ;This Instruction places the value 255 Decimal
                                    ;(0xFF) (11111111b) into the W-Register.
            movwf   COUNT1          ;This Instruction places the value 255 Decimal
                                    ;(0xFF) (11111111b) into the Variable Register
                                    ;named COUNT1.

            movlw   0xFF            ;This Instruction places the value 255 Decimal
                                    ;(0xFF) (11111111b) into the W-Register.
            movwf   COUNT2          ;This Instruction places the value 255 Decimal
                                    ;(0xFF) (11111111b) into the Variable Register
                                    ;named COUNT2.

            movlw   0x01            ;This Instruction places the value (0x01)
                                    ;(00000001b) into the W-Register.
            movwf   PORTB           ;This Instruction places the value (0x01)
                                    ;(00000001b) into PortB. This Turns-On the LED.
            Call    Delay

            movlw   0x00            ;This Instruction places the value (0x00)
                                    ;(00000000b) into the W-Register.
            movwf   PORTB           ;This Instruction places the value (0x00)
                                    ;(00000000b) into PortB. This Turns-Off the LED.
            Call    Delay

            goto Start

Delay

            Loop
            decfsz COUNT1,1         ;This group of Instructions Creates a Delay
            goto Loop               ;The value of this Delay is "Approximated"
                                    ;as; (256 x 256) Times
                                    ;(the Time of One, One-Cycle Instruction)
                                    ;Times (3);or Approximately,
            decfsz COUNT2,1         ;(256 x 256 x (4/CPU_Osc_Freq) x 3)
            goto Loop
            return
```

The xor Command used to Toggle and Output

```
        movlw   0x01            ;This Instruction places 0x01 Hex (00000001b)
                                ;into the W-Register.

        xorwf   PORTB,0         ;This Instruction Exclusive OR's PortB, Bit-0, with
                                ;Bit-0 of the W-Register. If PortB, Bit-0 is a 1, it will
                                ;change to a 0. If PortB, Bit-0, is a 0, it will change
                                ;to a 1. The result is placed in PortB, Bit-0.
```

This following Assembly Code represents a more efficient use of memory for flashing the LED.

Flashing LED with a Subroutine, an On/Off Switch and the Exclusive OR'ing Toggle Method:

```
        TRISB equ 0x0F93        ;This Instruction equates the Hexadecimal address
                                ;(0x0F93) with its Name, which is TRISB.
        PORTB equ 0x0F81        ;This Instruction equates the Hexadecimal address
                                ;(0x0F81) with its Name, which is PortB.
        COUNTI equ 0x0200       ;This Instruction equates the General Purpose
                                ;Hexadecimal address (0x0200) with the Variable
                                ;Count1.
        COUNT2 equ 0x0201       ;This Instruction equates the General Purpose
                                ;Hexadecimal address (0x0201) with the Variable
                                ;Count2.
        movlw   0x02            ;This Instruction places the value 0 Decimal (0x02)
                                ;(00000010b) into the W-Register.
        movwf   TRISB           ;This Instruction moves the Hexadecimal value
                                ;(0x02) (00000010b) into TRISB, which is located
                                ;at address 0x0F93. This makes PortB, Bit-1 an
                                ;Input Pin and all other PortB Pins Outputs.
Start
        btfss PortB,1           ;This Command Performs a bit test on PortB,
                                ;Bit-1. If the result is a Logical "1" (High) (+5v),
                                ;then skip the next instruction.
        goto Start

        movlw   0xFF            ;This Instruction places the value 255 Decimal
                                ;(0xFF) (11111111b) into the W-Register.
        movwf   COUNT1          ;This Instruction places the value 255 Decimal
                                ;(0xFF) (11111111b) into the Variable Register
                                ;named COUNT1.
        movlw   0xFF            ;This Instruction places the value 255 Decimal
                                ;(0xFF) (11111111b) into the W-Register.
        movwf   COUNT2          ;This Instruction places the value 255 Decimal
                                ;(0xFF) (11111111b) into the Variable Register
                                ;named COUNT2.
        xorwf   PORTB,0         ;This Instruction Toggles the LED.
                                ;This Instruction Exclusive OR's PortB, Bit-0, with
                                ;Bit-0 of the W-Register. If PortB, Bit-0 is a 1, it will
                                ;change to a 0. If PortB, Bit-0, is a 0, it will change
                                ; to a 1. The result is placed in ;PortB, Bit-0, as
                                ;indicated by the number 1 after the comma.
        Call Delay

        goto Start

Delay
        Loop
        decfsz COUNT1,1         ;This group of Instructions Creates a Delay
        goto Loop               ;The value of this Delay is "Approximated"
                                ;as; (256 x 256) Times
                                ;(the Time of One, One-Cycle Instruction)
                                ;Times (3);or Approximately,
        decfsz COUNT2,1         ;(256 x 256 x (4/CPU_Osc_Freq) x 3)
        goto Loop
        return
```

External Interrupt

In the previous example, one can see that the looping program which flashes the LED can only be stopped by moving the input switch to the zero position. This means that, regardless when the stop single is initiated, the program must execute all of code up until the btfss line of code is encountered. Often times this is acceptable. However, whenever this is not acceptable, one can invoke the use of an external or peripheral interrupt.

When an external or peripheral interrupt is used, the program can be made to halt almost immediately. This means that the balance of looping program code, which occurs after the interrupt, will not be executed.

This is shown in the following example.

Flashing LED with a Subroutine, with an External Interrupt, NO On/Off Switch and the Exclusive OR'ing Toggle Method:

```
            TRISB equ 0x0F93        ;This Instruction equates the Hexadecimal address
                                    ;(0x0F93) with its Name, which is TRISB.
            PORTB equ 0x0F81        ;This Instruction equates the Hexadecimal address
                                    ;(0x0F81) with its Name, which is PortB.
            COUNTI equ 0x0200       ;This Instruction equates the General Purpose
                                    ; Hexadecimal address (0x0200) with the Variable
                                    ;Count1.
            COUNT2 equ 0x0201       ;This Instruction equates the General Purpose
                                    ;Hexadecimal address (0x0201) with the Variable
                                    ;Count2.
            INTCount equ 0x0202     ;This Instruction equates the General Purpose
                                    ;Hexadecimal address (0x0202) with the Variable
                                    ;INTCount.
            INTCON equ 0x0FF2       ;This Instruction equates the Hexadecimal address
                                    ;(0x0FF2) with its Name, which INTCON
                                    ;(Interrupt Control Register)
            INTCON2 equ 0x0FF1      ;This Instruction equates the Hexadecimal address
                                    ;(0x0FF1) with its Name, which is INTCON2
                                    ;(Interrupt Control Register).
            INTCON3 equ 0x0FF0      ;This Instruction equates the Hexadecimal address
                                    ;(0x0FF0) with its Name, which is INTCON3
                                    ;(Interrupt Control Register).

            Org 0x0000              ;Memory Location of the Main Program.
            goto Start

            Org 0x0008              ;Memory Location of the Interrupt Service Routine.
            goto IServ

Start
            bsf     Intcon,7        ;This Instruction Enables Global ;Interrupts
                                    ;(INTCON, Bit-7)
            bsf     Intcon,6        ;This Instruction Enables Peripheral Interrupts
                                    ;(INTCON, Bit-6)
            bsf     Intcon3,4       ;This Instruction Enables Interrupt on RB2 (INT2)
                                    ;(INTCON3, Bit-4)
            bcf     Intcon3,1       ;This Instruction Clears the INT2 External Interrupt
                                    ;Flag (INTCON3, Bit-1)
            bsf     Intcon2,4       ;This Instruction Enables an Interrupt on the
                                    ;Rising Edge of the signal on RB2 (INT2)
                                    ;(INTCON2, Bit-4)
            movlw   0x02            ;This Instruction places the Hexadecimal value
                                    ;(0x02) (00000010b) into the W-Register.
            movwf   TRISB           ;This Instruction moves the Hexadecimal value
                                    ;(0x02) (00000010b) into TRISB, which is located
                                    ;at address 0x0F93 This makes PortB, Bit-1 an
                                    ;Input Pin and all other PortB Pins Outputs.
            movlw   0xFF            ;This Instruction places the value 255 Decimal
                                    ;(0xFF) (11111111b) into the W-Register.
            movwf   COUNT1          ;This Instruction places the value 255 Decimal
                                    ;(0xFF) (11111111b) into the Variable Register
                                    ;named COUNT1.
```

External Interrupt (continued)

```
        movlw   0xFF            ;This Instruction places the value 255 Decimal
                                ;(0xFF) (11111111b) into the W-Register.

        movwf   COUNT2          ;This Instruction places the value 255 Decimal
                                ; (0xFF) (11111111b) into the Variable Register
                                ;named COUNT2.

        xorwf   PORTB,0         ;This Instruction Toggles the LED. This Instruction
                                ;Exclusive OR's PortB, Bit-0, with Bit-0 of the
                                ;W-Register. If PortB, Bit-0 is a 1, it will change to
                                ;a 0. If PortB, Bit-0, is a 0, it will change to a 1.The
                                ;result is placed in PortB, Bit-0, as indicated by the
                                ;number 1 after the comma.
Call Delay

        goto Start

Delay
    Loop
        decfsz  COUNT1,1        ;This group of Instructions Creates a Delay
        goto Loop               ;The value of this Delay is "Approximated"
                                ;as; (256 x 256) Times
                                ;(the Time of One, One-Cycle Instruction)
                                ;Times (3);or Approximately,
        decfsz  COUNT2,1        ;(256 x 256 x (4/CPU_Osc_Freq) x 3)
        goto Loop
        return

IServ
        bcf     Intcon,1        ;This Instruction Clears the INT0 External Interrupt
                                ;Flag (INTCON, Bit-1), which Enables further
                                ;Rising Edge Interrupts on RB0 (INT0).
        incf    INTCount        ;This Instruction Increments the value of the
                                ; Interrupt Counter by one each the interrupt
                                ;occurs.

        retfie
```

Timer-0

The Timer-0 Module in the PIC18F4685 is an 8-Bit or 16-Bit software configurable high speed counter/timer. In addition to setting the bit size of this Timer, one may also select from 8 different pre-scale frequency divide values which are available within this module. The Timer itself and the pre-scale frequency divide can be enable or disabled thru software. Additionally, one can decide to trigger Timer-0 on a High-to-Low or a Low-to-High signal transition from either an external source, thru Port-A, Bit-4, or the internal clock. An input signal conditioning circuit must be added to protect the microcontroller from voltage which would damage it. The following example depicts how one can create a portion of code that would form the basis of a frequency counter.

```
                TRISA equ 0x0F92        ;This Instruction equates the Hexadecimal address
                                        ;(0x0F92) with its Name, which is TRISA.
                PORTA equ 0x0F80        ;This Instruction equates the Hexadecimal address
                                        ;(0x0F80) with its Name, which is PORTA.
                T0CON equ 0x0FD5        ;This Instruction equates the Hexadecimal address
                                        ;(0x0FD5) with its Name, which is T0CON
                                        ;(Interrupt Control Register).
                TMR0L equ 0x0FD6        ;This Instruction equates the Hexadecimal address
                                        ;(0x0FD6) with its Name, which is TMR0L
                                        ;(Interrupt Control Register).
                TMR0H equ 0x0FD7        ;This Instruction equates the Hexadecimal address
                                        ;(0x0FD7) with its Name, which is TMR0H
                                        ;(Interrupt Control Register).
                CountValue equ 0x0203   ;This Instruction equates the General Purpose
                                        ;Hexadecimal address (0x0203) with the Variable
Start                                   ;CountValue.
        bsf     t0con,7                 ;This Instruction Enables Timer-0
                                        ; (T0CON, Bit-7 = 1)
        bsf     t0con,6                 ;This Instruction Enables Timer-0, 8-Bit Mode
                                        ; (T0CON, Bit-6 = 1)
        bsf     t0con,5                 ;This Instruction Selects the External Signal on
                                        ;Port-A, Bit-4 as the Input Trigger to Timer-0
                                        ; (T0CON, Bit-5 = 1)
        bcf     t0con,4                 ;This Instruction Enables Timer-0 to Trigger on
                                        ;Low-to-High input Signal Transitions
                                        ;(T0CON, Bit-4 = 0)
        cf      t0con,3                 ;This Instruction Enables the Prescale Freq..Divide
                                        ;Function of Timer-0 (T0CON, Bit-3 = 0)
        bcf     t0con,2                 ;The Bits 2,1 & 0 determine the Value of the
                                        ;Pre-scale Frequency Divide
        bcf     t0con,1                 ; (T0CON, Bit-2 = 0, T0CON, Bit-1 = 0, T0CON,
                                        ;Bit-0 = 0)
        bcf     t0con,0                 ;The above values will create a Pre-scale
                                        ;Frequency Divide = 2
        movlw   0x10                    ;This Instruction places the Hexadecimal value
                                        ;(0x10, (00010000b) into the W-Register.
        movwf   TRISA                   ;This Instruction moves the Hexadecimal value
                                        ;(0x10); (00010000b) into TRISA, which is located
                                        ;at address 0x0F92 This makes PortA, Bit-4 ;an
Loop                                    ;Input Pin and all other PortA Pins Outputs.
        movlw   0x00                    ;This Instruction places the value 0 Decimal
                                        ; (0x00); (11111111b) into the W-Register.
        movwf   TMR0L                   ;This Instruction places the value 0 Decimal
                                        ;(0x00); (11111111b) into the Variable Register
                                        ; named TMR0L.
        decfsz COUNT1,1                 ;This group of Instructions Creates a Delay
        movff  TMR0L,COUNTVALUE         ;This Instruction places the 8-Bit value of Timer-0
                                        ;into the Variable Register named COUNTVALUE
        goto Loop                       ;The value of this Delay is "Approximated"
                                        ;as; (256 x 256) Times
                                        ;(the Time of One, One-Cycle Instruction)
                                        ;Times (3);or Approximately,
        decfsz COUNT2,1                 ;(256 x 256 x (4/CPU_Osc_Freq) x 3)
        goto Loop
```

The decimal value of COUNTVALUE divided by the time of the delay is equal to the frequency of the incoming signal.

Assembly Code Schematic

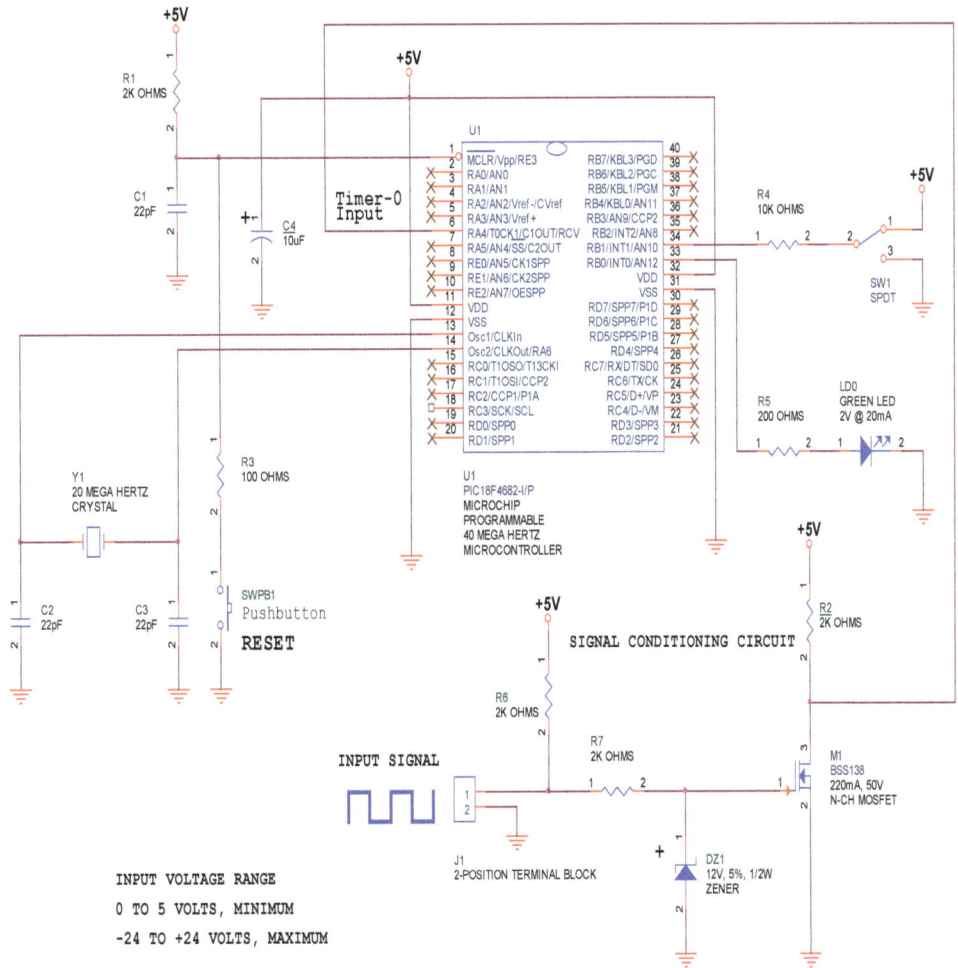

Figure 1.1: Assembly Code Demonstration Circuit

Engineering Practices for the PIC Microcontroller

By: Prof. Sal R. Riggio Jr., PhD, PE

Chapter - 2

Flash Memory Programming with the PICKit-2 Programmer

This circuit board contains an In-Circuit Flash Memory Programmer, which is called the **Microchip PICKit2 In-Circuit Flash Memory Programmer**. This programming circuit is compatible with the **PIC18F2550 Microcontroller** and the **Microchip PICKit-2 Software Package**, only. The PICKit-2 programming circuit has a USB circuit, which interfaces to any personal computer USB Port with a specific piece of Human Interface Device (HID) software, in the form of a Hex File. This software is provided to the user free of charge, by Microchip, and can be downloaded from the www.microchip.com website.

The PICKIt-2 Programming Circuit is also capable of programming a virgin part, but only after the USB HID interface software Hex File is loaded into the PIC18F2550 device, within the PICKit2 programming circuit, by way of an external programmer. This **PICKit-2 PIC18F2550 Hex File** is called **PK2VO23200.hex.** This file is provided to the user free of charge, by Microchip, and can be downloaded from the www.microchip.com website.

The Microchip **PICKit2 In-Circuit Flash Memory Programming Circuit** is shown in **Figure: 2.1.**

The PIC18Fx Family of Microcontrollers requires a Programming Voltage of +7 to +15 volts. This voltage range is provided by the MAX232 integrated circuit. The MAX232 integrated circuit also generates -7 to -15 volts, which is required by the Serial Port Communication circuits, along with +7 to +15 volts.

The Microchip **MAX232 RS232 Interface Circuit and Voltage Generator** is shown in **Figure: 2.2.**

Please see the **PICKit-2 Programmer/Debugger User's Guide Publication DS51553E**, which is provided by Microchip and can be downloaded from the www.microchip.com website.

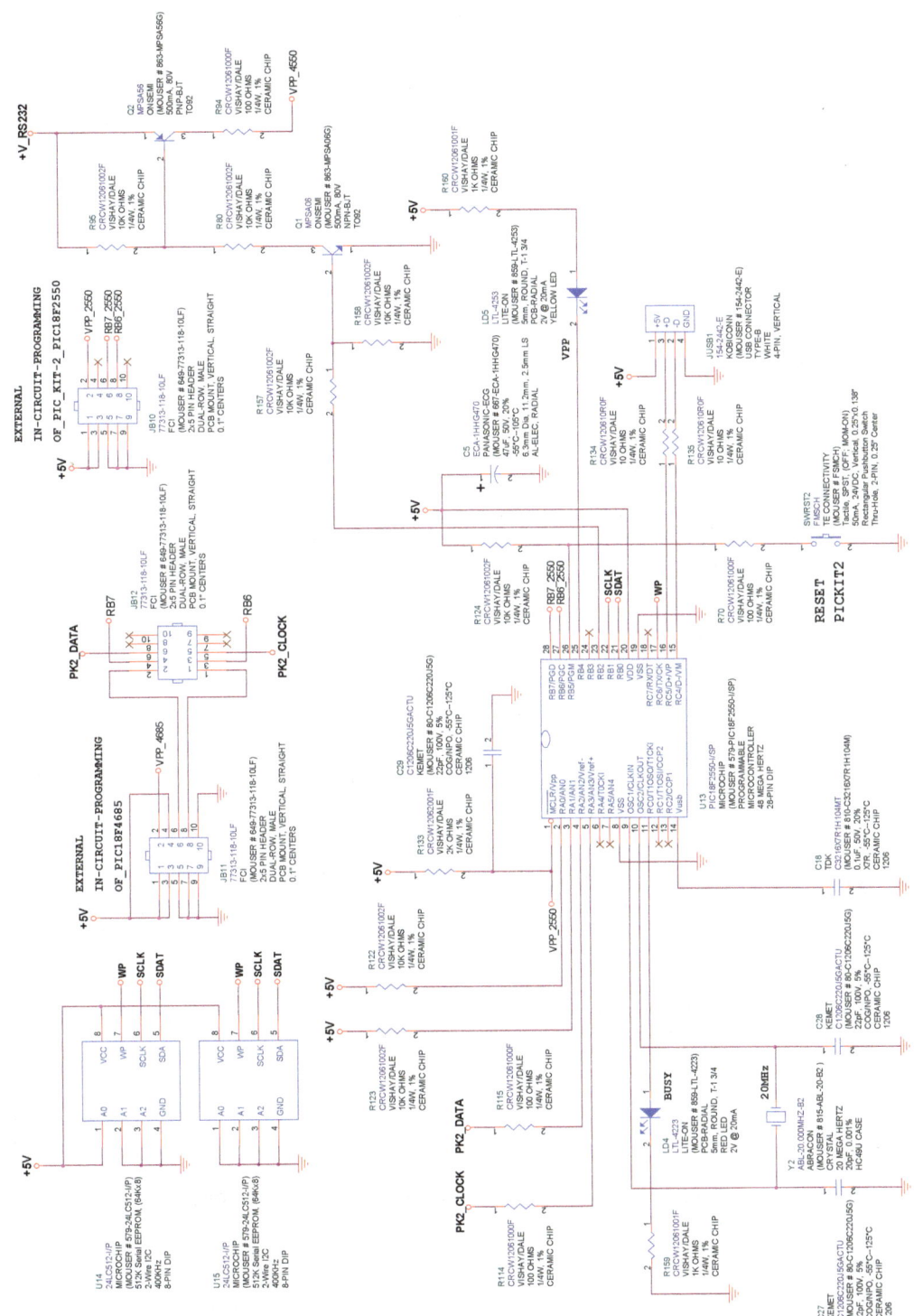

Figure-2.1: PICKit-2 Flash Memory Programmer

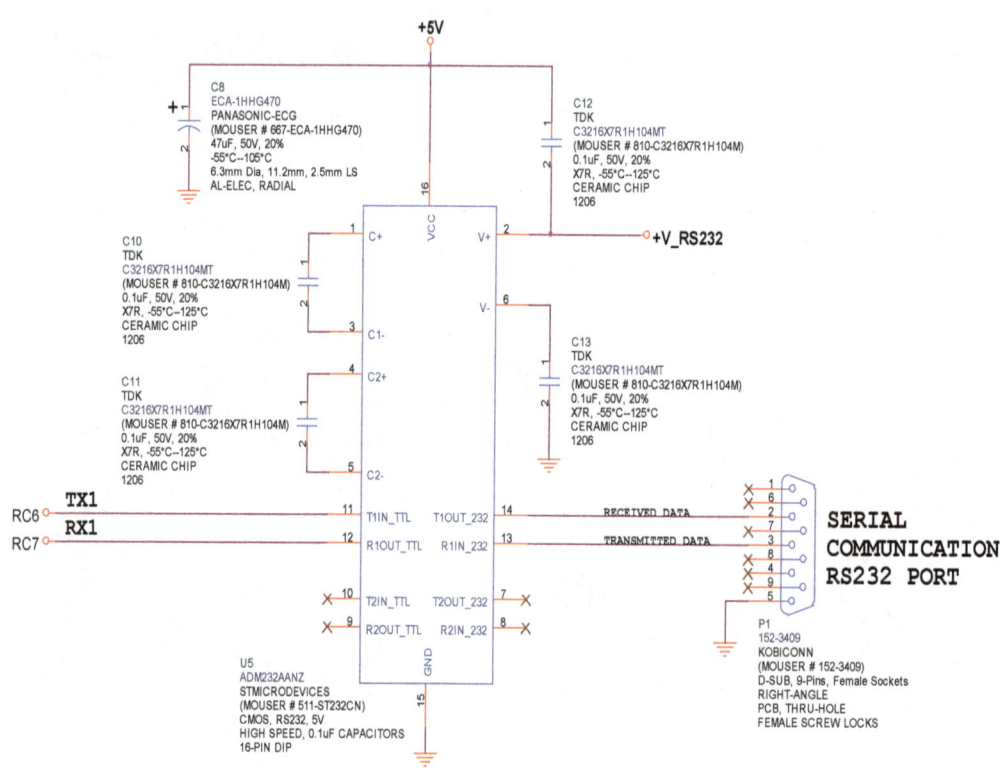

Figure-2.2: MAX232 RS232 Interface Circuit and Voltage Generator

Engineering Practices for the PIC Microcontroller

By: Prof. Sal R. Riggio Jr., PhD, PE

Chapter - 3

Laboratory Experiment - 1

*Delay Loop Timing,
Computer-to-PIC Serial-Communications,
The Keypad, The Light-Emitting-Diode,
The Liquid-Crystal-Display,
Relay Control,
Sound Generation*

Introduction

The purpose of this lab experiment is for the student to develop basic skills in making practical use of the Microchip **PIC18F4685** Microcontroller and become familiar with the importance, operation and hardware implementation of Serial Communications, Light-Emitting-Diodes (LEDs), the Liquid-Crystal-Display ((LCD), the Relay, the Delay Loop and Sound Generation.

The student is also expected to analyze PICBasic-Pro/Assembly & CCS C-code/Assembly, which will produce Delay-Loops and other Routines that will generate the desired electrical waveforms. The student is to analyze these waveforms to determine if the edge-to-edge timing of each waveform matches the code that was written to generate this waveform. Additionally, the student is expected to analyze PICBasic-Pro Code and CCS C-code, which will allow one to communicate with a Personal Computer over the RS232 Serial Port.

The student is expected to analyze PICBasic-Pro Code and CCS C-code, which will exercise the Light-Emitting-Diodes (LEDs), the Keypad, the Relay circuit and Liquid-Crystal-Display (LCD). The student is expected to analyze PICBasic-Pro Code and CCS C-code, which will generate sound through a small Speaker.

For the **Lab Written Report,** the student must <u>Record</u> the **Frequency Values** at which the **Red, Green & Yellow LEDs**, the **Orange, Green & Yellow LED Cluster Lamps**, the **Relay** and the **Speaker** <u>Start Operating</u> and <u>Stop Operating</u>. In other words, the student is to record the frequencies at which the lights start and stop flashing, the relay starts and stops clicking and the speaker starts and stops producing sound.

It is strongly suggested that the student read the PICBasic-Pro code comments, which are written at the end of each line, to learn how to write code for the PIC microcontrollers. It has been proven, many times, to be the most effective way to learn how to generate effective code for the PIC microcontrollers. This code is shown at the end of this chapter.

Portions of the Experiment Board Schematics, which are relevant to this experiment, are shown below in **Figures 3.1, 3.2, 3.3, 3.4, 3.5 & 3.6**. Refer to **Appendix-A** for the complete set of **Experiment Board Schematics** and refer to **Appendix-B** for the complete **Lab Board Schematic**.

Hardware Setup

1.) Connect the Square end of the <u>USB Type-A to Type-B Cable</u> to connector <u>JUSB1</u> of the Experiment Board. Then, connect the other end of this Cable to one of the <u>USB</u> Inputs of the Computer.
2.) Connect the DB9 end of the <u>DB9 Serial-to-USB Cable</u> to connector <u>P1</u> of the Experiment Board. Then, connect the USB end of this Cable to one of the other <u>USB</u> Inputs of the Computer. The DB9 Serial-to-USB Cable is used to communicate between the Serial HyperTerminal of the Computer and the Experiment Board.
3.) The Jumpers for **JB12** are to be placed in **Positions 3-5 & 4-6.**
4.) Connect the Channel-A Input of an oscilloscope to <u>Pin-1</u> of connector (J4), of the Experiment Board.
5.) Connect the Ground clip of the oscilloscope to Pin-2 (GND) of J8, of the Experiment Board.
6.) Connect a Wire between Pin-2 (GND) of J1, of the Lab Board, to Pin-2 (GND) of J8, of the Experiment Board.
7.) Connect a Wire between Pin-1 (SPK) of J1, of the Lab Board, to Pin-1 (SPK) of J12, of the Experiment Board.
8.) Connect a Wire between Pin-1 (Orange) of J8, of the Lab Board, to Pin-2 (Drain) of J12, of the Experiment Board.
9.) Connect a Wire between Pin-1 (Yellow) of J9, of the Lab Board, to Pin-2 (N.O.), of J13, of the Experiment Board.
10.) Connect a Wire between Pin-1 (Green) of J10, of the Lab Board, to Pin-2 (N.O.), of J13, of the Experiment Board.
11.) Connect a Wire between Pin-1 (Common) of connector (J13), of the Experiment Board, and Pin-2 (-V2), of connector (J10), of the Experiment Board.

Assignment #1

Analyze the **PICBasic-Pro Code** and the **CCS C-code** below, which uses the **SEROUT Command**, to **Send** Messages from your **PIC Experiment Board** to the Serial **HyperTerminal** of the Computer. Communications are transmitted at **2400 BAUD,** with **8-Bits, No Parity, 1 Stop Bit** and is sent in **Non-Inverted Mode**. You may change the code to send and receive whatever messages you wish.

Use the **20MHz Crystal Divide by 5 Oscillator Selection**, the **(HS Oscillator, HS used by USB) Oscillator** setting and the **No Divide CPU System Clock** setting.

Assignment #2

Analyze the **Assembly Code,** which has been added to the **PICBasic Code**, after **ASM** and before **ENDASM**. This code will cause the **Red, Yellow** and **Green LED's** to **Flash On and Off, simultaneously**. You will have to add more Delay (your choice) to actually see the LED's Flash. In addition, you must look up the **Register Addresses for TRISE & PORTE** & use **equ** statements to give these addresses names. Be sure that **DIP Switches 2, 3 & 4** of **Switch SW1** are placed in the **Grounded** Position. This assignment does not involve
CCS C-code.

Use the **20MHz Crystal Divide by 5 Oscillator Selection**, the **(HS Oscillator, HS used by USB) Oscillator** setting and the **No Divide CPU System Clock** setting.

Assignment #3

Compile and **Program** the PIC18F4685 microcontroller, using the **20MHz Crystal Divide by 5 Oscillator Selection**, the **(HS Oscillator, HS used by USB) Oscillator** setting and the **Divide by 2 CPU System Clock** setting. This setup will generate a **10 MHz** CPU System Clock.

Measure and **Capture** a **Picture** of the time of the "**high-to-low**" to "**low-to-high**" **pulse widths** on Pin-1 of connector (J4), of the Experiment Board, by using the Oscilloscope and the Computer display. One of these pulse widths represents the time use by the Assembly Code and the other represents the time use by the PICBasic-Pro Code.

Re-compile and **Re-program** the PIC18F4685 microcontroller, using the **20MHz Crystal Divide by 5 Oscillator Selection**, the **(HS Oscillator, PLL Enabled, HS used by USB) Oscillator** setting and the **Divide by 2 CPU System Clock** setting. This setup will generate a **48 MHz** CPU System Clock.

Measure and **Capture** a **Picture** of the time of the "**high-to-low**" to "**low-to-high**" **pulse widths** on Pin-1 of connector (J4), of the Experiment Board, using the Oscilloscope and the Computer display. One of these pulse widths represents the time use by the Assembly Code and the other represents the time use by the PICBasic-Pro Code.

Calculate (Very Important for the Lab Written Report) the **Time** of the **Assembly Code Delay Loop** and **Verify** that it **Matches** the above **Measured Values** for each **Oscillator Frequency**. **Measure** and **Capture** a **Picture** of the **Time** used by the **PICBasic-Pro Code**.

Assignment #4

Write a **PICBasic-Pro Code** routine and a **CCS C-code** routine, which will generate the following **Waveform**.

Use the **20MHz Crystal** <u>**Divide by 5**</u> **Oscillator Selection**, the **(HS Oscillator, PLL Enabled, HS used by USB) Oscillator** setting and the <u>**Divide by 6**</u> **CPU System Clock** setting. This setup will generate a **16 MHz** CPU System Clock.

PICBasic-Pro Oscillator Setting: Code set to **DEFINE OSC 16**
CCS C-code Oscillator Setting: Code set to **#use delay(clock=16000000)**

High Pulse Width = **200us**, Low Pulse Width = **100us**, High Pulse Width = **100us**
Then, place the output in a Low Pulse Width State for **600us**, and **continuously repeat** this entire process.

Display these **High** and **Low Signals** using the Oscilloscope and the Computer display.

Assignment #5:

Write a **PICBasic-Pro Code** routine and a **CCS C-code** routine, which will invoke the following sequence.

First, **Turn-On** the <u>**Red**</u> **LED** and the <u>**Relay**</u> of **your Experiment Board** for **0.8 seconds** and then **Off** for **1.6 seconds**. This process will also **Turn-On** the <u>**Green & Yellow**</u> **LED Cluster Lamps** of the **Lab Board**.
Repeat this Operation <u>**2 more times**</u>.

Second, **Turn-On** the <u>**Yellow**</u> **LED** of **your Experiment Board** for **0.4 seconds** and then **Off** for **0.8 seconds**.
Repeat this Operation <u>**3 more times.**</u>

Third, **Turn-On** the <u>**Green**</u> **LED** of **your Experiment Board** for **0.2 seconds** and then **Off** for **0.4 seconds**.
Repeat this Operation <u>**4 more times.**</u>

Repeat this **Entire Process 1** more time, and then return to the **Start Loop**.

Use the **"A" Key** of the **Keypad** to **Start** this sequence.

Use the **"0" Key** of the **Keypad** to **Stop** this sequence.

Use the **20MHz Crystal** <u>**Divide by 5**</u> **Oscillator Selection**, the **(HS Oscillator, HS used by USB) Oscillator** setting and the <u>**No Divide**</u> **CPU System Clock** setting.

Assignment #6:

Write a **PICBasic-Pro Code** routine and a **CCS C-code** routine, which will do the following.

Use the **"B" Key** of the **Keypad** to produce the following **Sequence-of-Sounds, 1 Time**.

250 Hertz for **1.5 seconds**, then **1K Hertz** for **2.5 seconds**, and then **4K Hertz** for **3.5 seconds**

This code must also **Display** the **Frequency** of the **Produced Sound**, in the **Liquid-Crystal-Display (LCD)** and in the **HyperTerminal** of the **Computer,** during the time that this sound is being heard.

Verify the **Value** of **Each Frequency** with the Oscilloscope connected across the **Speaker Terminals**.

Use the **20MHz Crystal** <u>**Divide by 5**</u> **Oscillator Selection**, the **(HS Oscillator, HS used by USB) Oscillator** setting and the <u>**No Divide**</u> **CPU System Clock** setting.

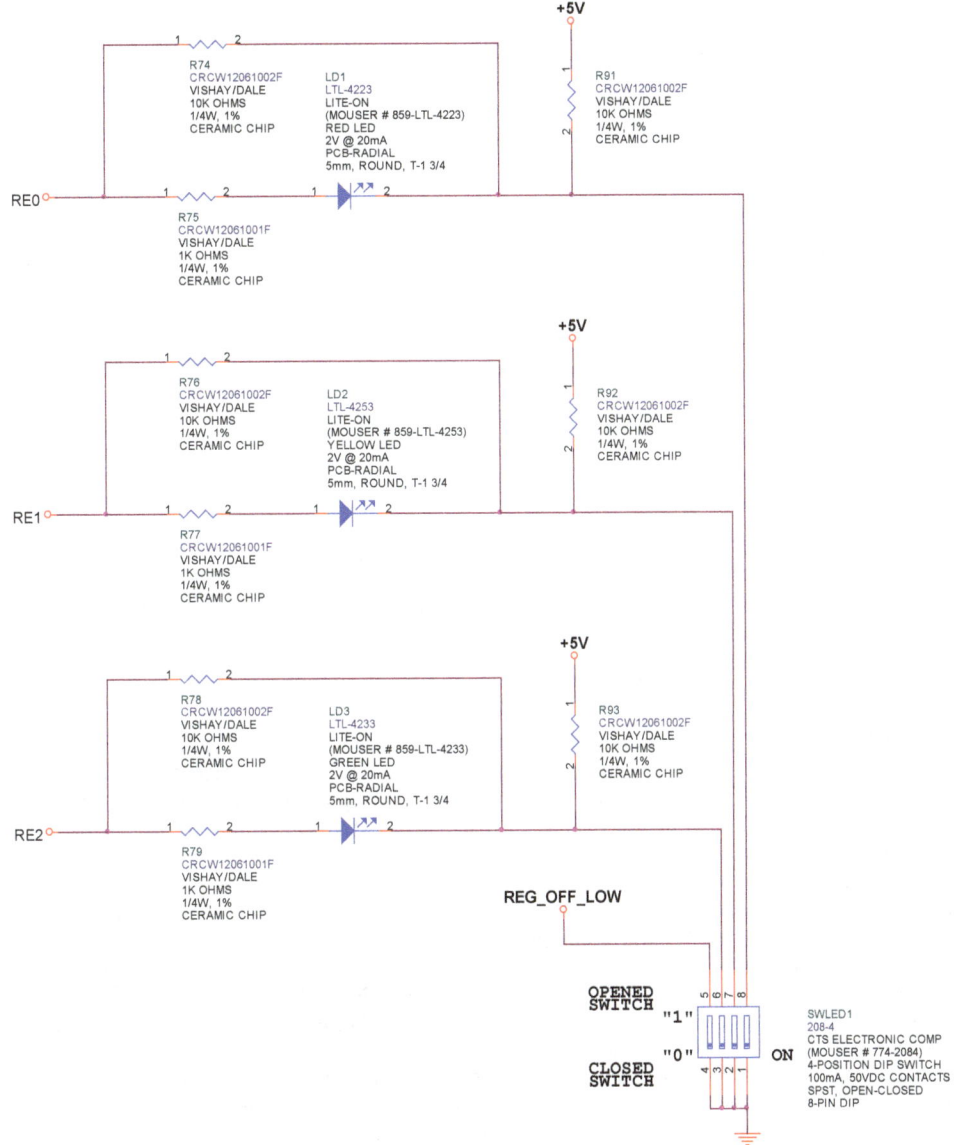

Figure 3.1:   Input Control Switches and LED Driver Circuits

Figure 3.1 shows a circuit which allows the designer to make use of Port-E, Bits 0,1 & 2 as either a input pin, which will detect the position of a manually operated Dip-Switch, or as an output pin which will light an LED. Whether these pins are set to be an input pin or output pin is determined in the code and can be changed from line to line within this code. This feature, which is known as a common I/O pin, represents one of the most important advances in the development of the microcontroller. If a Dip Switch is set in the logical "0" position and the corresponding Port-E bit is configured to be an output pin, the LED connected to this pin can be lighted. If this pin is configured to be an output pin, then the code can read the position of the Dip Switch as a logical "1" or a Logical "0". Because an input pin is high resistance, the LED cannot be lighted and the resistors provide a path for voltage and current between the Dip Switch and the input pin.

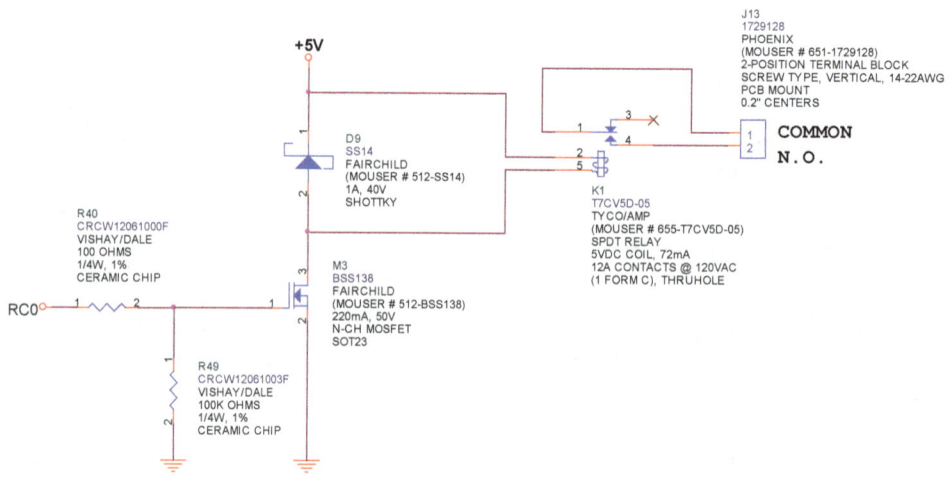

Figure 3.2: **Relay Driver Circuit**

Figure 3.2 shows a circuit, which allows the designer to actuate a relay directly from a microcontroller. This circuit is of great importance to the function of a microcontroller, because it permits the microcontroller to control high voltage and high power equipment with low voltage circuits and microcontroller code. Transistor (M3) in turned on into its triode region by a logic "1" high state voltage signal from the microcontroller. This action allows current to flow through the relay coil, which in turn, causes the high voltage and high power relay contacts to close by way of the generated magnetic field. The closing of these contacts will permit the high power equipment to operate. Diode (D9) is present to provide a path for the current, which is stored in the inductance of the relay coil, to continue to flow when Transistor (M3) is turned off by a logical "0" low state voltage signal from the microcontroller. If this path of current through Diode (D9) were not provided, the voltage at the Drain of Transistor (M3) will rise until the transistor is destroyed. This would occur because the off resistance of Transistor (M3) is extremely high and any amount of current flowing through this large resistance for any amount of time will cause a destructive high voltage to appear across Transistor (M3). Resistors (R40) and (R49) are used to limit transient current flow from the microcontroller.

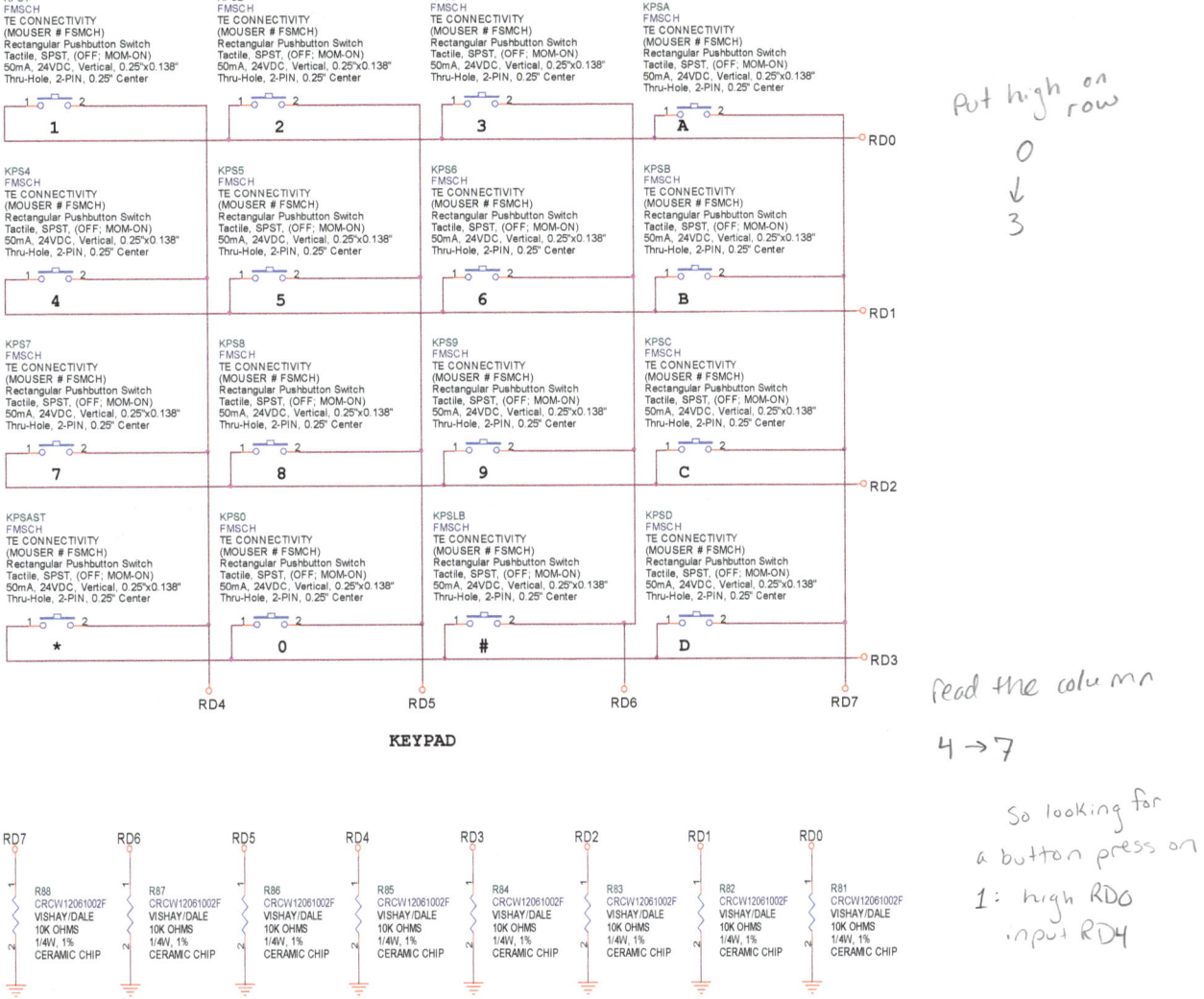

Figure 3.3: Keypad Circuit

Figure 3.3 shows a Keypad circuit which is made up of 16 momentary push-button switches and 8 pull-down resistors. This circuit is connected in the form of a 4 row by 4 column matrix with a resistor connected between each of these 8 points and ground. The microcontroller drives each keypad row (RD0, RD1, RD2 & RD3), one at a time, with a logical "1" (+5 volts) and scans for the presents of a logical "1" (+5 volts) at each keypad column (RD4, RD5, RD6 & RD7), one at a time. If only one keypad button is pressed at any given time, only one logical "1" (+5 volts) level can be present at any keypad column. Thus, the microcontroller code can easily determine which button has been pressed. If no button is pressed, no voltage will appear at any of the 4 column points.

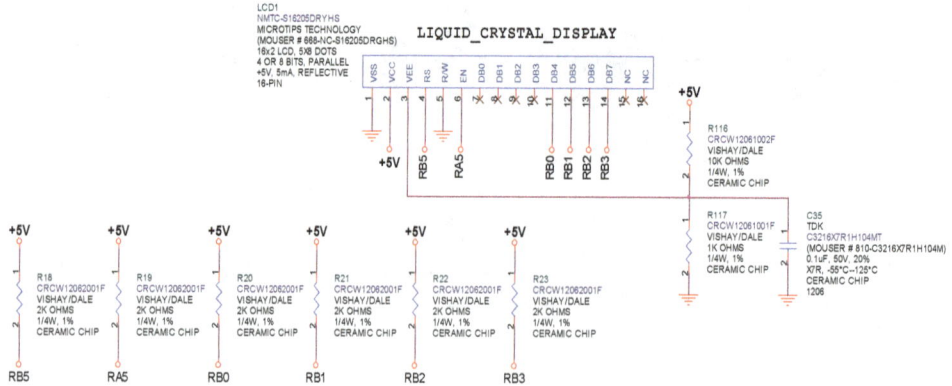

Figure 3.4: LED Display Circuit

Figure 3.4 shows an LCD Display circuit which, on this experiment board, is set to run in 4-Bit Mode. There are a total of 6 control lines between the microcontroller and this display. In addition to the 4 Data Bit lines, these is a line for determining which data register line is active and a line for enabling the operation of this display. The enable line is of particular importance, because the LCD Display must be disabled during the time the data bit lines are used to control the Atmel CPLD. The operation of this display is largely under the control of the operational PICBasic code. Therefore, one must refer to this code to gain a full understanding of the operation of this device.

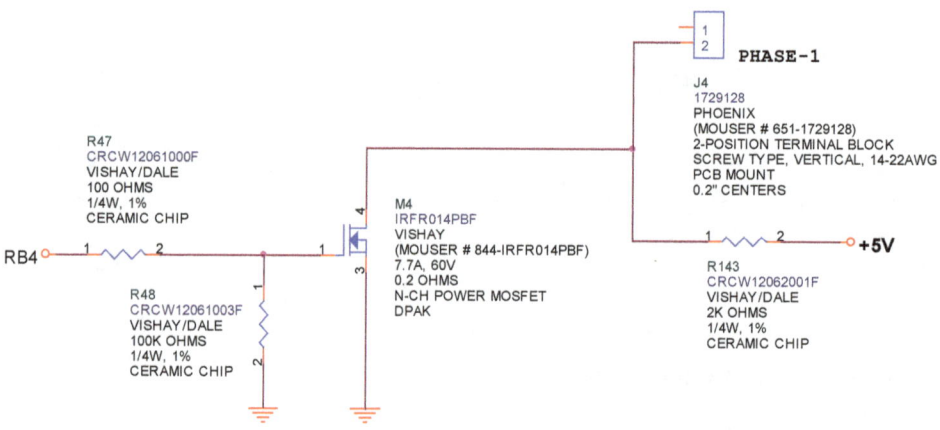

Figure 3.5: LED Display Circuit

Figure 3.5 shows a MOSFET Inverter Circuit that is used to buffer PortB, Bit-4 (Pin RB4) and make its signal available to the oscilloscope through connector (J4) Pin-2, of the experiment board. Additionally, this circuit is used as the Phase-1 drive signal for the 4-Phase Stepper-Motor-Controller.

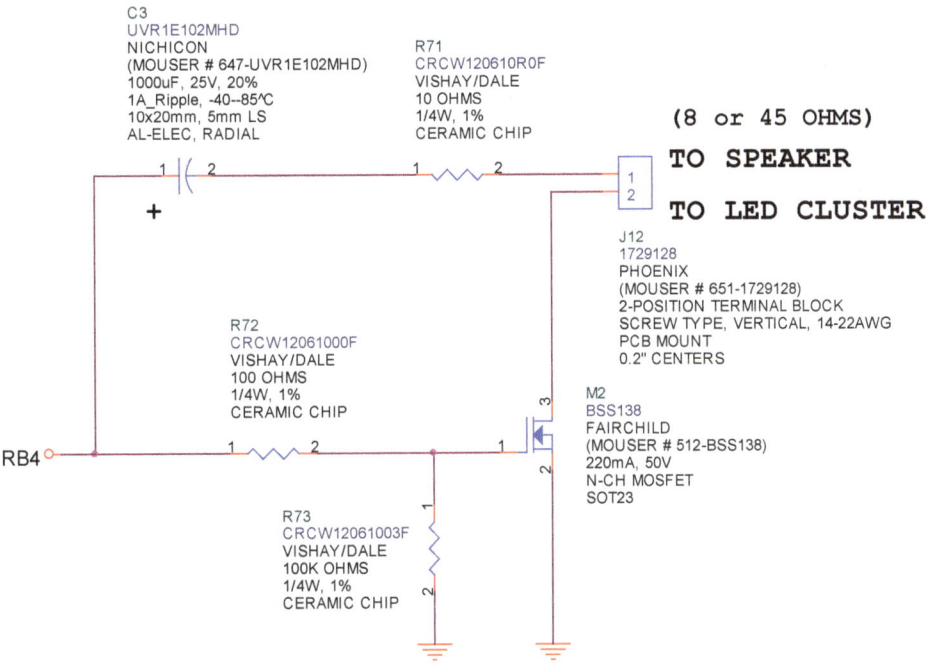

Figure 3.6: Speaker and LED Cluster Lamp Driver Circuit

Figure 3.6 shows a circuit, which allows the designer to drive a speaker of between 8 and 45 ohms, an LED Cluster Lamp, or any other device, which can be driven from an Open-Drain MOSFET Transistor (M2). This transistor is turned on by placing a logical "1" (+5 volts) on Port-B, Bit-4, thru the microcontroller and It is turned off by placing a logical "0" (0 volts) on Port-B, Bit-4. The speaker is driven through a large 1000 microfarad capacitor to pass low frequencies without a large amount of signal attenuation and to block the logical "1" +5v state DC voltage from possibly damaging the microcontroller or the speaker. This circuit is only capable of passing audible tones, which are generated from square-wave pulsed signals by the microcontroller code, to the speaker. True sound wave are passed from the Digital-to-Analog converter, of this experiment board, to the speaker.

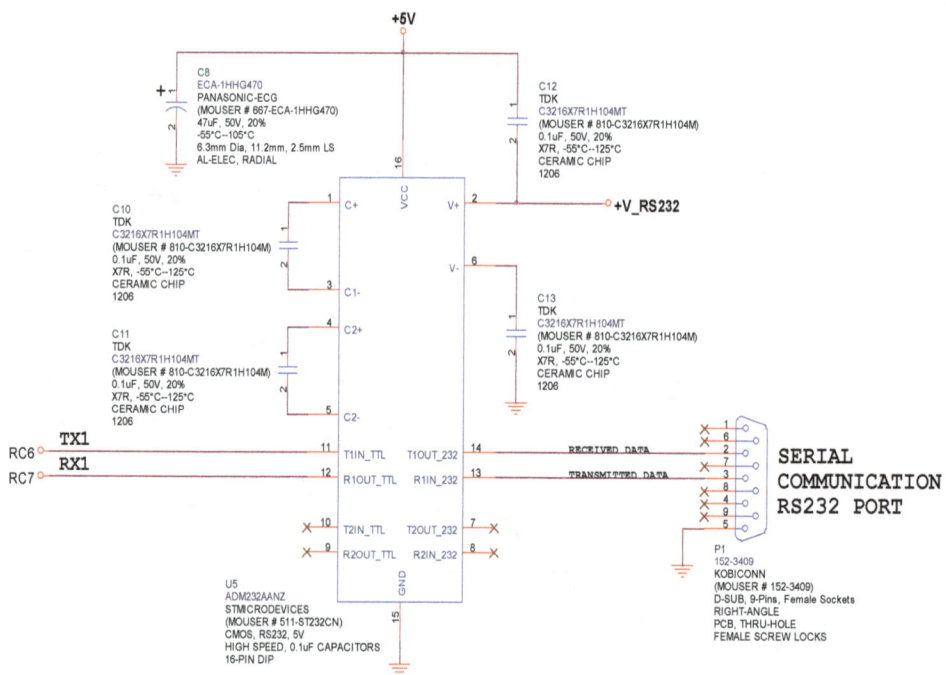

Figure 3.7: Serial Communications Circuit

Figure 3.7 shows a circuit, which provides a way for serial communications to take place between the microcontroller and serial communications port of any personal computer. Because the serial communications port of a personal computer uses the RS232 asynchronous protocol, a signal level-shifting circuit must be provided to allow this communication to take place. The signal voltage levels on the computer side of this communication channel are between +7 and +13 volts for a logical "1" signal and between -7 and -13 volts for a logical "0" signal. The signal voltage levels on the microcontroller side of this communication channel are +5 volts for a logical "1" signal and 0 volts for a logical "0" signal. Analog Devices and other supplies have made a chip that provides for the needs of the RS232 protocol. This not only provides the necessary signal level-shifting, it also provides a dual power supply feature which produces and supplies +/- 7 to 13 volts at 6 milli-amperes, which can be used to fill other needs such as the programming voltage for the PIC microcontrollers. Transmitted data is sent out on pin-3 of the DB9 connector interface of the computer to Port-C, Bit-7 of the microcontroller. Received data is sent from Port-C, Bit-6 of the microcontroller to pin-2 of the DB9 connector interface of the computer.

Laboratory Experiment #1

'Lab-1A_PICBasic-PRO_and_Assembly-Code_Loop-Time
INCLUDE "modedefs.bas"   'Identifies the Command File

'The following assembly code is written to configure the oscillator structure 'of the PIC18F4685.

'To change the oscillator function from HS to HSPLL replace _HS_ with _HSPLL_HS_

```
@   __CONFIG    _CONFIG1L, _PLLDIV_5_1L & _CPUDIV_OSC1_PLL2_1L & _USBDIV_2_1L
@   __CONFIG    _CONFIG1H, _FOSC_HS_1H & _FCMEN_OFF_1H & _IESO_OFF_1H

DEFINE OSC 20                   'Set PIC-Pro Oscillator Value in Mega Hertz
DEFINE ADC_BITS 10              'Set for 10-Bit Internal ADC

DEFINE ADC_CLOCK 5              'DEFINE ADC_CLOCK 1, for OSC=10, MAX ADC
                                'Clock=10MHz/8) =1.25MHz
                                'DEFINE ADC_CLOCK 5, for OSC=16, MAX ADC
                                'Clock=16MHz/16) =1.00MHz
                                'DEFINE ADC_CLOCK 5, for OSC=20, MAX ADC
                                'Clock=20MHz/16) =1.25MHz
                                'DEFINE ADC_CLOCK 2, for OSC=24, MAX ADC
                                'Clock=24MHz/32) =1.25MHz
                                'DEFINE ADC_CLOCK 2, for OSC=32, MAX ADC
                                'Clock=32MHz/32) =1.00MHz
                                'DEFINE ADC_CLOCK 6, for OSC=48, MAX ADC
                                'Clock=48MHz/64) = 0.75MHz

DEFINE CCP1_BIT 2               'Setup Hardware PWM Timer
DEFINE LCD_BITS 4               'Setup LCD for 4-Bit Mode
DEFINE LCD_DREG PORTB           'Drive LCD from Port-B
DEFINE LCD_DBIT 0               'Starting with Bit-0
DEFINE LCD_RSREG PORTA          'Drive LCD Register-Select from
DEFINE LCD_RSBIT 4              'Port-A, Bit-4
DEFINE LCD_EREG PORTA           'Drive LCD Enable from
DEFINE LCD_EBIT 5               'Port-A, Bit-5
DEFINE LCD_LINES 2              'LCD can hold 2-Lines of Characters
DEFINE LCD_COMMANDSUS 4000      'Delay between Commands
DEFINE LCD_DATAUS 200           'Delay between Data Transfers

ADCON0 = %00001111              'ADC Enabled, Analog Input 3 Enabled
ADCON1 = %00001011              'ADC Voltage Range is 0 to 5 Volts &
                                'Analog Inputs 0 Thru 3 are available
ADCON2 = %10101101              'Right Justify to use Upper 8-Bits, only,
                                '12*Tad, & ADC Clock=20MHz/16=1.25MHz

W0 VAR WORD                     'Set 16-Bit Word Variable,W0
W1 VAR WORD                     ' "    "     "      "    ,W1
W2 VAR WORD                     ' "    "     "      "    ,W2
W3 VAR WORD                     ' "    "     "      "    ,W3

B0 VAR BYTE                     'Set 8-Bit Byte Variable,B0
B1 VAR BYTE                     ' "   "    "       "    ,B1

A1 VAR BYTE[16]                 'Set a 16 Character, 8-Bit Byte per Character,
                                'Array Variable
A2 VAR BYTE[16]                 'Set a 16 Character, 8-Bit Byte per Character,
                                'Array Variable

TRISB=$FF                       'Make PortB an Input Port to Enable
                                'In-Circuit-Programming
PAUSE 1000                      'Pause or Delay for 1 Seconds to Enable
                                'In-Circuit-Programming
LCDOUT $FE,1                    'Clear the LCD
LCDOUT $FE, $80                 'Set the Cursor to Line 1, Position 0
LCDOUT "Power Up"               'Send, Power Up, to the LCD

SERout PORTC.6,T2400,["Power Up",13,10,10]  'Send "Power Up" to the HyperTerminal
                                'with a Carriage Return & two Line Feeds,
                                'at 2400 Bits/Second, 8-Bits, 1-Stop Bit,
                                'No Parity, Driven True (T2400).

PAUSE 2000                      'Pause or Delay for 2 Seconds
                                'to Read "Power Up" in the LCD
LCDOUT $FE,1                    'Clear the LCD
TRISB=$00                       'Make PortB an Output Port to Allow for Normal Operation
TRISD=$F0                       'Make PortD, Bits-4,5,6,7 Inputs, & Bits-3
```

```
START:  HIGH PORTB.4            'Make PortB, Bit-4 +5 Volts
                                '(Output Signal is inverted on the Oscilloscope)

ASM

MOVLW   255                     ;Set Variable B1 to 255 Decimal
MOVWF   _B1
MOVLW   255                     ;Set Variable B0 to 255 Decimal
MOVWF   _B0
                                ;Inner Delay Loop
LOOP    DECFSZ _B0,1            ;Decrement File Register B0 by 1
        GOTO LOOP               ;and skip the next instruction
                                ;(GOTO LOOP) if B0=0

                                ;Outer Delay Loop
DECFSZ _B1,1                    ;Decrement File Register B1 by 1
GOTO LOOP                       ;and skip the next instruction
                                ;(GOTO LOOP) if B1=0

ENDASM


LOW PORTB.4                     'Make PortB, Bit-4 Zero Volts
                                '(Output Signal is inverted on the Oscilloscope)
PAUSE 20                        'Pause or Delay for 20 milli-seconds
LCDOUT $FE,1                    'Clear the LCD
LCDOUT $FE, $80                 'Set Cursor to Line 1, Position 1
LCDOUT "Looping"                'Send Looping, to the LCD

SERout2 PORTC.6,396,["Looping",13,10,10]
                                'Send "Looping" to the HyperTerminal
                                'with a Carriage Return & two Line Feeds,
                                'at 2400 Bits/Second, 8-Bits, 1-Stop Bit,
                                'No Parity, Driven True (396).
GOTO START

END
```

Assembly Code (annotation bracketing the ASM...ENDASM block)

Laboratory Experiment #1

'Lab-1B_PICBasic-PRO_Computer_to_PIC Communications
INCLUDE "modedefs.bas"   'Identifies the Command File

'The following assembly code is written to configure the oscillator structure 'of the PIC18F4685.

'To change the oscillator function from HS to HSPLL replace _HS_ with _HSPLL_HS_

```
@   __CONFIG    _CONFIG1L, _PLLDIV_5_1L & _CPUDIV_OSC1_PLL2_1L & _USBDIV_2_1L
@   __CONFIG    _CONFIG1H, _FOSC_HS_1H & _FCMEN_OFF_1H & _IESO_OFF_1H

    DEFINE OSC 20                   'Set PIC-Pro Oscillator Value in Mega Hertz
    DEFINE ADC_BITS 10              'Set for 10-Bit Internal ADC
    DEFINE ADC_CLOCK 5              'DEFINE ADC_CLOCK 1, for OSC=10, MAX ADC
                                    'Clock=10MHz/8) = 1.25MHz
                                    'DEFINE ADC_CLOCK 5, for OSC=16, MAX ADC
                                    'Clock=16MHz/16) =1.00MHz
                                    'DEFINE ADC_CLOCK 5, for OSC=20, MAX ADC
                                    'Clock=20MHz/16) =1.25MHz
                                    'DEFINE ADC_CLOCK 2, for OSC=24, MAX ADC
                                    'Clock=24MHz/32) =1.25MHz
                                    'DEFINE ADC_CLOCK 2, for OSC=32, MAX ADC
                                    'Clock=32MHz/32) =1.00MHz
                                    'DEFINE ADC_CLOCK 6, for OSC=48, MAX ADC
                                    'Clock=48MHz/64) = 0.75MHz

    DEFINE CCP1_BIT 2               'Setup Hardware PWM Timer
    DEFINE LCD_BITS 4               'Setup LCD for 4-Bit Mode
    DEFINE LCD_DREG PORTB           'Drive LCD from Port-B
    DEFINE LCD_DBIT 0               'Starting with Bit-0
    DEFINE LCD_RSREG PORTA          'Drive LCD Register-Select from
    DEFINE LCD_RSBIT 4              'Port-A, Bit-4
    DEFINE LCD_EREG PORTA           'Drive LCD Enable from
    DEFINE LCD_EBIT 5               'Port-A, Bit-5
    DEFINE LCD_LINES 2              'LCD can hold 2-Lines of Characters
    DEFINE LCD_COMMANDSUS 4000      'Delay between Commands
    DEFINE LCD_DATAUS 200           'Delay between Data Transfers

    ADCON0 = %00001111              'ADC Enabled, Analog Input 3 Enabled
    ADCON1 = %00001011              'ADC Voltage Range is 0 to 5 Volts &
                                    'Analog Inputs 0 Thru 3 are available
    ADCON2 = %10101101              'Right Justify to use Upper 8-Bits, only,
                                    '12*Tad, & ADC Clock=20MHz/16=1.25MHz

    W0 VAR WORD                     'Set 16-Bit Word Variable,W0
    W1 VAR WORD                     '"   "     "      "    ,W1
    W2 VAR WORD                     '"   "     "      "    ,W2
    W3 VAR WORD                     '"   "     "      "    ,W3

    B0 VAR BYTE                     'Set 8-Bit Byte Variable,B0
    B1 VAR BYTE                     '"   "    "      "    ,B1

    A1 VAR BYTE[16]                 'Set a 16 Character, 8-Bit Byte per Character,
                                    'Array Variable
    A2 VAR BYTE[16]                 'Set a 16 Character, 8-Bit Byte per Character,
                                    'Array Variable

    TRISB=$FF                       'Make PortB an Input Port to Enable
                                    'In-Circuit-Programming
    PAUSE 1000                      'Pause or Delay for 1 Seconds to Enable
                                    'In-Circuit-Programming
    LCDOUT $FE,1                    'Clear the LCD
    LCDOUT $FE, $80                 'Set the Cursor to Line 1, Position 0
    LCDOUT "Power Up"               'Send, Power Up, to the LCD

    SERout PORTC.6,T2400,["Power Up",13,10,10]
                                    'Send "Power Up" to the HyperTerminal
                                    'with a Carriage Return & two Line Feeds,
                                    'at 2400 Bits/Second, 8-Bits, 1-Stop Bit,
                                    'No Parity, Driven True (T2400).

    PAUSE 2000                      'Pause or Delay for 2 Seconds
                                    'to Read "Power Up" in the LCD
    LCDOUT $FE,1                    'Clear the LCD
    TRISB=$00                       'Make PortB an Output Port to Allow for Normal Operation
    TRISD=$F0                       'Make PortD, Bits-4,5,6,7 Inputs, & Bits-3
```

```
START:
    LCDOUT $FE,1                        'Clear the LCD
    LCDOUT $FE, $80                     'Set the Cursor to Line 1, Position 1
    LCDOUT "Press[A] to send"           'Send "Press[A] to send" to the LCD
    LCDOUT $FE, $C0                     'Set the Cursor to Line 2, Position 1
    LCDOUT "Arrays A1 & A2"             'Send "Arrays A1 & A2" to the LCD

    SERout PORTC.6,T2400,
    ["Press the Keyboard Button [A] to send the Message contained in Arrays A1 & A2",13,10,10]
                                        'Send "Press the Keyboard Button [A]
                                        'to send the Message contained
                                        'in Arrays A1 & A2" to the HyperTerminal,
                                        'followed by a Carriage-Return, & Two Line Feeds,
                                        'at 2400 Bits/Second, 8-Bits, 1-Stop Bit,
                                        'No Parity, Driven True (T2400).

    A1[0]="M"
    A1[1]="E"
    A1[2]="S"
    A1[3]="S"
    A1[4]="A"
    A1[5]="G"
    A1[6]="E"
    A1[7]=" "
    A1[8]="F"
    A1[9]="R"
    A1[10]="O"
    A1[11]="M"
    A1[12]=" "
    A1[13]="A"
    A1[14]="1"
    A1[15]=" "


    A2[0]="&"
    A2[1]=" "
    A2[2]="A"
    A2[3]="2"
    A2[4]=" "
    A2[5]="A"
    A2[6]="R"
    A2[7]="R"
    A2[8]="A"
    A2[9]="Y"
    A2[10]="S"
    A2[11]=" "
    A2[12]="G"
    A2[13]="O"
    A2[14]="O"
    A2[15]="D"


    HIGH PORTD.0                        'Detect Keypad Button "A"
    INPUT PORTD.7                       '"   "    "    "  "
    IF (PORTD.7=1)                      '"   "    "    "  "
    LOW PORTD.0                         '"   "    "    "  "
    PAUSE 2000                          'Pause or Delay for 2 Seconds
    GOTO START

SEND:   SERout2 PORTC.6,396,[str a1\16,13,10,str a2\16,13,10,10]
                                        'Send the Contents of the Array Variable A1,
                                        'a Carriage-Return & One Line Feed
                                        'to the HyperTerminal. Then,
                                        'Send the Contents of the Array Variable A2,
                                        'a Carriage-Return, & Two Line Feeds
                                        'to the HyperTerminal.
                                        'Use 2400 Bits/Second, 8-Bits, 1-Stop Bit,
                                        'No Parity, Driven True (396).

    PAUSE 2000                          'Pause or Delay for 2 Seconds
    GOTO START

    END
```

Laboratory Experiment #1

'Lab-1C_PICBasic-PRO _and_Keypad_Sound_Relay_LCD_and_LEDs
INCLUDE "modedefs.bas"   'Identifies the Command File

'The following assembly code is written to configure the oscillator structure 'of the PIC18F4685.

'To change the oscillator function from HS to HSPLL replace _HS_ with _HSPLL_HS_

```
@ __CONFIG   _CONFIG1L, _PLLDIV_5_1L & _CPUDIV_OSC1_PLL2_1L & _USBDIV_2_1L
@ __CONFIG   _CONFIG1H, _FOSC_HS_1H & _FCMEN_OFF_1H & _IESO_OFF_1H

DEFINE OSC 20                   'Set PIC-Pro Oscillator Value in Mega Hertz
DEFINE ADC_BITS 10              'Set for 10-Bit Internal ADC
DEFINE ADC_CLOCK 5              'DEFINE ADC_CLOCK 1, for OSC=10, MAX ADC
                                'Clock=10MHz/8) = 1.25MHz
                                'DEFINE ADC_CLOCK 5, for OSC=16, MAX ADC
                                'Clock=16MHz/16) = 1.00MHz
                                'DEFINE ADC_CLOCK 5, for OSC=20, MAX ADC
                                'Clock=20MHz/16) = 1.25MHz
                                'DEFINE ADC_CLOCK 2, for OSC=24, MAX ADC
                                'Clock=24MHz/32) = 1.25MHz
                                'DEFINE ADC_CLOCK 2, for OSC=32, MAX ADC
                                'Clock=32MHz/32) = 1.00MHz
                                'DEFINE ADC_CLOCK 6, for OSC=48, MAX ADC
                                'Clock=48MHz/64) = 0.75MHz

DEFINE CCP1_BIT 2               'Setup Hardware PWM Timer
DEFINE LCD_BITS 4               'Setup LCD for 4-Bit Mode
DEFINE LCD_DREG PORTB           'Drive LCD from Port-B
DEFINE LCD_DBIT 0               'Starting with Bit-0
DEFINE LCD_RSREG PORTA          'Drive LCD Register-Select from
DEFINE LCD_RSBIT 4              'Port-A, Bit-4
DEFINE LCD_EREG PORTA           'Drive LCD Enable from
DEFINE LCD_EBIT 5               'Port-A, Bit-5
DEFINE LCD_LINES 2              'LCD can hold 2-Lines of Characters
DEFINE LCD_COMMANDSUS 4000      'Delay between Commands
DEFINE LCD_DATAUS 200           'Delay between Data Transfers

ADCON0 = %00001111              'ADC Enabled, Analog Input 3 Enabled
ADCON1 = %00001011              'ADC Voltage Range is 0 to 5 Volts &
                                'Analog Inputs 0 Thru 3 are available
ADCON2 = %10101101              'Right Justify to use Upper 8-Bits, only,
                                '12*Tad, & ADC Clock=20MHz/16=1.25MHz

W0 VAR WORD                     'Set 16-Bit Word Variable,W0
W1 VAR WORD                     '"    "     "     "     ,W1
W2 VAR WORD                     '"    "     "     "     ,W2
W3 VAR WORD                     '"    "     "     "     ,W3

B0 VAR BYTE                     'Set 8-Bit Byte Variable,B0
B1 VAR BYTE                     '"    "    "     "     ,B1

A1 VAR BYTE[16]                 'Set a 16 Character, 8-Bit Byte per Character,
                                'Array Variable

A2 VAR BYTE[16]                 'Set a 16 Character, 8-Bit Byte per Character,
                                'Array Variable

TRISB=$FF                       'Make PortB an Input Port to Enable
                                'In-Circuit-Programming

PAUSE 1000                      'Pause or Delay for 1 Seconds to Enable
                                'In-Circuit-Programming

LCDOUT $FE,1                    'Clear the LCD
LCDOUT $FE,$80                  'Set the Cursor to Line 1, Position 0
LCDOUT "Power Up"               'Send, Power Up, to the LCD

SERout PORTC.6,T2400,["Power Up",13,10,10]
                                'Send "Power Up" to the HyperTerminal
                                'with a Carriage Return & two Line Feeds,
                                'at 2400 Bits/Second, 8-Bits, 1-Stop Bit,
                                'No Parity, Driven True (T2400).

PAUSE 2000                      'Pause or Delay for 2 Seconds
                                'to Read "Power Up" in the LCD
LCDOUT $FE,1                    'Clear the LCD
TRISB=$00                       'Make PortB an Output Port to Allow for Normal Operation
TRISD=$F0                       'Make PortD, Bits-4,5,6,7 Inputs, & Bits-3
```

```
KSCAN:
    LCDOUT $FE, 1                       'Clear the LCD
    LCDOUT $FE, $80                     'Set Cursor to Line 1, Position 1
    LCDOUT "Scanning"                   'Send, Scanning, to the LCD
    SERout2 PORTC.6,396,["Scanning",13,10,10]
                                        'Send, Scanning, to the HyperTerminal
                                        'with a Carriage Return & 2 Line Feeds.
                                        'at 2400 Bits/Second, 8-Bits, 1-Stop Bit,
                                        'No Parity, Driven True (396).

    HIGH PORTD.0                        'Make 0th Row High to Scan for the Following:
    PAUSE 500
    INPUT PORTD.4
    IF (PORTD.4=1) THEN MES1            'Check for Keypad Key "1" , Column-4
    PAUSE 100
    INPUT PORTD.5
    IF (PORTD.5=1) THEN MES2            'Check for Keypad Key "2" , Column-5
    PAUSE 100
    INPUT PORTD.6
    IF (PORTD.6=1) THEN MES3            'Check for Keypad Key "3" , Column-6
    PAUSE 100
    INPUT PORTD.7
    IF (PORTD.7=1) THEN MESA            'Check for Keypad Key "A" , Column-7
    PAUSE 100
    LOW PORTD.0

    HIGH PORTD.1                        'Make 1st Row High to Scan for the Following:
    PAUSE 500
    INPUT PORTD.4
    IF (PORTD.4=1) THEN MES4            'Check for Keypad Key "4" , Column-4
    PAUSE 100
    INPUT PORTD.5
    IF (PORTD.5=1) THEN MES5            'Check for Keypad Key "5" , Column-5
    PAUSE 100
    INPUT PORTD.6
    IF (PORTD.6=1) THEN MES6            'Check for Keypad Key "6" , Column-6
    PAUSE 100
    INPUT PORTD.7
    IF (PORTD.7=1) THEN MESB            'Check for Keypad Key "B" , Column-7
    PAUSE 100
    LOW PORTD.1

    HIGH PORTD.2                        'Make 2nd Row High to Scan for the Following
    PAUSE 500
    INPUT PORTD.4
    IF (PORTD.4=1) THEN MES7            'Check for Keypad Key "7" , Column-4
    PAUSE 100
    INPUT PORTD.5
    IF (PORTD.5=1) THEN MES8            'Check for Keypad Key "8" , Column-5
    PAUSE 100
    INPUT PORTD.6
    IF (PORTD.6=1) THEN MES9            'Check for Keypad Key "9" , Column-6
    PAUSE 100
    INPUT PORTD.7
    IF (PORTD.7=1) THEN MESC            'Check for Keypad Key "C" , Column-7
    PAUSE 100
    LOW PORTD.2

    HIGH PORTD.3                        'Make 3rd Row High to Scan for the Following:
    PAUSE 500
    INPUT PORTD.4
    IF (PORTD.4=1) THEN MESAT           'Check for Keypad Key "AT" , Column-4
    PAUSE 100
    INPUT PORTD.5
    IF (PORTD.5=1) THEN MES0            'Check for Keypad Key "0" , Column-5
    PAUSE 100
    INPUT PORTD.6
    IF (PORTD.6=1) THEN MESLB           'Check for Keypad Key "LB" , Column-6
    PAUSE 100
    INPUT PORTD.7
    IF (PORTD.7=1) THEN MESD            'Check for Keypad Key "D" , Column-7
    PAUSE 100
    LOW PORTD.3
```

```
HIGH PORTE.0                        'Turn-On the Red LED
PAUSE 300
LOW PORTE.0                         'Turn-Off the Red LED
PAUSE 300
HIGH PORTE.1                        'Turn-On the Yellow LED
PAUSE 300
LOW PORTE.1                         'Turn-Off the Yellow LED
PAUSE 300
HIGH PORTE.2                        'Turn-On the Green LED
PAUSE 300
LOW PORTE.2                         'Turn-On the Green LED
PAUSE 300
GOTO KSCAN
```

MES0:
```
LCDOUT $FE,1                        'Clear LCD
LCDOUT $FE, $89                     'Set Cursor to Line 1, Position 10
LCDOUT "Key 0"                      'Send, Key 0, to the LCD
LCDOUT $FE, $C0                     'Set Cursor to Line 2, Position 1
LCDOUT "0.5 Hertz"                  'Send, 0.5 Hertz, to the LCD
SERout2 PORTC.6,396,["Key 0",13,10,10]
                                    'Send, Key 0, to the HyperTerminal
SERout2 PORTC.6,396,["0.5 Hertz",13,10,10]
                                    'Send, 0.5 Hertz, to the HyperTerminal
SD=2                                'Set the Number of Times to Repeat the Waveform
```
LOOP0:
```
High PORTC.0                        'Turn-On the Relay
High PORTB.4                        'Turn-On Off-Board LED Cluster Lamps
PAUSE 1024
LOW PORTC.0                         'Turn-Off the Relay
LOW PORTB.4                         'Turn-Off Off-Board LED Cluster Lamps
PAUSE 1024
SD=SD-1                             'Check for Return to Keypad Scan
IF SD!=0 THEN LOOP0
GOTO KSCAN
```

MES1:
```
LCDOUT $FE,1
LCDOUT $FE, $8A
LCDOUT "Key 1"
LCDOUT $FE, $C1
LCDOUT "1 Hertz"
SERout2 PORTC.6,396,["Key 1",13,10,10]
SERout2 PORTC.6,396,["1 Hertz",13,10,10]
SD=4
```
LOOP1:
```
High PORTC.0                        'Turn-On the Relay
High PORTB.4                        'Turn-On Off-Board LED Cluster Lamps
PAUSE 512
LOW PORTC.0                         'Turn-Off the Relay
LOW PORTB.4                         'Turn-Off Off-Board LED Cluster Lamps
PAUSE 512
SD=SD-1
IF SD!=0 THEN LOOP1
GOTO KSCAN
```

MES2:
```
LCDOUT $FE,1
LCDOUT $FE, $8B
LCDOUT "Key 2"
LCDOUT $FE, $C2
LCDOUT "2 Hertz"
SERout PORTC.6,T2400,["Key 2",13,10,10]
SERout PORTC.6,T2400,["2 Hertz",13,10,10]
SD=8
```
LOOP2:
```
High PORTC.0                        'Turn-On the Relay
High PORTB.4                        'Turn-On Off-Board LED Cluster Lamps
PAUSE 256
LOW PORTC.0                         'Turn-Off the Relay
LOW PORTB.4                         'Turn-Off Off-Board LED Cluster Lamps
PAUSE 256
SD=SD-1
IF SD!=0 THEN LOOP2
GOTO KSCAN
```

MES3:
```
LCDOUT $FE,1
LCDOUT $FE, $89
LCDOUT "Key 3"
LCDOUT $FE, $C0
LCDOUT "4 Hertz"
SERout PORTC.6,T2400,["Key 3",13,10,10]
SERout PORTC.6,T2400,["4 Hertz",13,10,10]
SD=16
```
LOOP3:
```
High PORTC.0            'Turn-On the Relay
High PORTB.4            'Turn-On Off-Board LED Cluster Lamps
PAUSE 128
LOW PORTC.0             'Turn-Off the Relay
LOW PORTB.4             'Turn-Off Off-Board LED Cluster Lamps
PAUSE 128
SD=SD-1
IF SD!=0 THEN LOOP3
GOTO KSCAN
```

MES4:
```
LCDOUT $FE,1
LCDOUT $FE, $8A
LCDOUT "Key 4"
LCDOUT $FE, $C1
LCDOUT "8 Hertz"
SERout PORTC.6,T2400,["Key 4",13,10,10]
SERout PORTC.6,T2400,["8 Hertz",13,10,10]
SD=32
```
LOOP4:
```
High PORTC.0            'Turn-On the Relay
High PORTB.4            'Turn-On Off-Board LED Cluster Lamps
PAUSEUS 65535
LOW PORTC.0             'Turn-Off the Relay
LOW PORTB.4             'Turn-Off Off-Board LED Cluster Lamps
PAUSEUS 65535
SD=SD-1
IF SD!=0 THEN LOOP4
GOTO KSCAN
```

MES5:
```
LCDOUT $FE,1
LCDOUT $FE, $8B
LCDOUT "Key 5"
LCDOUT $FE, $C2
LCDOUT "16 Hertz"
SERout PORTC.6,T2400,["Key 5",13,10,10]
SERout PORTC.6,T2400,["16 Hertz",13,10,10]
SD=64
```
LOOP5:
```
High PORTC.0            'Turn-On the Relay
High PORTB.4            'Turn-On Off-Board LED Cluster Lamps
PAUSEUS 32768
LOW PORTC.0             'Turn-Off the Relay
LOW PORTB.4             'Turn-Off Off-Board LED Cluster Lamps
PAUSEUS 32768
SD=SD-1
IF SD!=0 THEN LOOP5
GOTO KSCAN
```

MES6:
```
LCDOUT $FE,1
LCDOUT $FE, $89
LCDOUT "Key 6"
LCDOUT $FE, $C0
LCDOUT "32 Hertz"
SERout PORTC.6,T2400,["Key 6",13,10,10]
SERout PORTC.6,T2400,["32 Hertz",13,10,10]
SD=128
```
LOOP6:
```
High PORTC.0            'Turn-On the Relay
High PORTB.4            'Turn-On Off-Board LED Cluster Lamps
PAUSEUS 16384
LOW PORTC.0             'Turn-Off the Relay
LOW PORTB.4             'Turn-Off Off-Board LED Cluster Lamps
PAUSEUS 16384
SD=SD-1
IF SD!=0 THEN LOOP6
GOTO KSCAN
```

MES7:
```
LCDOUT $FE,1
LCDOUT $FE, $8A
LCDOUT "Key 7"
LCDOUT $FE, $C1
LCDOUT "64 Hertz"
SERout PORTC.6,T2400,["Key 7",13,10,10]
SERout PORTC.6,T2400,["64 Hertz",13,10,10]
SD=256
```
LOOP7:
```
High PORTC.0              'Turn-On the Relay
High PORTB.4              'Turn-On Off-Board LED Cluster Lamps
PAUSEUS 8192
LOW PORTC.0               'Turn-Off the Relay
LOW PORTB.4               'Turn-Off Off-Board LED Cluster Lamps
PAUSEUS 8192
SD=SD-1
IF SD!=0 THEN LOOP7
GOTO KSCAN
```

MES8:
```
LCDOUT $FE,1
LCDOUT $FE, $8B
LCDOUT "Key 8"
LCDOUT $FE, $C2
LCDOUT "128 Hertz"
SERout PORTC.6,T2400,["Key 8",13,10,10]
SERout PORTC.6,T2400,["128 Hertz",13,10,10]
SD=512
```
LOOP8:
```
High PORTC.0              'Turn-On the Relay
High PORTB.4              'Turn-On Off-Board LED Cluster Lamps
PAUSEUS 4096
LOW PORTC.0               'Turn-Off the Relay
LOW PORTB.4               'Turn-Off Off-Board LED Cluster Lamps
PAUSEUS 4096
SD=SD-1
IF SD!=0 THEN LOOP8
GOTO KSCAN
```

MES9:
```
LCDOUT $FE,1
LCDOUT $FE, $89
LCDOUT "Key 9"
LCDOUT $FE, $C0
LCDOUT "256 Hertz"
SERout PORTC.6,T2400,["Key 9",13,10,10]
SERout PORTC.6,T2400,["256 Hertz",13,10,10]
SD=1024
```
LOOP9:
```
High PORTC.0              'Turn-On the Relay
High PORTB.4              'Turn-On Off-Board LED Cluster Lamps
PAUSEUS 2048
LOW PORTC.0               'Turn-Off the Relay
LOW PORTB.4               'Turn-Off Off-Board LED Cluster Lamps
PAUSEUS 2048
SD=SD-1
IF SD!=0 THEN LOOP9
GOTO KSCAN
```

MESA:
```
 LCDOUT $FE,1
 LCDOUT $FE, $8A
 LCDOUT "Key A"
 LCDOUT $FE, $C1
 LCDOUT "512 Hertz"
 SERout PORTC.6,T2400,["Key A",13,10,10]
 SERout PORTC.6,T2400,["512 Hertz",13,10,10]
 SD=2048
```
LOOP10:
```
 High PORTC.0                    'Turn-On the Relay
 High PORTB.4                    'Turn-On Off-Board LED Cluster Lamps
 PAUSEUS 1024
 LOW PORTC.0                     'Turn-Off the Relay
 LOW PORTB.4                     'Turn-Off Off-Board LED Cluster Lamps
 PAUSEUS 1024
 SD=SD-1
 IF SD!=0 THEN LOOP10
 GOTO KSCAN
```

MESB:
```
 LCDOUT $FE,1
 LCDOUT $FE, $8B
 LCDOUT "Key B"
 LCDOUT $FE, $C2
 LCDOUT "1024 Hertz"
 SERout PORTC.6,T2400,["Key B",13,10,10]
 SERout PORTC.6,T2400,["1024 Hertz",13,10,10]
 SD=4096
```
LOOP11:
```
 High PORTC.0                    'Turn-On the Relay
 High PORTB.4                    'Turn-On Off-Board LED Cluster Lamps
 PAUSEUS 512
 LOW PORTC.0                     'Turn-Off the Relay
 LOW PORTB.4                     'Turn-Off Off-Board LED Cluster Lamps
 PAUSEUS 512
 SD=SD-1
 IF SD!=0 THEN LOOP11
 GOTO KSCAN
```

MESC:
```
 LCDOUT $FE,1
 LCDOUT $FE, $89
 LCDOUT "Key C"
 LCDOUT $FE, $C0
 LCDOUT "2048 Hertz"
 SERout PORTC.6,T2400,["Key C",13,10,10]
 SERout PORTC.6,T2400,["2048 Hertz",13,10,10]
 SD=8192
```
LOOP12:
```
 High PORTC.0                    'Turn-On the Relay
 High PORTB.4                    'Turn-On Off-Board LED Cluster Lamps
 PAUSEUS 256
 LOW PORTC.0                     'Turn-Off the Relay
 LOW PORTB.4                     'Turn-Off Off-Board LED Cluster Lamps
 PAUSEUS 256
 SD=SD-1
 IF SD!=0 THEN LOOP12
 GOTO KSCAN
```

MESD:
```
 LCDOUT $FE,1
 LCDOUT $FE, $8A
 LCDOUT "Key D"
 LCDOUT $FE, $C1
 LCDOUT "4096 Hertz"
 SERout PORTC.6,T2400,["Key D",13,10,10]
 SERout PORTC.6,T2400,["4096 Hertz",13,10,10]
 SD=16384
```
LOOP13:
```
 High PORTC.0                    'Turn-On the Relay
 High PORTB.4                    'Turn-On Off-Board LED Cluster Lamps
 PAUSEUS 128
 LOW PORTC.0                     'Turn-Off the Relay
 LOW PORTB.4                     'Turn-Off Off-Board LED Cluster Lamps
 PAUSEUS 128
 SD=SD-1
 IF SD!=0 THEN LOOP13
 GOTO KSCAN
```

MESAT:
```
LCDOUT $FE,1
LCDOUT $FE, $8B
LCDOUT "Key *"
LCDOUT $FE, $C2
LCDOUT "8192 Hertz"
SERout PORTC.6,T2400,["Key *",13,10,10]
SERout PORTC.6,T2400,["8192 Hertz",13,10,10]
SD=32768
```
LOOP14:
```
High PORTC.0             'Turn-On the Relay
High PORTB.4             'Turn-On Off-Board LED Cluster Lamps
PAUSEUS 64
LOW PORTC.0              'Turn-Off the Relay
LOW PORTB.4              'Turn-Off Off-Board LED Cluster Lamps
PAUSEUS 64
SD=SD-1
IF SD!=0 THEN LOOP14
GOTO KSCAN
```

MESLB:
```
LCDOUT $FE,1
LCDOUT $FE, $89
LCDOUT "Key #"
LCDOUT $FE, $C0
LCDOUT "16535 Hertz"
SERout PORTC.6,T2400,["Key #",13,10,10]
SERout PORTC.6,T2400,["16535 Hertz",13,10,10]
SD=65535
```
LOOP15:
```
High PORTC.0             'Turn-On the Relay
High PORTB.4             'Turn-On Off-Board LED Cluster Lamps
PAUSEUS 32
LOW PORTC.0              'Turn-Off the Relay
LOW PORTB.4              'Turn-Off Off-Board LED Cluster Lamps
PAUSEUS 32
SD=SD-1
IF SD!=0 THEN LOOP15
GOTO KSCAN

END
```

Laboratory Experiment #1

//Lab-1A_CCS_C-code_and_Assembly-Code_Loop-Time

```c
#include <18F4685.h>                //Identify Microcontroller
#device adc=10                      //Identify ADC Bit Width
#fuses HS,NOWDT,NOPROTECT,NOLVP,MCLR //Setup Programmer Oscillator Value
                                    //and make Master Clear Pin an
                                    //Enable Master Clear=MCLR
                                    //Disable Master Clear=NOMCLR
#use delay(clock=20000000)          //Setup C-Code Oscillator Value
#use fast_io(d)                     //Leave the State of Port-D (all bits)
                                    //the same until changed, again

#use rs232(baud=2400, xmit=PIN_C6, rcv=PIN_C7, stream=HT)
                                    //Set RS232 HyperTerminal
                                    //Communication Parameters

#use rs232(baud=2400, xmit=PIN_B4, rcv=PIN_B5, invert, stream=BB)
                                    //Set RS232 Board-to-Board
                                    //Communication Parameters
#use I2C(master, SCL=PIN_B6, SDA=PIN_B7)
                                    //Set I2C Communication Parameters
#include <lcd_flex.c>                //Identify LCD Driver File
#include <math.h>                    //Include Math Functions

Short B0,B1;                         //1-Bit Variable

void main()
  {
  set_tris_b(0xFF);                  //Make PortB an Input Port to
                                     //Enable In-Circuit-Programming
  delay_ms(1000);                    //Pause or Delay for 1 second

  lcd_init();                        //Initialize the LCD
  lcd_putc(0x0c);                    //Clear the LCD
  lcd_gotoxy(1,1);                   //Set the Cursor to Position-1 of Line-1
  printf(lcd_putc,"Power Up");       //Send, Power Up, to the LCD

  fprintf(HT, "\n\rPower Up\n\r");   //Send, Power Up, to the HyperTerminal with
                                     //a next Line & a Carriage Return
  delay_ms(1000);                    //Pause or Delay for 1 second

  setup_port_a(AN0_to_AN3);          //Identify all of the Analog Inputs
  setup_adc(VSS_VDD);                //Setup the ADC Voltage Range to be
                                     //0 Volts to 5 Volts
                                     //5v-0v)/1024~=.005 Volts Per Bit

  setup_adc(ADC_CLOCK_DIV_32);       //for delay(clock=10000000),
                                     //setup_adc(ADC_CLOCK_DIV_8),
                                     //10MHz/8=1.25MHz
                                     //for delay(clock=16000000),
                                     //setup_adc(ADC_CLOCK_DIV_16),
                                     //16MHz/16=1.00MHz
                                     //for delay(clock=20000000),
                                     //setup_adc(ADC_CLOCK_DIV_16),
                                     //20MHz/16=1.25MHz
                                     //for delay(clock=24000000),
                                     //setup_adc(ADC_CLOCK_DIV_32),
                                     //24MHz/32=1.25MHz
                                     //for delay(clock=32000000),
                                     //setup_adc(ADC_CLOCK_DIV_32),
                                     //32MHz/32=1.00MHz
                                     //for delay(clock=48000000),
                                     //setup_adc(ADC_CLOCK_DIV_64),
                                     //48MHz/64=0.75MHz

  set_adc_channel(0);                //Identify the Analog Input to be
                                     //Sampled (Channel Input Zero)

  setup_ccp1(CCP_PWM);               //Setup Capture/Compare/PWM Pin-1 to PWM Mode

  setup_ccp2(CCP_CAPTURE_RE);        //Setup Capture/Compare/PWM Pin-2 to
                                     //Capture Mode

  setup_timer_1(T1_INTERNAL);        //Setup Timer-1, which controls the
                                     //Capture/Compare portion of
                                     //Capture/Compare/PWM Pin-2, only,
                                     //to Internal Mode

  setup_timer_2(T2_DIV_BY_4, 19, 1); //Setup Timer-2, which controls the PWM portion //of
                                     //Capture/Compare/PWM Pin-1, only.
                                     //DIV_BY_Mode, Prescale (Period), Postscale
                                     //The Following Limits are for the
                                     //PIC18F4685 Microcontroller
                                     //Mode may equal 1 or 4 or 16
                                     //Prescale (Period) may equal any integer
                                     //between 0 and 255
                                     //Postscale may equal any integer between
                                     //1 and 16 (Post = 1 to set PWM Frequency)
                                     //T2Freq=OSC/[(4)*(Mode)*(Prescale+1)]=
                                     //24MegaHz/[(4)*(4)*(PS+1)] = 75KHz
```

```c
   enable_interrupts(INT_CCP2);           //Enable the Internal Interrupt Input
                                          //of Capture/Compare/PWM Pin-2

   enable_interrupts(GLOBAL);             //Enable Global Interrupts for the
                                          //PIC18F4685 Microcontroller

   set_tris_b(0x00);                      //Make PortB an Output Port to Allow for
                                          //Normal Operation
   tris_d(0xF0);                          //Make PortD, Bits-4,5,6,7 Inputs
                                          //and Bits-3,2,1,0 Outputs

   LOOP:

   output_high(PIN_B4);                   //Make PortB, Bit-4 +5 Volts (Output Signal
                                          //is Inverted on the Oscilloscope)
   #ASM

   MOVLW  255                             //Set Variable B1 to 255 Decimal
   MOVWF  B1
   MOVLW  255                             //Set Variable B0 to 255 Decimal
   MOVWF  B0

   LOOP:                                  //Inner Delay Loop
   DECFSZ B0,1                            //Decrement File Register B0 by 1 & skip the
                                          //next instruction if B0=0

   GOTO LOOP                              //Outer Delay Loop
   DECFSZ B1,1                            //Decrement File Register B1 by 1 & skip the
                                          //next instruction if B1=0

   GOTO LOOP

   #ENDASM

   output_low(PIN_B4);                    //Make PortB, Bit-4 Zero Volts
                                          //(Output Signal is Inverted on the Oscilloscope)

   delay_ms(20);                          //Delay or Pause for 20 milli-seconds
   lcd_putc(0x0c);                        //Clear the LCD
   lcd_gotoxy(1,1);                       //Set Cursor to the 1st Position of the 1st Line
   printf(lcd_putc,"Looping");            //Send Looping, to the LCD
   fprintf(HT, "\n\rLooping\n\r");        //Send Looping, to the HyperTerminal

   goto start;

    }
```

Laboratory Experiment #1

```c
//Lab-1C_CCS_C-code_and_Keypad_Relay_Sound_LCD_and_LEDs
#include <18F4685.h>                                     //Identify Microcontroller
#device adc=10                                           //Identify ADC Bit Width
#fuses HS,NOWDT,NOPROTECT,NOLVP,MCLR                     //Setup Programmer Oscillator Value
                                                         //and make Master Clear Pin an
                                                         //Enable Master Clear=MCLR
                                                         //Disable Master Clear=NOMCLR
#use delay(clock=20000000)                               //Setup C-Code Oscillator Value
#use fast_io(d)                                          //Leave the State of Port-D (all bits)
                                                         //the same until changed, again

#use rs232(baud=2400, xmit=PIN_C6, rcv=PIN_C7, stream=HT)
                                                         //Set RS232 HyperTerminal
                                                         //Communication Parameters

#use rs232(baud=2400, xmit=PIN_B4, rcv=PIN_B5, invert, stream=BB)
                                                         //Set RS232 Board-to-Board
                                                         //Communication Parameters
#use I2C(master, SCL=PIN_B6, SDA=PIN_B7)
                                                         //Set I2C Communication Parameters
#include <lcd_flex.c>                                    //Identify LCD Driver File
#include <math.h>                                        //Include Math Functions

long int SD;                                             //16-Bit Word Variable
short D4,D5,D6,D7;                                       //1-Bit Variable

void main()
     {
set_tris_b(0xFF);                                        //Make PortB an Input Port to
                                                         //Enable In-Circuit-Programming
delay_ms(1000);                                          //Pause or Delay for 1 second

lcd_init();                                              //Initialize the LCD
lcd_putc(0x0c);                                          //Clear the LCD
lcd_gotoxy(1,1);                                         //Set the Cursor to Position-1 of Line-1
printf(lcd_putc,"Power Up");                             //Send, Power Up, to the LCD

fprintf(HT, "\n\rPower Up\n\r");                         //Send, Power Up, to the HyperTerminal with
                                                         //a next Line & a Carriage Return
delay_ms(1000);                                          //Pause or Delay for 1 second

setup_port_a(AN0_to_AN3);                                //Identify all of the Analog Inputs
setup_adc(VSS_VDD);                                      //Setup the ADC Voltage Range to be
                                                         //0 Volts to 5 Volts
                                                         //5v-0v)/1024~=.005 Volts Per Bit

setup_adc(ADC_CLOCK_DIV_32);                             //for delay(clock=10000000),
                                                         //setup_adc(ADC_CLOCK_DIV_8),
                                                         //10MHz/8=1.25MHz
                                                         //for delay(clock=16000000),
                                                         //setup_adc(ADC_CLOCK_DIV_16),
                                                         //16MHz/16=1.00MHz
                                                         //for delay(clock=20000000),
                                                         //setup_adc(ADC_CLOCK_DIV_16),
                                                         //20MHz/16=1.25MHz
                                                         //for delay(clock=24000000),
                                                         //setup_adc(ADC_CLOCK_DIV_32),
                                                         //24MHz/32=1.25MHz
                                                         //for delay(clock=32000000),
                                                         //setup_adc(ADC_CLOCK_DIV_32),
                                                         //32MHz/32=1.00MHz
                                                         //for delay(clock=48000000),
                                                         //setup_adc(ADC_CLOCK_DIV_64),
                                                         //48MHz/64=0.75MHz

set_adc_channel(0);                                      //Identify the Analog Input to be
                                                         //Sampled (Channel Input Zero)

setup_ccp1(CCP_PWM);                                     //Setup Capture/Compare/PWM Pin-1 to PWM Mode

setup_ccp2(CCP_CAPTURE_RE);                              //Setup Capture/Compare/PWM Pin-2 to
                                                         //Capture Mode

setup_timer_1(T1_INTERNAL);                              //Setup Timer-1, which controls the
                                                         //Capture/Compare portion of
                                                         //Capture/Compare/PWM Pin-2, only,
                                                         //to Internal Mode

setup_timer_2(T2_DIV_BY_4, 19, 1);                       //Setup Timer-2, which controls the PWM portion //of
                                                         //Capture/Compare/PWM Pin-1, only.
                                                         //DIV_BY_Mode, Prescale (Period), Postscale
                                                         //The Following Limits are for the
                                                         //PIC18F4685 Microcontroller
                                                         //Mode may equal 1 or 4 or 16
                                                         //Prescale (Period) may equal any integer
                                                         //between 0 and 255
                                                         //Postscale may equal any integer between
                                                         //1 and 16 (Post = 1 to set PWM Frequency)
                                                         //T2Freq=OSC/[(4)*(Mode)*(Prescale+1)]=
                                                         //24MegaHz/[(4)*(4)*(PS+1)] = 75KHz
```

```c
   enable_interrupts(INT_CCP2);              //Enable the Internal Interrupt Input
                                             //of Capture/Compare/PWM Pin-2

   enable_interrupts(GLOBAL);                //Enable Global Interrupts for the
                                             //PIC18F4685 Microcontroller

   set_tris_b(0x00);                         //Make PortB an Output Port to Allow for
                                             //Normal Operation
   tris_d(0xF0);                             //Make PortD, Bits-4,5,6,7 Inputs
                                             //and Bits-3,2,1,0 Outputs

kscan:

   lcd_putc(0x0c);                           //Clear LCD
   lcd_gotoxy(1,1);                          //Set Cursor to Position 1, Line 1
   printf(lcd_putc,"Scanning");              //Send Scanning, to the LCD
   printf("\n\rScanning\n\r");               //Send Scanning, to the HyperTerminal with
                                             //a Next Line Feed & a Carriage Return

   output_high(PIN_D0);                      //Make 0th Row High to Scan for the Following:
   delay_ms(500);
   D4= input(PIN_D4);                        //Check for Keypad Key "1" , Column-4
   if (D4) goto mes1;
   delay_ms(100);
   D5 = input(PIN_D5);                       //Check for Keypad Key "2" , Column-5
   if (D5)  goto mes2;
   delay_ms(100);
   D6 = input(PIN_D6);                       //Check for Keypad Key "3" , Column-6
   if (D6) goto mes3;
   delay_ms(100);
   D7 = input(PIN_D7);                       //Check for Keypad Key "A" , Column-7
   if (D7) goto mesa;
   delay_ms(100);
   output_low(PIN_D0);                       //Turn-Off 0th Row with Low

   output_high(PIN_D1);                      //Make 1st Row High to Scan for the Following:
   delay_ms(500);
   D4 = input(PIN_D4);                       //Check for Keypad Key "4" , Column-4
   if (D4) goto mes4;
   delay_ms(100);
   D5 = input(PIN_D5);                       //Check for Keypad Key "5" , Column-5
   if (D5) goto mes5;
   delay_ms(100);
   D6 = input(PIN_D6);                       //Check for Keypad Key "6" , Column-6
   if (D6) goto mes6;
   delay_ms(100);
   D7 = input(PIN_D7);                       //Check for Keypad Key "B" , Column-7
   if (D7) goto mesb;
   delay_ms(100);
   output_low(PIN_D1);                       //Turn-Off 1th Row with Low

   output_high(PIN_D2);                      //Make 2nd Row High to Scan for the Following:
   delay_ms(500);
   D4 = input(PIN_D4);                       //Check for Keypad Key "7" , Column-4
   if (D4) mes7;
   delay_ms(100);
   D5 = input(PIN_D5);                       //Check for Keypad Key "8" , Column-5
   if (D5)  goto mes8;
   delay_ms(100);
   D6 = input(PIN_D6);                       //Check for Keypad Key "9" , Column-6
   if (D6) goto mes9;
   delay_ms(100);
   D7 = input(PIN_D7);                       //Check for Keypad Key "C" , Column-7
   if (D7) goto mesc;
   delay_ms(100);
   output_low(PIN_D2);                       //Turn-Off 2nd  Row with Low

   output_high(PIN_D3);                      //Make 3rd Row High to Scan for the Following:
   delay_ms(500);
   D4 = input(PIN_D4);                       //Check for Keypad Key "*" , Column-4
   if (D4)goto mesat;
   delay_ms(100);
   D5 = input(PIN_D5);                       //Check for Keypad Key "0" , Column-5
   if (D5) goto mes0;
   delay_ms(100);
   D6 = input(PIN_D6);                       //Check for Keypad Key "#" , Column-6
   if (D6) goto meslb;
   delay_ms(100);
   D7 = input(PIN_D7);                       //Check for Keypad Key "D" , Column-7
   if (D7) goto mesd;
   delay_ms(100);
   output_low(PIN_D3);                       //Turn-Off 3rd Row with Low
```

```c
output_high(PIN_E0);              //Turn-On the Red LED
delay_ms(200);
output_low(PIN_E0);               //Turn-Off the Red LED
delay_ms(200);
output_high(PIN_E1);              //Turn-On the Yellow LED
delay_ms(200);
output_low(PIN_E1);               //Turn-Off the Yellow LED
delay_ms(200);
output_high(PIN_E2);              //Turn-On the Green LED
delay_ms(200);
output_low(PIN_E2);               //Turn-Off the Green LED
goto kscan;

mes0:
lcd_putc(0x0c);                   //Clear the LCD
lcd_gotoxy(9,1);                  //Set Cursor to Position 9, Line 1
printf(lcd_putc,"Key 0");         //Send Key 0, to the LCD
lcd_gotoxy(1,2);                  //Set Cursor to Position 1, Line 2
printf(lcd_putc,"0.5 Hertz");     //Send 0.5 Hertz, to the LCD
printf("\n\rKey 0\n\r");          //Send Key 0, to the HyperTerminal with
                                  //a Next Line Feed & a Carriage Return
printf("\n\r0.5 Hertz\n\r");      //Send 0.5 Hertz, to the HyperTerminal with
                                  //a Next Line Feed & a Carriage Return
SD=2;                             //Set the Number of Times to Repeat the Waveform
loop0:
output_high(PIN_C0);              //Turn-On the Relay
output_high(PIN_B4);              //Turn-On Off-Board LED Cluster Lamps
delay_ms(1024);
output_low(PIN_C0);               //Turn-Off the Relay
output_low(PIN_B4);               //Turn-Off Off-Board LED Cluster Lamps
delay_ms(1024);
SD=SD-1;                          //Decrement SD by 1
if (SD!=0) goto loop0;            //Check for Return to Keypad Scan
goto kscan;

mes1:
lcd_putc(0x0c);
lcd_gotoxy(10,1);
printf(lcd_putc,"Key 1");
lcd_gotoxy(1,2);
printf(lcd_putc,"1 Hertz");
printf("\n\rKey 1\n\r");
printf("\n\r1 Hertz\n\r");
SD=4;
loop1:
output_high(PIN_C0);              //Turn-On the Relay
output_high(PIN_B4);              //Turn-On Off-Board LED Cluster Lamps
delay_ms(512);
output_low(PIN_C0);               //Turn-Off the Relay
output_low(PIN_B4);               //Turn-Off Off-Board LED Cluster Lamps
delay_ms(512);
SD=SD-1;                          //Decrement SD by 1
if (SD!=0) goto loop1;
goto kscan;

mes2:
lcd_putc(0x0c);
gotoxy(11,1);
printf(lcd_putc,"Key 2");
lcd_gotoxy(1,2);
printf(lcd_putc,"2 Hertz");
printf("\n\rKey 2\n\r");
printf("\n\r2 Hertz\n\r");
SD=8;
loop2:
output_high(PIN_C0);              //Turn-On the Relay
output_high(PIN_B4);              //Turn-On Off-Board LED Cluster Lamps
delay_ms(256);
output_low(PIN_C0);               //Turn-Off the Relay
output_low(PIN_B4);               //Turn-Off Off-Board LED Cluster Lamps
delay_ms(256);
SD=SD-1;
if (SD!=0) goto loop2;
goto kscan;
```

mes3:
```
lcd_putc(0x0c);
lcd_gotoxy(12,1);
printf(lcd_putc,"Key 3");
lcd_gotoxy(1,2);
printf(lcd_putc,"4 Hertz");
printf("\n\rKey 3\n\r");
printf("\n\r4 Hertz\n\r");
SD=16;
```
loop3:
```
output_high(PIN_C0);      //Turn-On the Relay
output_high(PIN_B4);      //Turn-On Off-Board LED Cluster Lamps
delay_ms(128);
output_low(PIN_C0);       //Turn-Off the Relay
output_low(PIN_B4);       //Turn-Off Off-Board LED Cluster Lamps
delay_ms(128);
SD=SD-1;
if (SD!=0) goto loop3;
goto kscan;
```

mes4:
```
lcd_putc(0x0c);
lcd_gotoxy(1,1);
printf(lcd_putc,"Key 4");
lcd_gotoxy(1,2);
printf(lcd_putc,"8 Hertz");
printf("\n\rKey 4\n\r");
printf("\n\r8 Hertz\n\r");
SD=32;
```
loop4:
```
output_high(PIN_C0);      //Turn-On the Relay
output_high(PIN_B4);      //Turn-On Off-Board LED Cluster Lamps
delay_us(65535);
output_low(PIN_C0);       //Turn-Off the Relay
output_low(PIN_B4);       //Turn-Off Off-Board LED Cluster Lamps
delay_us(65535);
SD=SD-1;
if (SD!=0) goto loop4;
goto kscan;
```

mes5:
```
lcd_putc(0x0c);
lcd_gotoxy(2,1);
printf(lcd_putc,"Key 5");
lcd_gotoxy(1,2);
printf(lcd_putc,"16 Hertz");
printf("\n\rKey 5\n\r");
printf("\n\r16 Hertz\n\r");
SD=64;
```
loop5:
```
output_high(PIN_C0);      //Turn-On the Relay
output_high(PIN_B4);      //Turn-On Off-Board LED Cluster Lamps
delay_us(32768);
output_low(PIN_C0);       //Turn-Off the Relay
output_low(PIN_B4);       //Turn-Off Off-Board LED Cluster Lamps
delay_us(32768);
SD=SD-1;
if (SD!=0) goto loop5;
goto kscan;
```

mes6:
```
lcd_putc(0x0c);
lcd_gotoxy(3,1);
printf(lcd_putc,"Key 6");
lcd_gotoxy(1,2);
printf(lcd_putc,"32 Hertz");
printf("\n\rKey 6\n\r");
printf("\n\r32 Hertz\n\r");
SD=128;
```
loop6:
```
output_high(PIN_C0);      //Turn-On the Relay
output_high(PIN_B4);      //Turn-On Off-Board LED Cluster Lamps
delay_us(16384);
output_low(PIN_C0);       //Turn-Off the Relay
output_low(PIN_B4);       //Turn-Off Off-Board LED Cluster Lamps
delay_us(16384);
SD=SD-1;
if (SD!=0) goto loop6;
goto kscan;
```

mes7:
```
lcd_putc(0x0c);
lcd_gotoxy(4,1);
printf(lcd_putc,"Key 7");
lcd_gotoxy(1,2);
printf(lcd_putc,"64 Hertz");
printf("\n\rKey 7\n\r");
printf("\n\r64 Hertz\n\r");
SD=256;
```
loop7:
```
output_high(PIN_C0);            //Turn-On the Relay
output_high(PIN_B4);            //Turn-On Off-Board LED Cluster Lamps
delay_us(8192);
output_low(PIN_C0);             //Turn-Off the Relay
output_low(PIN_B4);             //Turn-Off Off-Board LED Cluster Lamps
delay_us(8192);
SD=SD-1;
if (SD!=0) goto loop7;
goto kscan;
```

mes8:
```
lcd_putc(0x0c);
lcd_gotoxy(5,1);
printf(lcd_putc,"Key 8");
lcd_gotoxy(1,2);
printf(lcd_putc,"128 Hertz");
printf("\n\rKey 8\n\r");
printf("\n\r128 Hertz\n\r");
SD=512;
```
loop8:
```
output_high(PIN_C0);            //Turn-On the Relay
output_high(PIN_B4);            //Turn-On Off-Board LED Cluster Lamps
delay_us(4096);
output_low(PIN_C0);             //Turn-Off the Relay
output_low(PIN_B4);             //Turn-Off Off-Board LED Cluster Lamps
delay_us(4096);
SD=SD-1;
if (SD!=0) goto loop8;
goto kscan;
```

mes9:
```
lcd_putc(0x0c);
lcd_gotoxy(6,1);
printf(lcd_putc,"Key 9");
lcd_gotoxy(1,2);
printf(lcd_putc,"256 Hertz");
printf("\n\rKey 9\n\r");
printf("\n\r256 Hertz\n\r");
SD=1024;
```
loop9:
```
output_high(PIN_C0);            //Turn-On the Relay
output_high(PIN_B4);            //Turn-On Off-Board LED Cluster Lamps
delay_us(2048);
output_low(PIN_C0);             Turn-Off the Relay
output_low(PIN_B4);             //Turn-Off Off-Board LED Cluster Lamps
delay_us(2048);
SD=SD-1;
if (SD!=0) goto loop9;
goto kscan;
```

mesa:
```
lcd_putc(0x0c);
lcd_gotoxy(7,1);
printf(lcd_putc,"Key A");
lcd_gotoxy(1,2);
printf(lcd_putc,"512 Hertz");
printf("\n\rKey A\n\r");
printf("\n\r512 Hertz\n\r");
SD=2048;
```
loopa:
```
output_high(PIN_C0);            //Turn-On the Relay
output_high(PIN_B4);            //Turn-On Off-Board LED Cluster Lamps
delay_us(1024);
output_low(PIN_C0);             //Turn-Off the Relay
output_low(PIN_B4);             //Turn-Off Off-Board LED Cluster Lamps
delay_us(1024);
SD=SD-1;
if (SD!=0) goto loopa;
goto kscan;
```

```c
mesb:
lcd_putc(0x0c);
lcd_gotoxy(8,1);
printf(lcd_putc,"Key B");
lcd_gotoxy(1,2);
printf(lcd_putc,"1024 Hertz");
printf("\n\rKey B\n\r");
printf("\n\r1024 Hertz\n\r");
SD=4096;
loopb:
output_high(PIN_C0);            //Turn-On the Relay
output_high(PIN_B4);            //Turn-On Off-Board LED Cluster Lamps
delay_us(512);
output_low(PIN_C0);             //Turn-Off the Relay
output_low(PIN_B4);             //Turn-Off Off-Board LED Cluster Lamps
delay_us(512);
SD=SD-1;
if (SD!=0) goto loopb;
goto kscan;

mesc:
lcd_putc(0x0c);
lcd_gotoxy(9,1);
printf(lcd_putc,"Key C");
lcd_gotoxy(1,2);
printf(lcd_putc,"2048 Hertz");
printf("\n\rKey C\n\r");
printf("\n\r2048 Hertz\n\r");
SD=8192;
loopc:
output_high(PIN_C0);            //Turn-On the Relay
output_high(PIN_B4);            //Turn-On Off-Board LED Cluster Lamps
delay_us(256);
output_low(PIN_C0);             //Turn-Off the Relay
output_low(PIN_B4);             //Turn-Off Off-Board LED Cluster Lamps
delay_us(256);
SD=SD-1;
if (SD!=0) goto loopc;
goto kscan;

mesd:
lcd_putc(0x0c);
lcd_gotoxy(10,1);
printf(lcd_putc,"Key D");
lcd_gotoxy(1,2);
printf(lcd_putc,"4096 Hertz");
printf("\n\rKey D\n\r");
printf("\n\r4096 Hertz\n\r");
SD=16384;
loopd:
output_high(PIN_C0);            //Turn-On the Relay
output_high(PIN_B4);            //Turn-On Off-Board LED Cluster Lamps
delay_us(128);
output_low(PIN_C0);             //Turn-Off the Relay
output_low(PIN_B4);             //Turn-Off Off-Board LED Cluster Lamps
delay_us(128);
SD=SD-1;
if (SD!=0) goto loopd;
goto kscan;

mesat:
lcd_putc(0x0c);
lcd_gotoxy(11,1);
printf(lcd_putc,"Key *");
lcd_gotoxy(1,2);
printf(lcd_putc,"8192 Hertz");
printf("\n\rKey *\n\r");
printf("\n\r8192 Hertz\n\r");
SD=32768;
loopat:
output_high(PIN_C0);            //Turn-On the Relay
output_high(PIN_B4);            //Turn-On Off-Board LED Cluster Lamps
delay_us(64);
output_low(PIN_C0);             //Turn-Off the Relay
output_low(PIN_B4);             //Turn-Off Off-Board LED Cluster Lamps
delay_us(64);
SD=SD-1;
if (SD!=0) goto loopat;
goto kscan;
```

meslb:
lcd_putc(0x0c);
lcd_gotoxy(12,1);
printf(lcd_putc,"Key #");
lcd_gotoxy(1,2);
printf(lcd_putc,"16535 Hertz");
printf("\n\rKey #\n\r");
printf("\n\r16535 Hertz\n\r");
SD=65535;
looplb:
output_high(PIN_C0);   //Turn-On the Relay
output_high(PIN_B4);   //Turn-On Off-Board LED Cluster Lamps
delay_us(32);
output_low(PIN_C0);   //Turn-Off the Relay
output_low(PIN_B4);   //Turn-Off Off-Board LED Cluster Lamps
delay_us(32);
SD=SD-1;
if (SD!=0) goto looplb;
goto kscan;

}

Engineering Practices for the PIC Microcontroller

By: Prof. Sal R. Riggio Jr., PhD, PE

Chapter - 4

Laboratory Experiment - 2

*The Digital Voltage Regulator,
The Brushless-DC Fan Speed Regulator*

Introduction

The purpose of this lab experiment is for the student to develop basic skills in making practical use of the Microchip **PIC18F4685** Microcontroller and become familiar with the importance, operation and hardware implementation of a Digital Voltage Regulator Control Loop and a Digital Motor Speed Control Loop.

The student is expected to analyze the **PICBasic-Pro Code** and the **CCS C-code,** below, which <u>Regulates</u> the <u>Output Voltage</u> of a **Digital Voltage Regulator**. This means that the output voltage is to be maintained at a constant value, while the **input voltage** is **varied**. The **input** voltage must be, **at least**, **1 volt** higher than the **output** voltage of the voltage regulator, at all times. The **Minimum** input voltage to this voltage regulator must be **+11VDC** because of the required turn-on voltage of the P-Channel Power MOSFET. The voltage regulator will operate from an Input Voltage Range of **+11VDC** to **+20VDC.**

The student is expected to analyze the **PICBasic-Pro Code** and the **CCS C-code,** below, which enables a <u>Digital Fan Speed Regulator</u> to maintain the <u>Rotational speed</u> of the fan. The fan speed is to be maintained while the **input voltage** and the **fan backpressure** are changed. The **input** voltage must be, **at least, 1 volt** higher than the **output voltage** of the voltage regulator that drives the fan, at all times. The **Minimum** input voltage to this voltage regulator must be **+11VDC** because of the required turn-on voltage of the P-Channel Power MOSFET. The fan will operate from **+10VDC (2400 RPM)** to **+15VDC (3200 RPM)**. This means that the **Minimum** input voltage must be **+11VDC**. The **Maximum** input voltage is limited to **+20VDC.**

It is strongly suggested that the student **read** the **PICBasic-Pro code** and the **CCS C-code**, particularly the **comments** written at the end of each line, to learn how to write code for the PIC microcontrollers. It has been proven, many times, to be the most effective way to learn how to generate effective code for the PIC microcontrollers. This code is shown at the end of this chapter

Portions of the Experiment Board Schematics, which are relevant to this experiment, are shown below in **Figure 4.1**. Refer to **Appendix-A** for the complete set of **Experiment Board Schematics** and refer to **Appendix-B** for the complete **Lab Board Schematic**.

Hardware Setup

1.) Connect the Square end of the <u>USB Type-A to Type-B Cable</u> to connector <u>JUSB1</u> of the Experiment Board. Then, connect the other end of this Cable to one of the <u>USB</u> Inputs of the Computer.
2.) Connect the DB9 end of the <u>DB9 Serial-to-USB Cable</u> to connector <u>P1</u> of the Experiment Board. Then, connect the USB end of this Cable to one of the other <u>USB</u> Inputs of the Computer. The DB9 Serial-to-USB Cable is used to communicate between the Serial HyperTerminal of the Computer and the Experiment Board.
3.) The Jumpers for **JB12** are to be placed in **Positions 3-5 & 4-6.**
4.) Connect the Channel-A Input of an oscilloscope to <u>Pin-1</u> of connector (J7), of the Experiment Board.
5.) Connect the Ground clip of the oscilloscope to <u>Pin-2</u> (GND) of connector (J7), of the Experiment Board.
6.) Connect the output of a <u>+20VDC, 0.5A Variable Power Supply</u> and a DMM to the terminals of Connector (J6), of the Experiment Board.
   Use <u>AWG #22 Solid Red and Black Wires</u> to make the two connections. Be sure to connect the <u>Black Lead</u> to the <u>Ground Terminal of (J6)</u>, and the <u>Red Lead</u> to the <u>+Vin-Reg Terminal of (J6)</u>, of the Experimentation Board.
7.) Connect the Channel-A Input of an oscilloscope to <u>Pin-1</u> (FSP) of connector (J8), of the Experiment Board.
8.) Connect the Ground clip of the oscilloscope to <u>Pin-2</u> (GND) of connector (J8), of the Experiment Board.
9.) Connect a <u>50 ohm, 20 Watt Load Resistor</u> between Pin-1 and Pin-2 of connector (J7), of the Experiment Board.
10.) Connect a <u>DMM</u> across the <u>50 ohm, 20 Watt Load Resistor</u>. Set this meter to the <u>20 volts DC Scale</u>.
11.) Place all 4 DIP Switches (SW) of the Experiment Board in the <u>Logic"1" Up Position</u>.
12.) Connect a Wire between Pin-2 (+12VF) of J3, of the Lab Board, to Pin-1 (+Vout-Reg) of J7, of the Experiment Board.
13,) Connect a Wire between Pin-1 (GND) of J3, of the Lab Board, to Pin-2 (GND) of J7, of the Experiment Board.
14.) Connect a Wire between Pin-3 (Speed) of J3, of the Lab Board, to Pin-1 (FSP) of J8, of the Experiment Board.

Assignment #1

Analyze the **PICBasic-Pro Code** and the **CCS C-code** below, which allows the **Output Voltage** of the **Digital Voltage Regulator** to be varied from **+5VDC** to **+12VDC**, according to the **Table** below. The **keypad** is to be used as the input device to **Set** the **Output Voltage**. A comparison variable called the **Target Value (TV)** must be used. This variable is to be updated through the keypad, using the **Keypad Input Key Sequence** below. The **Target Value (TV)** is a decimal number between **0 and 1023.** This variable is to be updated through the keypad, using the keypad input sequence below, through the keypad scan routine. This number represents the value of the output voltage and is to be compared to the value that is continuously recorded through the **ADC Routine (VS0)**, to maintain the output voltage. The Target Value must also be updated in the **Slow-Start Routine (SLOWST)**. The decimal number **0** represents **0VDC**, and the decimal number **1023** represents **+5VDC**. The actual output voltage is monitored thru a **6:1** Voltage Divider, which must be considered when writing the code. The **Keypad Scan Routine** must include an **LCD command line** to **output** the **character** of each **button** as it is **pressed**, in **sequence**.

Use **Switch (SW0) (PortC.3)** to turn the **Regulator** on and off. The **Regulator** must be **turned-off** and **turned-on** again, before the **Keypad Input Sequence** is recognized. Use **Switch (SW1) (PortC.4)** to allow the voltage regulator to produce an **Un-Regulated Output Voltage (Free-Run Mode)**, depending upon the **input voltage**.

In **Free-Run Mode {Switch (SW1) (PortC.4) = 1}**, **Measure** and **Record** the **Output Voltage** and the **DutyCycle** at VinDC = +15VDC, and at VinDC = +20VDC.

Input Voltage (J6)	Output Voltage (J7)	Keypad Sequence	Output Voltage	DutyCycle
+15 VDC	+ 5 VDC	0 5 #	_____	_____
+20 VDC	+ 5 VDC	0 5 #	_____	_____

{For PIC-Pro Code, Only}
TV = (Desired Output Voltage)(1024) / (5v)(6)

{For CCS C-Code, Only}
TV = The Actual Output Voltage

Use the Following Equations to Compute the Percent Efficiency:
%EFF = (100)(VoutDC)(IoutDC) / (VinDC)(IinAVG)   DutyCycle = Ton / T (To Be Measured)
IoutDC = VoutDC / 24 Ohms $T \rightarrow$ The Switching Period
IinAVG = (Ton / T)(IoutDC)

Measure the Actual Value of the **24 Ohm Resistor** and use this **Value** in all **Calculations**. RL = _____

Measure and Record the following:

Input Voltage (J6) Minimum Maximum	Output Voltage (J7)	Keypad Sequence	Min Input Voltage Output Voltage	Max Input Voltage Output Voltage
+10 VDC +20 VDC	+ 5 VDC	0 5 #	_____	_____
+10 VDC +20 VDC	+ 6 VDC	0 6 #	_____	_____
+13 VDC +20 VDC	+ 9 VDC	0 9 #	_____	_____
+16 VDC +20 VDC	+12 VDC	1 2 #	_____	_____

Measure, Calculate and Record the following   Also, Capture each Waveform with the Oscilloscope

Input Voltage (J6) Minimum Maximum	Output Voltage (J7)	Keypad Sequence	Min Input Voltage DutyCycle	Max Input Voltage DutyCycle
+10 VDC +20 VDC	+ 5 VDC	0 5 #	_____	_____
+10 VDC +20 VDC	+ 6 VDC	0 6 #	_____	_____
+13 VDC +20 VDC	+ 9 VDC	0 9 #	_____	_____
+16 VDC +20 VDC	+12 VDC	1 2 #	_____	_____

Input Voltage (J6) Minimum Maximum	Output Voltage (J7)	Keypad Sequence	Min Input Voltage % Efficiency	Max Input Voltage % Efficiency
+10 VDC +20 VDC	+ 5 VDC	0 5 #	_____	_____
+10 VDC +20 VDC	+ 6 VDC	0 6 #	_____	_____
+13 VDC +20 VDC	+ 9 VDC	0 9 #	_____	_____
+16 VDC +20 VDC	+12 VDC	1 2 #	_____	_____

Input Voltage (J6) Minimum Maximum	Output Voltage (J7)	Keypad Sequence	Min Input Voltage Output Voltage Slow-Start Rise-Time	Max Input Voltage Output Voltage Slow-Start Rise-Time
+10 VDC +20 VDC	+ 5 VDC	0 5 #	_____	_____
+10 VDC +20 VDC	+ 6 VDC	0 6 #	_____	_____
+13 VDC +20 VDC	+ 9 VDC	0 9 #	_____	_____
+16 VDC +20 VDC	+12 VDC	1 2 #	_____	_____

Input Voltage (J6) Minimum Maximum	Output Voltage (J7)	Keypad Sequence	Min Input Voltage Output Voltage Fall-Time	Max Input Voltage Output Voltage Fall-Time
+10 VDC +20 VDC	+ 5 VDC	0 5 #	_____	_____
+10 VDC +20 VDC	+ 6 VDC	0 6 #	_____	_____
+13 VDC +20 VDC	+ 9 VDC	0 9 #	_____	_____
+16 VDC +20 VDC	+12 VDC	1 2 #	_____	_____

Input Voltage (J6) Minimum Maximum Voltage Voltage	Output Voltage (J7)	Keypad Sequence	Min Input Voltage Peak-to-Peak Output Noise	Max Input Voltage Peak-to-Peak Output Noise
+10 VDC +20 VDC	+ 5 VDC	0 5 #	_____	_____
+10 VDC +20 VDC	+ 6 VDC	0 6 #	_____	_____
+13 VDC +20 VDC	+ 9 VDC	0 9 #	_____	_____
+16 VDC +20 VDC	+12 VDC	1 2 #	_____	_____

Assignment #2

Analyze the **PICBasic-Pro Code** and the **CCS C-code** below, which allows the <u>Rotational Speed</u> of the Fan to be varied from <u>2400 RPM</u> to <u>3200 RPM</u>, according to the table below. The **Keypad** is to be used as the input device to **Set** the **Fan Speed**. A comparison variable called the <u>Target Value</u> **(TV)** must be used. This variable is to be updated through the keypad, using the **Keypad Input Key Sequence** below. The **Target Value (TV)** represents the value of the **Rotational Speed** of the fan in <u>RPM x100</u>, and is to be compared to the value that is continuously recorded through the **Pulse Counting Routine (FS)**, to maintain the fan speed. The Target Value must also be updated in the **Slow-Start Routine (SLOWST)**. The **Keypad Scan Routine** must include an **LCD command line** to **output** the **character** of each **button** as it is **Pressed**, in **sequence**. The following equations are to be used to compute the <u>Target Value</u> (TV).

Fan → PPR=Poles/2 Pulses/Revolution
<u>**PPS = Counter's Pulses per Second**</u>

<u>Poles=4</u>

TV & FS (in RPM) = (**PPS** Pulses/Sec)/(**PPR** Pulses/Rev)(60 Sec/Min)

Use the **Reset Switch** to turn the **Regulator** on and off. The **Regulator** must be **turned-off** and **turned-on**, again, before the **Keypad Input Sequence** is recognized. Use the **"Asterisk" Key** of the **Keypad** to allow the fan to **rotate freely**, at **any speed**, depending only upon the **input voltage**. Use the **"Pound-Sign" Key** of the **Keypad** to allow the fan to return to Speed Regulation Mode.

<u>Use the Following Table for the keypad scan routine:</u>

Rotational Fan Speed (RPM) (Set the Input Voltage to <u>+20VDC</u> for all <u>5 Speeds</u>)	Keypad Input Sequence	Record the Voltage Regulator Output Voltage with a DMM	Record the Duty-Cycle with an Oscilloscope
2400	2 4 #	_____	_____
2600	2 6 #	_____	_____
2800	2 8 #	_____	_____
3000	3 0 #	_____	_____
3200	3 2 #	_____	_____

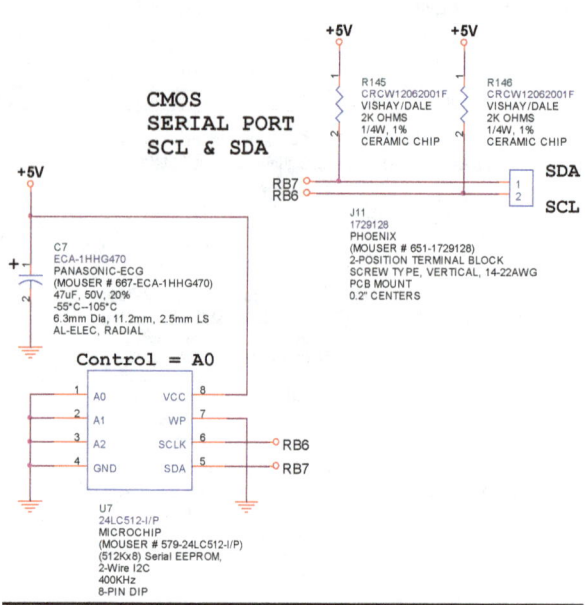

Figure 4.1: 512x8 EEPROM and interface connector

<u>Figure 4.1</u> shows a circuit which allows for Writing date to and and Reading Data from and EEPROM.

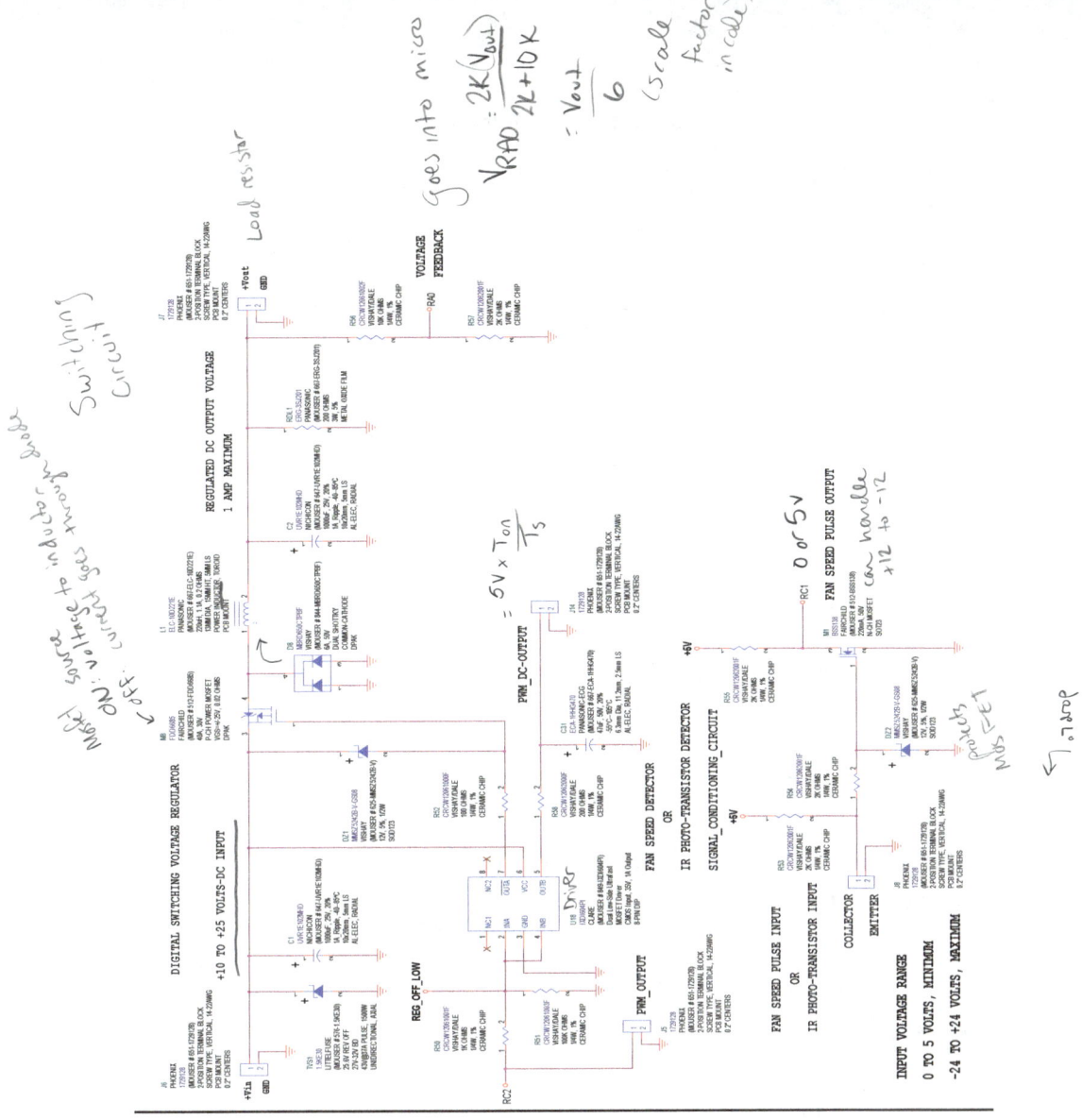

Figure 4.2: Digital Pulse-Width-Modulated Switching Voltage Regulator and Pulse Counting Interface Circuit

Figure 4.2 shows a circuit that allows one to write code to digitally regulate the output voltage of a switching voltage regulator under load and with a changing input voltage. The code writer can also use this regulator circuit to digitally regulate any given parameter of the load that is sensitive to changes in the voltage applied to the load. In this experiment, a DC-Brushless Fan is used as the load and this digital switching voltage regulator, along with code, will regulated the rotational speed of this fan to a constant value. The rotational speed can be set to a number of different values through a Keypad entry.

The regulation feedback uses to use resistors, (R56) and (R57), in a standard constant voltage divider arrangement to feedback an error voltage to PIC microcontroller's analog-to-digital converter.
This error voltage will alter the duty-cycle of the pulse-width-modulated (PWM) drive signal, which is delivered from Port-C, Bit-2 of the PIC microcontroller to the IXD604P MOSFET driver module and the P-Channel MOSFET transistor (M8). The High State of the PWM drive signal is called the On-Time. The Low State of the PWM drive signal is called the Off-Time. When the On-time is made larger, the output voltage becomes higher and vice versa.
Since the operating frequency and consequently period of this digital switching voltage regulator are both constant, when the On-time is increased the Off-time is decreased and vice versa.

When the P-Channel MOSFET (M8) is placed in its triode region (On-Time), it will conduct current to the load from the input voltage source. When the P-Channel MOSFET (M8) is placed in its cutoff region (Off-Time), the load current is maintained through the inductor (L1) by way of the power diode (D8). The capacitor (C2) and the inductor (L1) form a filter, which removes the switching frequency and its harmonics. The Zener diode (DZ1) protects the gate-to-source voltage of the P-Channel MOSFET (M8) from voltages greater than 12 volts. This (M8) has a maximum gate-to-source voltage of +/- 20 volts. The Transient Voltage Suppressor (TVS1) protects the input of this voltage regulator from temporary voltages greater than +27 volts and -0.7 volts.

$$V_{out} = V_{in}\left[\frac{T_{on}}{T_s}\right] \qquad T_s = \frac{1}{f_s} \qquad f_s = 30\,kHz$$

Laboratory Experiment #2

'Lab-2A_PICBasic-Pro_KeyPad_Digital_Voltage_Regulator

```
INCLUDE "modedefs.bas"            'Identifies the Command File

DEFINE OSC 20                     'Set PIC-Pro Oscillator Value in Mega Hertz
DEFINE ADC_BITS 10                'Set for 10-Bit Internal ADC

DEFINE ADC_CLOCK 5                'DEFINE ADC_CLOCK 1, for OSC=10, MAX ADC
                                  'Clock=10MHz/8) = 1.25MHz
                                  'DEFINE ADC_CLOCK 5, for OSC=16, MAX ADC
                                  'Clock=16MHz/16) =1.00MHz
                                  'DEFINE ADC_CLOCK 5, for OSC=20, MAX ADC
                                  'Clock=20MHz/16) =1.25MHz
                                  'DEFINE ADC_CLOCK 2, for OSC=24, MAX ADC
                                  'Clock=24MHz/32) =1.25MHz
                                  'DEFINE ADC_CLOCK 2, for OSC=32, MAX ADC
                                  'Clock=32MHz/32) =1.00MHz
                                  'DEFINE ADC_CLOCK 6, for OSC=48, MAX ADC
                                  'Clock=48MHz/64) = 0.75MHz

DEFINE CCP1_BIT 2                 'Setup Hardware PWM Timer
DEFINE LCD_BITS 4                 'Setup LCD for 4-Bit Mode
DEFINE LCD_DREG PORTB             'Drive LCD from Port-B
DEFINE LCD_DBIT 0                 'Starting with Bit-0
DEFINE LCD_RSREG PORTA            'Drive LCD Register-Select from
DEFINE LCD_RSBIT 4                'Port-A, Bit-4
DEFINE LCD_EREG PORTA             'Drive LCD Enable from
DEFINE LCD_EBIT 5                 'Port-A, Bit-5
DEFINE LCD_LINES 2                'LCD can hold 2-Lines of Characters
DEFINE LCD_COMMANDSUS 4000        'Delay between Commands
DEFINE LCD_DATAUS 200             'Delay between Data Transfers

ADCON0 = %00001111                'ADC Enabled, Analog Input 3 Enabled
ADCON1 = %00001011                'ADC Voltage Range is 0 to 5 Volts &
                                  'Analog Inputs 0 Thru 3 are available
ADCON2 = %10101101                'Right Justify to use Upper 8-Bits, only,
                                  '12*Tad, & ADC Clock=20MHz/16=1.25MHz

W0 VAR WORD                       'Set 16-Bit Word Variable,W0
W1 VAR WORD                       ' "    "   "    "     "    ,W1
W2 VAR WORD                       ' "    "   "    "     "    ,W2
W3 VAR WORD                       ' "    "   "    "     "    ,W3
VS0 VAR WORD                      ' "    "   "    "     "    ,VS0
VS0R VAR WORD                     ' "    "   "    "     "    ,VS0R
VS0WHOLE VAR WORD                 ' "    "   "    "     "    ,VS0WHOLE
VS0DECIMAL VAR WORD               ' "    "   "    "     "    ,VS0DECIMAL
MAXVOLTRANGE VAR WORD             ' "    "   "    "     "    ,MAXVOLTRANGE
OUTVOLTDIVRATIO VAR WORD          ' "    "   "    "     "    ,OUTVOLTDIVRATIO
LCDOC VAR WORD                    ' "    "   "    "     "    ,LCDOC

B0 VAR BYTE                       'Set 8-Bit Byte Variable,B0
B1 VAR BYTE                       ' "   "   "    "     "  ,B1
DutyCycle VAR BYTE                ' "   "   "    "     "  ,DutyCycle
PASSCODE VAR BYTE                 ' "   "   "    "     "  ,PASSCODE

TRISB=$FF                         'Make PortB an Input Port to Enable
                                  'In-Circuit-Programming
PAUSE 1000                        'Pause or Delay for 1 Seconds to Enable
                                  'In-Circuit-Programming
LCDOUT $FE,1                      'Clear the LCD
LCDOUT $FE, $80                   'Set the Cursor to Line 1, Position 0
LCDOUT "Power Up"                 'Send, Power Up, to the LCD

SEROUT2 PORTC.6,396,["Power Up",13,10,10] 'Send "Power Up" to the HyperTerminal
                                  'with a Carriage Return & two Line Feeds,
                                  'at 2400 Bits/Second, 8-Bits, 1-Stop Bit,
                                  'No Parity, Driven True (T2400).
PAUSE 2000                        'Pause or Delay for 2 Seconds
                                  'to Read "Power Up" in the LCD
LCDOUT $FE,1                      'Clear the LCD
TRISB=$00                         'Make PortB an Output Port to Allow for Normal Operation
TRISD=$F0                         'Make PortD, Bits-4,5,6,7 Inputs, & Bits-3
```

```
MAXVOLTRANGE=5                              'Set the Maximum ADC Voltage Range

SFACTOR=1024/MAXVOLTRANGE                   'Determine the Scale Factor

OUTVOLTDIVRATIO=(2000+10000)/2000           'The Output Voltage, Voltage Divider Ratio

LCDOC=10                                    'Initialize the LCD Output Counter
                                            'Send the Output Voltage Value to
                                            'the LCD and the HyperTerminal every
                                            '250 Samples
EEPROM:
PAUSE 10
I2CWRITE PORTB.7,PORTB.6,$A0,$00,[OUTVOLTDIVRATIO]
                                            'PORTB.7 is The Data, PORTB.6 is the Clock,
                                            'A0 is the Hex Address of the 24LC512 EEPROM
                                            ,00 is the starting memory HEX address for the write.
PAUSE 10
I2CREAD PORTB.7,PORTB.6,$A0,$00,[PASSCODE]
                                            'PORTB.7 is The Data, PORTB.6 is the Clock,
                                            'A0 is the Hex Address of the 24LC512 EEPROM
                                            ,00 is the starting memory HEX address for the write.
PAUSE 10
IF (PASSCODE=$06) THEN CONTINUE             'Check EEPROM PASSCODE = OUTVOLTDIVRATIO
LCDOUT $FE, 1                               'Clear the LCD
LCDOUT $FE, $80                             'Set Cursor to Line 1, Position 0
LCDOUT "WRONG"                              'Send, WRONG PASSCODE, to the LCD
LCDOUT $FE, $C0                             'Set Cursor to Line 2, Position 0
LCDOUT "PASSCODE"                           'Send, PASSCODE = 06, to the LCD
SEROUT2 PORTC.6,396,["WRONG PASSCODE",13,10,10]
                                            'Send, WRONG PASSCODE, to the HyperTerminal
GOTO EEPROM

CONTINUE:
LCDOUT $FE, 1                               'Clear the LCD
LCDOUT $FE, $80                             'Set Cursor to Line 1, Position 0
LCDOUT "CORRECT"                            'Send, CORRECT, to the LCD
LCDOUT $FE, $C0                             'Set Cursor to Line 2, Position 0
LCDOUT "PASSCODE = 06"                      'Send, PASSCODE = 06, to the LCD
SEROUT2 PORTC.6,396,["CORRECT, PASSCODE = 06",13,10,10]
                                            'Send, CORRECT, PASSCODE = 06,
                                            'to the HyperTerminal
PAUSE 2000
LCDOUT $FE, 1                               'Clear the LCD


KSCAN:
DutyCycle=0                                 'Initialize the DutyCycle to Zero
HPWM 1,DutyCycle,30000                      'Setup Capture/Compare/PWM Pin-1 to
                                            'PWM Mode, Set Frequency to 30KHz &
                                            'Set the DutyCycle Variable equal to
                                            'any value between 0 and 255
                                            'example: DutyCycle = 128
                                            '(128/256)*100% = 50%

LCDOUT $FE, 1                               'Clear the LCD
LCDOUT $FE, $80                             'Set Cursor to Line 1, Position 0
LCDOUT "The Reg. is Off"                    'Send, The Reg. is Off, to the LCD
LCDOUT $FE, $C0                             'Set Cursor to Line 2, Position 0
LCDOUT "Enter Voltage"                      'Send, Enter Voltage, to the LCD

SEROUT2 PORTC.6,396,["The Regulator is Off",13,10,10]
                                            'Send, The Regulator is Off, to the HyperTerminal
                                            'with a Carriage Return & 2 Line Feeds,
                                            'at 2400 Bits/Second, 8-Bits, 1-Stop Bit,
                                            'No Parity, Driven True (396).

HIGH PORTD.3                                'Make 3rd Row High to Scan for the Following:
PAUSE 500
INPUT PORTD.5
IF (PORTD.5=1) THEN L569                    'Check for Keypad Key "0" , Column-5
PAUSE 100
LOW PORTD.3

HIGH PORTD.0                                'Make 0th Row High to Scan for the Following:
PAUSE 500
INPUT PORTD.4
IF (PORTD.4=1) THEN L2                      'Check for Keypad Key "1" , Column-4
PAUSE 100
LOW PORTD.0
GOTO KSCAN
```

```
L569:
LCDOUT $FE, 1                           'Clear the LCD
LCDOUT $FE, $80                         'Set Cursor to Line 1, Position 0
LCDOUT "0"                              'Send 0 to the LCD
SERout2 PORTC.6,396,["0",13,10,10]      'Send 0 to the HyperTerminal with
                                        'a Carriage Return & 2 Line Feeds

HIGH PORTD.1                            'Make 1st Row High to Scan for the Following:
PAUSE 500
INPUT PORTD.5
IF (PORTD.5=1) THEN LB05                'Check for Keypad Key "5" , Column 5
PAUSE 100
LOW PORTD.1

HIGH PORTD.1                            'Make 1st Row High to Scan for the Following:
PAUSE 500
INPUT PORTD.6
IF (PORTD.6=1) THEN LB06                'Check for Keypad Key "6" , Column 6
PAUSE 100
LOW PORTD.1


HIGH PORTD.2                            'Make 2nd Row High to Scan for the Following:
PAUSE 500
INPUT PORTD.6
IF (PORTD.6=1) THEN LB09                'Check for Keypad Key "9", Column 6
PAUSE 100
LOW PORTD.2
GOTO L569

L2:
LCDOUT $FE, 1                           'Clear the LCD
LCDOUT $FE, $80                         'Set Cursor to Line 1, Position 0
LCDOUT "1"                              'Send 1 to the LCD
SERout2 PORTC.6,396,["1",13,10,10]      'Send 1 to the HyperTerminal with
                                        'a Carriage Return & 2 Line Feeds

HIGH PORTD.0                            'Make 0th Row High to Scan for the Following:
PAUSE 500
INPUT PORTD.5
IF (PORTD.5=1) THEN LB12                'Check for Keypad Key "2", Column 5
PAUSE 100
LOW PORTD.0
GOTO L2

LB05:
LCDOUT $FE, $80                         'Set Cursor to Line 1, Position 0
LCDOUT "05"                             'Send 05 to the LCD
SERout2 PORTC.6,396,["05",13,10,10]     'Send 05 to the HyperTerminal with
                                        'a Carriage Return & 2 Line Feeds

HIGH PORTD.3                            'Make 3rd Row High to Scan for the Following:
PAUSE 500
INPUT PORTD.6
IF (PORTD.6=1) THEN TV05                'Check for Keypad Key "#", Column 6
PAUSE 100
LOW PORTD.3
GOTO LB05

LB06:
LCDOUT $FE, $80                         'Set Cursor to Line 1, Position 0
LCDOUT "06"                             'Send 06 to the LCD
SERout2 PORTC.6,396,["06",13,10,10]     'Send 06 to the HyperTerminal with
                                        'a Carriage Return & 2 Line Feeds

HIGH PORTD.3                            'Make 3rd Row High to Scan for the Following:
PAUSE 500
INPUT PORTD.6
IF (PORTD.6=1) THEN TV06                'Check for Keypad Key "#", Column 6
PAUSE 100
LOW PORTD.3
GOTO LB06

LB09:
LCDOUT $FE, $80                         'Set Cursor to Line 1, Position 0
LCDOUT "09"                             'Send 09 to the LCD
SERout2 PORTC.6,396,["09",13,10,10]     'Send 09 to the HyperTerminal with
                                        'a Carriage Return & 2 Line Feeds
```

```
                HIGH PORTD.3                           'Make 3rd Row High to Scan for the Following:
                PAUSE 500
                INPUT PORTD.6
                IF (PORTD.6=1) THEN TV09               'Check for Keypad Key "#", Column 6
                PAUSE 100
                LOW PORTD.3
                GOTO LB09

                LB12:
                LCDOUT $FE, $80                        'Set Cursor to Line 1, Position 0
                LCDOUT "12"                            'Send 12 to the LCD
                SERout2 PORTC.6,396,["12",13,10,10]    'Send 12 to the HyperTerminal with
                                                       'a Carriage Return & 2 Line Feeds

                HIGH PORTD.3                           'Make 3rd Row High to Scan for the Following:
                PAUSE 500
                INPUT PORTD.6
                IF (PORTD.6=1) THEN TV12               'Check for Keypad Key "#", Column 6
                PAUSE 100
                LOW PORTD.3
                GOTO LB12

                TV05:
                LCDOUT $FE, $80                        'Set Cursor to Line 1, Position 0
                LCDOUT "05#"                           'Send 05# to the LCD
                SERout2 PORTC.6,396,["05#",13,10,10]   'Send 05# to the HyperTerminal with
                                                       'a Carriage Return & 2 Line Feeds

                TV=170                                 'Set the Output Voltage Target Value to 170
                PAUSE 500
                GOTO SLOWST

                TV06:
                LCDOUT $FE, $80                        'Set Cursor to Line 1, Position 0
                LCDOUT "06#"                           'Send 06# to the LCD
                SERout2 PORTC.6,396,["06#",13,10,10]   'Send #06 to the HyperTerminal with
                                                       'a Carriage Return & 2 Line Feeds

                TV=204                                 'Set the Output Voltage Target Value to 204
                PAUSE 500
                GOTO SLOWST

                TV09:
                LCDOUT $FE, $80                        'Set Cursor to Line 1, Position 0
                LCDOUT "09#"                           'Send 09# to the LCD
                SERout2 PORTC.6,396,["09#",13,10,10]   'Send #09 to the HyperTerminal with
                                                       'a Carriage Return & 2 Line Feeds

                TV=306                                 'Set the Output Voltage Target Value to 306
                PAUSE 500
                GOTO SLOWST

                TV12:
                LCDOUT $FE, $80                        'Set Cursor to Line 1, Position 0
                LCDOUT "12#"                           'Send 12# to the LCD
                SERout2 PORTC.6,396,["12#",13,10,10]   'Send 12# to the HyperTerminal with
                                                       'a Carriage Return & 2 Line Feeds

                TV=408                                 'Set the Output Voltage Target Value to 408
                PAUSE 500
                GOTO SLOWST

                SLOWST:
                LCDOUT $FE, 1                          'Clear the LCD
                LCDOUT $FE, $80                        'Set Cursor to Line 1, Position 0
                LCDOUT "Regulator is in"               'Send "Regulator is in" to the LCD
                LCDOUT $FE, $C0                        'Set Cursor to Line 2, Position 0
                LCDOUT "Slow-Start"                    'Send "Reg. is Off" to the LCD
                SEROUT2 PORTC.6,396,["The Regulator is in Slow-Start",13,10,10,10,10]
                                                       'Send "The Regulator is in
                                                       'Slow-Start" to the HyperTerminal with
                                                       'a Carriage Return & 2 Line Feeds,
                                                       'at 2400 Bits/Second, 8-Bits, 1-Stop Bit,
                                                       'No Parity, Driven True (396).
```

```
        DutyCycle=10                          'Initialize DutyCycle to: (25/256)*100%

    AGAIN:
        DutyCycle=DutyCycle+5                 ' Increment the DutyCycle by (13/256)*100%

        HPWM 1,DutyCycle,30000                'Setup Capture/Compare/PWM Pin-1 to
                                              'PWM Mode, Set Frequency to 30KHz &
                                              'Set the DutyCycle Variable equal to
                                              'any value between 0 and 255
                                              'example: DutyCycle = 128
                                              '(128/256)*100% = 50%

        PAUSE 100                             'Increase the DutyCycle every
                                              '100 milli-seconds by (13/256)*100%
        ADCIN 0,VS0
        IF VS0>=((TV/10)*9) THEN REG          'Set Slow-Start Output Voltage Value
                                              'to be 90% of the Regulated Output
        GOTO AGAIN                            'Voltage Target Value


    REG:
        ADCIN 0,VS0                           'Sample ADC Input Voltage-0, and place the
                                              '10-Bit result in Variable VS0

        VS0R=OUTVOLTDIVRATIO*VS0              'Convert VS0 to VS0R by Multiplying by the
                                              'Output Voltage Divider Ratio
        VS0WHOLE=VS0R/SFACTOR                 'Scale VS0R to Determine the WHOLE Portion of VS0R

        W3=VS0R*10                            'Expand VS0R to 16-Bits
        W2=W3/SFACTOR                         'Scale the 16-Bit Expanded Value of VS0R
        W1=VS0WHOLE*10                        'Expand the WHOLE Portion of VS0R to 16-Bits
        VS0DECIMAL=(W2-W1)                    'Determine the Decimal Portion of VS0R by
                                              'Subtracting the Expanded 16-Bit WHOLE Portion
                                              'of VS0R from the Scaled
                                              '16-Bit Expanded value of VS0R

        IF VS0>TV THEN DDUTY                  'Go to Decrement the DutyCycle
        IF VS0<TV THEN IDUTY                  'Go to Increment the DutyCycle
        GOTO DNCDC                            'Go to Do Not Change the DutyCycle

    DDUTY:
        DutyCycle=DutyCycle-1                 'Decrement the DutyCycle by
                                              '(1/256)*100% = 0.4%
        HPWM 1,DutyCycle,30000                'Setup Capture/Compare/PWM Pin-1 to
                                              'PWM Mode, Set Frequency to 30KHz &
                                              'Set the DutyCycle Variable equal to
                                              'any value between 0 and 255
                                              'example: DutyCycle = 128
                                              '(128/256)*100% = 50%
        GOTO DNCDC                            'Go to Do Not Change the DutyCycle

    IDUTY:
        DutyCycle=DutyCycle+1                 'Increment the DutyCycle by
                                              '(1/256)*100% = 0.4%
        HPWM 1,DutyCycle,30000                'Setup Capture/Compare/PWM Pin-1 to
                                              'PWM Mode, Set Frequency to 30KHz &
                                              'Set the DutyCycle Variable equal to
                                              'any value between 0 and 255
                                              'example: DutyCycle = 128
                                              '(128/256)*100% = 50%

    DNCDC:

        HIGH PORTD.3                          'Scan the Keypad for an Asterisk
        PAUSE 500
        INPUT PORTD.4
        if (PORTD.4=1) THEN FRERUN            'If *, Then GOTO FRERUN
        PAUSE 100
        LOW PORTD.3

        LCDOC=LCDOC-1                         'Decrement LCD Output Counter
        IF LCDOC=0 THEN LCDR                  'Send the Output Voltage Value to
                                              'the LCD and the HyperTerminal every
        GOTO REG                              'LCDOC Samples
```

```
LCDR:
    LCDOUT $FE, 1                                    'Clear the LCD
    LCDOUT $FE, $80                                  'Set Cursor to Line 1, Position 0
    LCDOUT "Vo = ",#VS0WHOLE,".",#VS0DECIMAL," Volts "
                                                     'Send "Vo = .  Volts" to the LCD
    LCDOUT $FE, $C0                                  'Set Cursor to Line 2, Position 0
    LCDOUT "VS0(BCD) = ",#vS0                        'Send "VS0(BCD) = " to the LCD

    SEROUT2 PORTC.6,396,["Vout =  ",#VS0WHOLE,".",#VS0DECIMAL," Volts ",13,10,10]
                                                     'Send "Vout = .  Volts" to the
                                                     'HyperTerminal with Carriage
                                                     'Return & Two Line Feeds

    SEROUT2 PORTC.6,396,["VS0(BCD) = ",#vS0,13,10,10,10,10]
                                                     'Send "VS0(BCD) = " to the
                                                     'HyperTerminal with Carriage
                                                     'Return & Two Line Feeds

    LCDOC=10                                         'Initialize the LCD Output Counter
                                                     'Send the Output Voltage Value to
                                                     'the LCD and the HyperTerminal every
                                                     'LCDOC Samples

    GOTO REG


FRERUN:
    HPWM 1,128,30000                                 'Setup Capture/Compare/PWM Pin-1 to
                                                     'PWM Mode, Set Frequency to 30KHz &
                                                     'Set the DutyCycle to:
                                                     '(128/256)*100% = 50%
    ADCIN 0,VS0                                      'Sample ADC Input Voltage-0, and place the
                                                     '10-Bit result in Variable VS0

    VS0R=OUTVOLTDIVRATIO*VS0                         'Convert VS0 to VS0R by Multiplying by the
                                                     'Output Voltage Divider Ratio
    VS0WHOLE=VS0R/SFACTOR                            'Scale VS0R to Determine the WHOLE Portion of VS0R

    W3=VS0R*10                                       'Expand VS0R to 16-Bits
    W2=W3/SFACTOR                                    'Scale the 16-Bit Expanded Value of VS0R
    W1=VS0WHOLE*10                                   'Expand the WHOLE Portion of VS0R to 16-Bits
    VS0DECIMAL=(W2-W1)                               'Determine the Decimal Portion of VS0R by
                                                     'Subtracting the Expanded 16-Bit WHOLE Portion

    HIGH PORTD.3                                     'Scan the Keypad for a Pound Sign
    PAUSE 500
    INPUT PORTD.6
    if (PORTD.6=1) THEN REG                          'If #, Then GOTO REG
    PAUSE 100
    LOW PORTD.3

    LCDOC=LCDOC-1                                    'Decrement LCD Output Counter
    IF LCDOC=0 THEN LCDF                             'Send the Output Voltage Value to
                                                     'the LCD and the HyperTerminal every
                                                     'LCDOC Samples
    GOTO FRERUN

LCDF:  LCDOUT $FE, 1                                 'Clear the LCD
    LCDOUT $FE, $80                                  'Set Cursor to Line 1, Position 0
    LCDOUT "Vo = ",#VS0WHOLE,".",#VS0DECIMAL," Volts "
                                                     'Send "Vo = .  Volts" to the LCD
    LCDOUT $FE, $C0                                  'Set Cursor to Line 2, Position 0
    LCDOUT "VS0(BCD) = ",#vS0                        'Send "VS0(BCD) = " to the LCD

    SEROUT2 PORTC.6,396,["Vout =  ",#VS0WHOLE,".",#VS0DECIMAL," Volts ",13,10,10]
                                                     'Send "Vout = .  Volts" to the
                                                     'HyperTerminal with Carriage
                                                     'Return & Two Line Feeds

    SEROUT2 PORTC.6,396,["VS0(BCD) = ",#vS0,13,10,10,10,10]
                                                     'Send "VS0(BCD) = " to the
                                                     'HyperTerminal with Carriage
                                                     'Return & Two Line Feeds

    LCDOC=10                                         'Initialize the LCD Output Counter
                                                     'Send the Output Voltage Value to
                                                     'the LCD and the HyperTerminal every
                                                     'LCDOC Samples

    GOTO FRERUN

    END
```

Laboratory Experiment #2

```
'Lab-2B_PICBasic-Pro_KeyPad_Digital_Brushless_Fan_Speed_Regulator
INCLUDE "modedefs.bas"           'Identifies the Command File

DEFINE OSC 20                    'Set PIC-Pro Oscillator Value in Mega Hertz
DEFINE ADC_BITS 10               'Set for 10-Bit Internal ADC

DEFINE ADC_CLOCK 5               'DEFINE ADC_CLOCK 1, for OSC=10, MAX ADC
                                 'Clock=10MHz/8) = 1.25MHz
                                 'DEFINE ADC_CLOCK 5, for OSC=16, MAX ADC
                                 'Clock=16MHz/16) =1.00MHz
                                 'DEFINE ADC_CLOCK 5, for OSC=20, MAX ADC
                                 'Clock=20MHz/16) =1.25MHz
                                 'DEFINE ADC_CLOCK 2, for OSC=24, MAX ADC
                                 'Clock=24MHz/32) =1.25MHz
                                 'DEFINE ADC_CLOCK 2, for OSC=32, MAX ADC
                                 'Clock=32MHz/32) =1.00MHz
                                 'DEFINE ADC_CLOCK 6, for OSC=48, MAX ADC
                                 'Clock=48MHz/64) = 0.75MHz

DEFINE CCP1_BIT 2                'Setup Hardware PWM Timer
DEFINE LCD_BITS 4                'Setup LCD for 4-Bit Mode
DEFINE LCD_DREG PORTB            'Drive LCD from Port-B
DEFINE LCD_DBIT 0                'Starting with Bit-0
DEFINE LCD_RSREG PORTA           'Drive LCD Register-Select from
DEFINE LCD_RSBIT 4               'Port-A, Bit-4
DEFINE LCD_EREG PORTA            'Drive LCD Enable from
DEFINE LCD_EBIT 5                'Port-A, Bit-5
DEFINE LCD_LINES 2               'LCD can hold 2-Lines of Characters
DEFINE LCD_COMMANDSUS 4000       'Delay between Commands
DEFINE LCD_DATAUS 200            'Delay between Data Transfers

ADCON0 = %00001111               'ADC Enabled, Analog Input 3 Enabled
ADCON1 = %00001011               'ADC Voltage Range is 0 to 5 Volts &
                                 'Analog Inputs 0 Thru 3 are available
ADCON2 = %10101101               'Right Justify to use Upper 8-Bits, only,
                                 '12*Tad, & ADC Clock=20MHz/16=1.25MHz

W0 VAR WORD                      'Set 16-Bit Word Variable,W0
W1 VAR WORD                      ' "     "      "      "   ,W1
W2 VAR WORD                      ' "     "      "      "   ,W2
W3 VAR WORD                      ' "     "      "      "   ,W3
VS0 VAR WORD                     ' "     "      "      "   ,VS0
VS0R VAR WORD                    ' "     "      "      "   ,VS0R
VS0WHOLE VAR WORD                ' "     "      "      "   ,VS0WHOLE
VS0DECIMAL VAR WORD              ' "     "      "      "   ,VS0DECIMAL
MAXVOLTRANGE VAR WORD            ' "     "      "      "   ,MAXVOLTRANGE
OUTVOLTDIVRATIO VAR WORD         ' "     "      "      "   ,OUTVOLTDIVRATIO
LCDOC VAR WORD                   ' "     "      "      "   ,LCDOC
TV VAR WORD                      ' "     "      "      "   ,TV
PPS VAR WORD                     ' "     "      "      "   ,Pulses per Second
FS VAR WORD                      ' "     "      "      "   ,Fan Speed, Revolutions per Minute

B0 VAR BYTE                      'Set 8-Bit Byte Variable,B0
B1 VAR BYTE                      ' "    "    "     "     ,B1
DutyCycle VAR BYTE               ' "    "    "     "     ,DutyCycle
PPR VAR BYTE                     ' "    "    "     "     ,Pulses per Revolution
PASSCODE VAR BYTE                ' "    "    "     "     ,PASSCODE

TRISB=$FF                        'Make PortB an Input Port to Enable
                                 'In-Circuit-Programming
PAUSE 1000                       'Pause or Delay for 1 Seconds to Enable
                                 'In-Circuit-Programming
LCDOUT $FE,1                     'Clear the LCD
LCDOUT $FE, $80                  'Set the Cursor to Line 1, Position 0
LCDOUT "Power Up"                'Send, Power Up, to the LCD

SEROUT2 PORTC.6,396,["Power Up",13,10,10]  'Send "Power Up" to the HyperTerminal
                                 'with a Carriage Return & two Line Feeds,
                                 'at 2400 Bits/Second, 8-Bits, 1-Stop Bit,
                                 'No Parity, Driven True (T2400).
PAUSE 2000                       'Pause or Delay for 2 Seconds
                                 'to Read "Power Up" in the LCD
LCDOUT $FE,1                     'Clear the LCD
TRISB=$00                        'Make PortB an Output Port to Allow for Normal Operation
TRISD=$F0                        'Make PortD, Bits-4,5,6,7 Inputs, & Bits-3
```

```
LCDOC=2                                 'Initialize the LCD Output Counter
                                        'Send the Output Voltage Value to
                                        'the LCD and the HyperTerminal every
                                        'LCDOC Samples

Poles=6                                 'Old Lab Board Fan, Poles=4
PPR=Poles/2    pulses per rev           'New Lab Board Fan, Poles=6

EEPROM:
PAUSE 10
I2CWRITE PORTB.7,PORTB.6,$A0,$00,[OUTVOLTDIVRATIO]
                                        'PORTB.7 is The Data, PORTB.6 is the Clock,
                                        'A0 is the Hex Address of the 24LC512 EEPROM
                                        ,00 is the starting memory HEX address for the write.
PAUSE 10
I2CREAD PORTB.7,PORTB.6,$A0,$00,[PASSCODE]
                                        'PORTB.7 is The Data, PORTB.6 is the Clock,
                                        'A0 is the Hex Address of the 24LC512 EEPROM
                                        ,00 is the starting memory HEX address for the write.
PAUSE 10
IF (PASSCODE=$06) THEN CONTINUE         'Check EEPROM PASSCODE = OUTVOLTDIVRATIO
LCDOUT $FE, 1                           'Clear the LCD
LCDOUT $FE, $80                         'Set Cursor to Line 1, Position 0
LCDOUT "WRONG"                          'Send, WRONG PASSCODE, to the LCD
LCDOUT $FE, $C0                         'Set Cursor to Line 2, Position 0
LCDOUT "PASSCODE"                       'Send, PASSCODE = 06, to the LCD
SEROUT2 PORTC.6,396,["WRONG PASSCODE",13,10,10]
                                        'Send, WRONG PASSCODE, to the HyperTerminal
GOTO EEPROM

CONTINUE:
LCDOUT $FE, 1                           'Clear the LCD
LCDOUT $FE, $80                         'Set Cursor to Line 1, Position 0
LCDOUT "CORRECT"                        'Send, CORRECT, to the LCD
LCDOUT $FE, $C0                         'Set Cursor to Line 2, Position 0
LCDOUT "PASSCODE = 06"                  'Send, PASSCODE = 06, to the LCD
SEROUT2 PORTC.6,396,["CORRECT, PASSCODE = 06",13,10,10]
                                        'Send, CORRECT, PASSCODE = 06,
                                        'to the HyperTerminal
PAUSE 2000
LCDOUT $FE, 1                           'Clear the LCD


KSCAN:
DutyCycle=0                             'Initialize the DutyCycle to Zero
HPWM 1,DutyCycle,30000                  'Setup Capture/Compare/PWM Pin-1 to
                                        'PWM Mode, Set Frequency to 30KHz &
                                        'Set the DutyCycle Variable equal to
                                        'any value between 0 and 255
                                        'Example: DutyCycle = 128
                                        '(128/256)*100% = 50%

LCDOUT $FE, 1                           'Clear the LCD
LCDOUT $FE, $80                         'Set Cursor to Line 1, Position 0
LCDOUT "Scanning"                       'Send, Scanning, to the LCD
LCDOUT $FE, $C0                         'Set Cursor to Line 2, Position 0
LCDOUT "The Fan is Off"                 'Send, The Fan is Off, to the LCD

SERout2 PORTC.6,396,["Scanning",13,10,10]
                                        'Send, Scanning, to the
                                        'HyperTerminal with a
                                        'Carriage Return & 2 Line Feeds,
                                        'at 2400 Bits/Second, 8-Bits, 1-Stop Bit,
                                        'No Parity, Driven True (396).

SERout2 PORTC.6,396,["The Fan is Off",13,10,10]
                                        'Send, The Fan is Off,
                                        'to the HyperTerminal with a
                                        'Carriage Return & 2 Line Feeds,
                                        'at 2400 Bits/Second, 8-Bits, 1-Stop Bit,
                                        'No Parity, Driven True (396).
```

```
L23:
HIGH PORTD.0                            'Make 0th Row High to Scan for the Following:
PAUSE 500
INPUT PORTD.5
IF PORTD.5=1 THEN L468                  'Check for Keypad Key "2", Column-5
PAUSE 100
INPUT PORTD.6
IF PORTD.6=1 THEN LZ2                   'Check for Keypad Key "3", Column-6
PAUSE 100
LOW PORTD.0

GOTO kscan

L468:                                                  2400, 2600, or 2800
LCDOUT $FE, 1                           'Clear the LCD
LCDOUT $FE, $80                         'Set Cursor to Line 1, Position 0
LCDOUT "2"                              'Send, 2, to the LCD

SERout2 PORTC.6,396,["2",13,10,10]      'Send, 2, to the HyperTerminal with
                                        'a Carriage Return & 2 Line Feeds,
                                        'at 2400 Bits/Second, 8-Bits, 1-Stop Bit,
                                        'No Parity, Driven True (396).

HIGH PORTD.1                            'Make 1st Row High to Scan for the Following:
PAUSE 500
INPUT PORTD.4
IF PORTD.4=1 THEN LB24                  'Check for Keypad Key "4", Column-4
PAUSE 100
INPUT PORTD.6
IF PORTD.6=1 THEN LB26                  'Check for Keypad Key "6", Column 6
PAUSE 100
LOW PORTD.1

HIGH PORTD.2                            'Make 2nd Row High to Scan for the Following:
PAUSE 500
INPUT PORTD.5
IF PORTD.5=1 THEN LB28                  'Check for Keypad Key "8", Column 5
PAUSE 100
LOW PORTD.2
GOTO L468

LZ2:
LCDOUT $FE, 1                           'Clear the LCD
LCDOUT $FE, $80                         'Set Cursor to Line 1, Position 0
LCDOUT "3"                              'Send, 3, to the LCD

SERout2 PORTC.6,396,["3",13,10,10]      'Send, 3, to the HyperTerminal with
                                        'a Carriage Return & 2 Line Feeds,
                                        'at 2400 Bits/Second, 8-Bits, 1-Stop Bit,
                                        'No Parity, Driven True (396).

HIGH PORTD.3                            'Make 3rd Row High to Scan for the Following:
PAUSE 500
INPUT PORTD.5
IF PORTD.5=1 THEN LB30                  'Check for Keypad Key "0", Column 5
PAUSE 100
LOW PORTD.3

HIGH PORTD.0                            'Make 0th Row High to Scan for the Following:
PAUSE 500
INPUT PORTD.5
IF PORTD.5=1 THEN LB32                  'Check for Keypad Key "2", Column-5
PAUSE 100
LOW PORTD.0
GOTO LZ2


LB24:
LCDOUT $FE, $80                         'Set Cursor to Line 1, Position 0
LCDOUT "24"                             'Send, 24, to the LCD
SERout2 PORTC.6,396,["24",13,10,10]     'Send, 24, to the HyperTerminal with
                                        'a Carriage Return & 2 Line Feeds

HIGH PORTD.3                            'Make 3rd Row High to Scan for the Following:
PAUSE 500
INPUT PORTD.6
IF PORTD.6=1 THEN TV24                  'Check for Keypad Key "#", Column 6
PAUSE 100
LOW PORTD.3
GOTO LB24
```

```
LB26:
    LCDOUT $FE, $80                         'Set Cursor to Line 1, Position 0
    LCDOUT "26"                             'Send, 26, to the LCD
    SERout2 PORTC.6,396,["26",13,10,10]     'Send, 26, to the HyperTerminal with
                                            'a Carriage Return & 2 Line Feeds

    HIGH PORTD.3                            'Make 3rd Row High to Scan for the Following:
    PAUSE 500
    INPUT PORTD.6
    IF PORTD.6=1 THEN TV26                  'Check for Keypad Key "#", Column 6
    PAUSE 100
    LOW PORTD.3
    GOTO LB26

LB28:
    LCDOUT $FE, $80                         'Set Cursor to Line 1, Position 0
    LCDOUT "28"                             'Send, 28, to the LCD
    SERout2 PORTC.6,396,["28",13,10,10]     'Send, 28, to the HyperTerminal with
                                            'a Carriage Return & 2 Line Feeds

    HIGH PORTD.3                            'Make 3rd Row High to Scan for the Following:
    PAUSE 500
    INPUT PORTD.6
    IF PORTD.6=1 THEN TV28                  'Check for Keypad Key "#", Column 6
    PAUSE 100
    LOW PORTD.3
    GOTO LB28

LB30:
    LCDOUT $FE, $80                         'Set Cursor to Line 1, Position 0
    LCDOUT "30"                             'Send, 30, to the LCD
    SERout2 PORTC.6,396,["30",13,10,10]     'Send, 30, to the HyperTerminal with
                                            'a Carriage Return & 2 Line Feeds

    HIGH PORTD.3                            'Make 3rd Row High to Scan for the Following:
    PAUSE 500
    INPUT PORTD.6
    IF PORTD.6=1 THEN TV30                  'Check for Keypad Key "#", Column 6
    PAUSE 100
    LOW PORTD.3
    GOTO LB30

LB32:
    LCDOUT $FE, $80                         'Set Cursor to Line 1, Position 0
    LCDOUT "32"                             'Send, 32, to the LCD
    SERout2 PORTC.6,396,["32",13,10,10]     'Send, 32, to the HyperTerminal with
                                            'a Carriage Return & 2 Line Feeds

    HIGH PORTD.3                            'Make 3rd Row High to Scan for the Following:
    PAUSE 500
    INPUT PORTD.6
    IF PORTD.6=1 THEN TV32                  'Check for Keypad Key "#", Column 6
    PAUSE 100
    LOW PORTD.3
    GOTO LB32


TV24:
    LCDOUT $FE, $80                         'Set Cursor to Line 1, Position 0
    LCDOUT "24#"                            'Send, 24#, to the LCD
    SERout2 PORTC.6,396,["24#",13,10,10]    'Send, 24#, to the HyperTerminal with
                                            'a Carriage Return & 2 Line Feeds

    TV=2400                                 'Set the Fan Speed Target Value in 2400 RPM
    PAUSE 500
    goto SLOWST

TV26:
    LCDOUT $FE, $80                         'Set Cursor to Line 1, Position 0
    LCDOUT "26#"                            'Send, 26#, to the LCD
    SERout2 PORTC.6,396,["26#",13,10,10]    'Send, 26#, to the HyperTerminal with
                                            'a Carriage Return & 2 Line Feeds
    TV=2600                                 'Set the Fan Speed Target Value in 2600 RPM
    PAUSE 500
    goto SLOWST
```

```
TV28:
    LCDOUT $FE, $80                         'Set Cursor to Line 1, Position 0
    LCDOUT "28#"                            'Send, 28#, to the LCD
    SERout2 PORTC.6,396,["28#",13,10,10]    'Send, 28#, to the HyperTerminal with
                                            'a Carriage Return & 2 Line Feeds

    TV=2800                                 'Set the Fan Speed Target Value in 2800 RPM
    PAUSE 500
    goto SLOWST

TV30:
    LCDOUT $FE, $80                         'Set Cursor to Line 1, Position 0
    LCDOUT "30#"                            'Send, 30#, to the LCD
    SERout2 PORTC.6,396,["30#",13,10,10]    'Send, 30#, to the HyperTerminal with
                                            'a Carriage Return & 2 Line Feeds

    TV=3000                                 'Set the Fan Speed Target Value in 3000 RPM
    PAUSE 500
    goto SLOWST

TV32:
    LCDOUT $FE, $80                         'Set Cursor to Line 1, Position 0
    LCDOUT "32#"                            'Send, 32#, to the LCD
    SERout2 PORTC.6,396,["32#",13,10,10]    'Send, 32#, to the HyperTerminal with
                                            'a Carriage Return & 2 Line Feeds

    TV=3200                                 'Set the Fan Speed Target Value in 3200 RPM
    PAUSE 500
    goto SLOWST

SLOWST:
    DutyCycle=10                            'Initialize DutyCycle to: (25/256)*100%    ≈ 10%

AGAIN:
    DutyCycle=DutyCycle+5                   'Increase DutyCycle to:
                                            '((Original Value + 5)/256)*100%

    HPWM 1,DutyCycle,30000                  'Setup Capture/Compare/PWM Pin-1 to
                                            'PWM Mode, Set Frequency to 30KHz &
                                            'Set the DutyCycle Variable equal to
                                            'any value between 0 and 255
                                            'example: DuyCycle = 128
                                            '(128/256)*100% = 50%

    PAUSE 100                               'Increase the DutyCycle every
                                            '100 milli-seconds by (25/256)*100%

    LCDOUT $FE,1                            'Clear the LCD
    LCDOUT $FE, $80                         'Set Cursor to Line 1, Position 0
    LCDOUT "The Fan is in"                  'Send, The Fan is in, to the LCD
    LCDOUT $FE, $C0                         'Set Cursor to Line 2, Position 0
    LCDOUT "Slow-Start"                     'Send, Slow-Start, to the LCD

    SEROUT2 PORTC.6,396,["The Fan is in Slow-Start",13,10,10]
                                            'Send, The Fan is in Slow-Start, to the HyperTerminal
                                            'with a Carriage Return & 2 Line Feeds,

    COUNT PORTC.1,1000,PPS                  'Count the Number of Pulses that
                                            'occur in one second & place the
                                            'result in the Variable PCT

    FS=(Pps/PPR)*60                         'Calculated Fan Speed in RPM

    IF FS=>((TV/10)*9) THEN REG             'Set Slow-Start RPM Value to be 90%
                                            'of the Regulated Target Value
    GOTO AGAIN

REG:
    COUNT PORTC.1,1000,PPS                  'Count the Number of Pulses that
                                            'occur in one second & place the
                                            'result in the Variable PCT

    FS=(Pps/PPR)*60                         'Calculated Fan Speed in RPM

    IF FS>TV THEN DDUTYC                    'If the Fan Speed is greater than the
                                            'Target Value, Decrement the DutyCycle
    IF FS<TV THEN IDUTYC                    'If the Fan Speed is less than the
                                            'Target Value, Increment the DutyCycle
    GOTO DNCDC
```

Handwritten annotations:
- "Duty cycle or freq - one can be variable, one must be fixed"
- Next to HPWM line: "freq 30 KHz", "hardware pwm", "CCP 1"
- "≥" next to IF FS=>((TV/10)*9) line

```
DDUTYC: DutyCycle=DutyCycle-3        'Decrement the DutyCycle by (8/256)*100%

       HPWM 1,DutyCycle,30000          'Setup Capture/Compare/PWM Pin-1 to
                                       'PWM Mode, Set Frequency to 30KHz &
                                       'Set the DutyCycle Variable equal to
                                       'any value between 0 and 255
                                       'example: DuyCycle = 128
                                       '(128/256)*100% = 50%

       GOTO DNCDC

IDUTYC: DutyCycle=DutyCycle+3          'Increment the DutyCycle by (8/256)*100%

       HPWM 1,DutyCycle,30000          'Setup Capture/Compare/PWM Pin-1 to
                                       'PWM Mode, Set Frequency to 30KHz &
                                       'Set the DutyCycle Variable equal to
                                       'any value between 0 and 255
                                       'example: DuyCycle = 128
                                       '(128/256)*100% = 50%

DNCDC:
       HIGH PORTD.3                    'Scan the Keypad for an Asterisk
       PAUSE 500
       INPUT PORTD.4
       if PORTD.4=1 THEN FRERUN        'If *, Then Goto FRERUN
       PAUSE 100
       LOW PORTD.3

       LCDOC=LCDOC-1                   'Decrement LCD Output Counter
       IF LCDOC=0 THEN LCDR            'Send the Fan Speed Value to
                                       'the LCD and the HyperTerminal every
                                       'LCDOC Samples

       GOTO REG

LCDR:
       LCDOUT $FE,1                    'Clear the LCD
       LCDOUT $FE, $80                 'Set Cursor to Line 1, Position 0
       LCDOUT "Pulses/Sec=",#PPS       'Send, Pulses/Sec, to the LCD
       LCDOUT $FE, $C0                 'Set Cursor to Line 2, Position 0
       LCDOUT "RPM=",#FS               'Send, RPM, to the LCD

       SEROUT2 PORTC.6,396,["Pulses/Second = ",#PPS,13,10,10]
                                       'Send, Pulses/Second, to the HyperTerminal
                                       'with a Carriage Return & 2 Line Feeds

       SEROUT2 PORTC.6,396,["RPM = ",#FS,13,10,10,10,10]
                                       'Send, RPM, to the HyperTerminal
                                       'with a Carriage Return & 4 Line Feeds

       LCDOC=2                         'Initialize the LCD Output Counter
                                       'Send the Output Voltage Value to
                                       'the LCD and the HyperTerminal every
                                       'LCDOC Samples

       GOTO REG


FRERUN:
       HPWM 1,192,30000                'Setup Capture/Compare/PWM Pin-1 to
                                       'PWM Mode, Set Frequency to 30KHz &
                                       'Set the DutyCycle to:
                                       '(192/256)*100% = 75%

       COUNT PORTC.1,1000,PPS          'Count the Number of Pulses that
                                       'occur in one second & place the
                                       'result in the Variable PCT

       FS=(Pps/PPR)*60                 'Calculated Fan Speed in RPM
       HIGH PORTD.3                    'Scan the Keypad for a Pound Sign
       PAUSE 500
       INPUT PORTD.6
       if PORTD.6=1 THEN REG           'If #, Then Goto REG
       PAUSE 100
       LOW PORTD.3

       LCDOC=LCDOC-1                   'Decrement LCD Output Counter
       IF LCDOC=0 THEN LCDF            'Send the Fan Speed Value to
                                       'the LCD and the HyperTerminal every
                                       'LCDOC Samples

       GOTO FRERUN
```

Handwritten annotations: "So you're not updating every time"; "* built in function"; "RC1"; "pulses per revolution"

```
LCDF:
    LCDOUT $FE,1                              'Clear the LCD
    LCDOUT $FE, $80                           'Set Cursor to Line 1, Position 0
    LCDOUT "Pulses/Sec=",#PPS                 'Send, Pulses/Sec, to the LCD
    LCDOUT $FE, $C0                           'Set Cursor to Line 2, Position 0
    LCDOUT "RPM=",#FS                         'Send, RPM, to the LCD

    SEROUT2 PORTC.6,396,["Pulses/Second = ",#PPS,13,10,10]
                                              'Send, Pulses/Second, to the HyperTerminal
                                              'with a Carriage Return & 2 Line Feeds

    SEROUT2 PORTC.6,396,["RPM = ",#FS,13,10,10,10,10]
                                              'Send, RPM, to the HyperTerminal
                                              'with a Carriage Return & 4 Line Feeds

    LCDOC=2                                   'Initialize the LCD Output Counter
                                              'Send the Output Voltage Value to
                                              'the LCD and the HyperTerminal every
                                              'LCDOC Samples
    GOTO FRERUN

    END
```

Laboratory Experiment #2

//Lab-2A_CCS_C-code_ Keypad_Digital_Voltage_Regulator

```c
#include <18F4685.h>                              //Identify Microcontroller
#device adc=10                                    //Identify ADC Bit Width
#fuses HS,NOWDT,NOPROTECT,NOLVP,MCLR              //Setup Programmer Oscillator Value
                                                  //and make Master Clear Pin an
                                                  //Enable Master Clear=MCLR
                                                  //Disable Master Clear=NOMCLR
#use delay(clock=20000000)                        //Setup C-Code Oscillator Value
#use fast_io(d)                                   //Leave the State of Port-D (all bits)
                                                  //the same until changed, again

#use rs232(baud=2400, xmit=PIN_C6, rcv=PIN_C7, stream=HT)
                                                  //Set RS232 HyperTerminal
                                                  //Communication Parameters
#use rs232(baud=2400, xmit=PIN_B4, rcv=PIN_B5, invert, stream=BB)
                                                  //Set RS232 Board-to-Board
                                                  //Communication Parameters
#use I2C(master, SCL=PIN_B6, SDA=PIN_B7)          //Set I2C Communication Parameters
#include <lcd_flex.c>                             //Identify LCD Driver File
#include <math.h>                                 //Include Math Functions

float Vout,SSV,TV,MaxVoltRange,SFactor,OutVoltDivRatio;
                                                  //16-Bit Floating Point Word Variable
unsigned long int VS0,DC;                         //16-Bit Word Variable
short D4,D5,D6;                                   //1-Bit Variable

void main()
     {
     set_tris_b(0xFF);                            //Make PortB an Input Port to
                                                  //Enable In-Circuit-Programming
     delay_ms(1000);                              //Pause or Delay for 1 second

     lcd_init();                                  //Initialize the LCD
     lcd_putc(0x0c);                              //Clear the LCD
     lcd_gotoxy(1,1);                             //Set the Cursor to Position-1 of Line-1
     printf(lcd_putc,"Power Up");                 //Send, Power Up, to the LCD

     fprintf(HT, "\n\rPower Up\n\r");             //Send, Power Up, to the HyperTerminal with
                                                  //a next Line & a Carriage Return
     delay_ms(1000);                              //Pause or Delay for 1 second

     setup_port_a(AN0_to_AN3);                    //Identify all of the Analog Inputs
     setup_adc(VSS_VDD);                          //Setup the ADC Voltage Range to be
                                                  //0 Volts to 5 Volts
                                                  //5v-0v)/1024~=.005 Volts Per Bit

     setup_adc(ADC_CLOCK_DIV_32);                 //for delay(clock=10000000),
                                                  //setup_adc(ADC_CLOCK_DIV_8),
                                                  //10MHz/8=1.25MHz
                                                  //for delay(clock=16000000),
                                                  //setup_adc(ADC_CLOCK_DIV_16),
                                                  //16MHz/16=1.00MHz
                                                  //for delay(clock=20000000),
                                                  //setup_adc(ADC_CLOCK_DIV_16),
                                                  //20MHz/16=1.25MHz
                                                  //for delay(clock=24000000),
                                                  //setup_adc(ADC_CLOCK_DIV_32),
                                                  //24MHz/32=1.25MHz
                                                  //for delay(clock=32000000),
                                                  //setup_adc(ADC_CLOCK_DIV_32),
                                                  //32MHz/32=1.00MHz
                                                  //for delay(clock=48000000),
                                                  //setup_adc(ADC_CLOCK_DIV_64),
                                                  //48MHz/64=0.75MHz

     set_adc_channel(0);                          //Identify the Analog Input to be
                                                  //Sampled (Channel Input Zero)

     setup_ccp1(CCP_PWM);                         //Setup Capture/Compare/PWM Pin-1 to PWM Mode

     setup_ccp2(CCP_CAPTURE_RE);                  //Setup Capture/Compare/PWM Pin-2 to
                                                  //Capture Mode

     setup_timer_1(T1_INTERNAL);                  //Setup Timer-1, which controls the
                                                  //Capture/Compare portion of
                                                  //Capture/Compare/PWM Pin-2, only,
                                                  //to Internal Mode

     setup_timer_2(T2_DIV_BY_4, 19, 1);           //Setup Timer-2, which controls the PWM portion //of
                                                  //Capture/Compare/PWM Pin-1, only.
                                                  //DIV_BY_Mode, Prescale (Period), Postscale
                                                  //The Following Limits are for the
                                                  //PIC18F4685 Microcontroller
                                                  //Mode may equal 1 or 4 or 16
                                                  //Prescale (Period) may equal any integer
                                                  //between 0 and 255
                                                  //Postscale may equal any integer between
                                                  //1 and 16 (Post = 1 to set PWM Frequency)
                                                  //T2Freq=OSC/[(4)*(Mode)*(Prescale+1)]=
                                                  //24MegaHz/[(4)*(4)*(PS+1)] = 75KHz
```

```c
enable_interrupts(INT_CCP2);            //Enable the Internal Interrupt Input
                                        //of Capture/Compare/PWM Pin-2

enable_interrupts(GLOBAL);              //Enable Global Interrupts for the
                                        //PIC18F4685 Microcontroller

set_tris_b(0x00);                       //Make PortB an Output Port to Allow for
                                        //Normal Operation
tris_d(0xF0);                           //Make PortD, Bits-4,5,6,7 Inputs

MaxVoltRange=5;                         //Set the Maximum ADC Voltage Range

SFactor=1024/MaxVoltRange;              //Determine the Scale Factor

OutVoltDivRatio=6;                      //The Output Voltage, Voltage Divider Ratio

Kscan:

DC=0;
set_pwm1_duty(DC);                      //Set the PWM DutyCycle to 0%, (Keep Fan Off)
lcd_putc(0x0c);                         //Clear the LCD
lcd_gotoxy(1,1);                        //Set the Cursor to Position-1 of Line-1
printf(lcd_putc,"Scanning");            //Send, Scanning, to the LCD
fprintf(HT, "\n\rScanning\n\r");        //Send, Scanning, to the HyperTerminal with
                                        //a next Line & a Carriage Return
lcd_gotoxy(1,2);                        //Set the Cursor to Position-1 of Line-2
printf(lcd_putc,"The Reg. is Off");     //Send, The Reg. is Off, to the LCD
fprintf(HT, "\n\rThe Regulator is Off\n\r"); //Send, The Regulator is Off, to the HyperTerminal
                                        //with a next Line & a Carriage Return

Output_high(PIN_D3);                    //Make 3rd Row High to Scan for the Following:
delay_ms(500);
D5 = input(PIN_D5);                     //Identify PortD, Bit-5 as an Input
if (D5) goto L569;                      //Check for Keypad Key "0", Column-5
delay_ms(100);
Output_low(PIN_D3);
Output_high(PIN_D0);                    //Make 0th Row High to Scan for the Following:
delay_ms(500);
D4 = input(PIN_D4);                     //Identify PortD, Bit-4 as an Input
if (D4) goto L2;                        //Check for Keypad Key "1", Column-4
delay_ms(100);
Output_low(PIN_D0);
goto Kscan;


L569:
lcd_putc(0x0c);                         //Clear the LCD
lcd_gotoxy(1,1);                        //Set the Cursor to Position-1 of Line-1
printf(lcd_putc,"0");                   //Send, 0, to the LCD
fprintf(HT, "\n\r0\n\r");               //Send, 0, to the HyperTerminal with
                                        //a next Line & a Carriage Return

Output_high(PIN_D1);                    //Make 1st Row High to Scan for the Following:
delay_ms(500);
D5 = input(PIN_D5);                     //Identify PortD, Bit-5 as an Input
if (D5) goto LB05;                      //Check for Keypad Key "5", Column 5
delay_ms(100);
Output_low(PIN_D1);
Output_high(PIN_D1);                    //Make 1st Row High to Scan for the Following:
delay_ms(500);
D6 = input(PIN_D6);                     //Identify PortD, Bit-6 as an Input
if (D6) goto LB06;                      //Check for Keypad Key "6", Column 6
delay_ms(100);
Output_low(PIN_D1);
Output_high(PIN_D2);                    //Make 2nd Row High to Scan for the Following:
delay_ms(500);
D6 = input(PIN_D6);                     //Identify PortD, Bit-6 as an Input
if (D6) goto LB09;                      //Check for Keypad Key "9", Column 6
delay_ms(100);
Output_low(PIN_D2);
goto L569;


L2:
lcd_putc(0x0c);                         //Clear the LCD
lcd_gotoxy(1,1);                        //Set the Cursor to Position-1 of Line-1
printf(lcd_putc,"1");                   //Send, 1, to the LCD
fprintf(HT, "\n\r1\n\r");               //Send, 1, to the HyperTerminal with
                                        //a next Line & a Carriage Return
Output_high(PIN_D0);                    //Make 0th Row High to Scan for the Following:
delay_ms(500);
D5 = input(PIN_D5);                     //Identify PortD, Bit-5 as an Input
if (D5) goto LB12;                      //Check for Keypad Key "2", Column 5
delay_ms(100);
Output_low(PIN_D0);
goto L2;
```

```c
LB05:
    lcd_putc(0x0c);                  //Clear the LCD
    lcd_gotoxy(1,1);                 //Set the Cursor to Position-1 of Line-1
    printf(lcd_putc,"05");           //Send, 05, to the LCD
    fprintf(HT, "\n\r05\n\r");       //Send, 05, to the HyperTerminal with
                                     //a next Line & a Carriage Return

    Output_high(PIN_D3);             //Make 3rd Row High to Scan for the Following:
    delay_ms(500);
    D6 = input(PIN_D6);              //Identify PortD, Bit-6 as an Input
    if (D6) goto TV05;               //Check for Keypad Key "#", Column 6
    delay_ms(100);
    Output_low(PIN_D3);
    goto LB05;

LB06:
    lcd_putc(0x0c);                  //Clear the LCD
    lcd_gotoxy(1,1);                 //Set the Cursor to Position-1 of Line-1
    printf(lcd_putc,"06");           //Send, 06, to the LCD
    fprintf(HT, "\n\r06\n\r");       //Send, 06, to the HyperTerminal with
                                     //a next Line & a Carriage Return
    Output_high(PIN_D3);             //Make 3rd Row High to Scan for the Following:
    delay_ms(500);
    D6 = input(PIN_D6);              ////Identify PortD, Bit-6 as an Input
    if (D6) goto TV06;               //Check for Keypad Key "#", Column 6
    delay_ms(100);
    Output_low(PIN_D3);
    goto LB06;

LB09:
    lcd_putc(0x0c);                  //Clear the LCD
    lcd_gotoxy(1,1);                 //Set the Cursor to Position-1 of Line-1
    printf(lcd_putc,"09");           //Send, 09, to the LCD
    fprintf(HT, "\n\r09\n\r");       //Send, 09, to the HyperTerminal with
                                     //a next Line & a Carriage Return
    Output_high(PIN_D3);             //Make 3rd Row High to Scan for the Following:
    delay_ms(500);
    D6 = input(PIN_D6);              //Identify PortD, Bit-6 as an Input
    if (D6) goto TV09;               //Check for Keypad Key "#", Column 6
    delay_ms(100);
    Output_low(PIN_D3);
    goto LB09;

LB12:
    lcd_putc(0x0c);                  //Clear the LCD
    lcd_gotoxy(1,1);                 //Set the Cursor to Position-1 of Line-1
    printf(lcd_putc,"12");           //Send, 12, to the LCD
    fprintf(HT, "\n\r12\n\r");       //Send, 12, to the HyperTerminal with
                                     //a next Line & a Carriage Return
    Output_high(PIN_D3);             //Make 3rd Row High to Scan for the Following:
    delay_ms(500);
    D6 = input(PIN_D6);              //Identify PortD, Bit-6 as an Input
    if (D6) goto TV12;               //Check for Keypad Key "#", Column 6
    delay_ms(100);
    Output_low(PIN_D3);
    goto LB12;


TV05:
    lcd_putc(0x0c);                  //Clear the LCD
    lcd_gotoxy(1,1);                 //Set the Cursor to Position-1 of Line-1
    printf(lcd_putc,"05#");          //Send, 05#, to the LCD
    fprintf(HT, "\n\r05#\n\r");      //Send, 05#, to the HyperTerminal with
                                     //a next Line & a Carriage Return

    TV=5;                            //Set the Output Voltage Target Value to 5
    delay_ms(500);
    goto SlowSt;

TV06:
    lcd_putc(0x0c);                  //Clear the LCD
    lcd_gotoxy(1,1);                 //Set the Cursor to Position-1 of Line-1
    printf(lcd_putc,"06#");          //Send, 06#, to the LCD
    fprintf(HT, "\n\r06#\n\r");      //Send, 06#, to the HyperTerminal with a next Line & a Carriage
Return

    TV=6;                            //Set the Output Voltage Target Value to 6
    delay_ms(500);
    goto SlowSt;

TV09:
    lcd_putc(0x0c);                  //Clear the LCD
    lcd_gotoxy(1,1);                 //Set the Cursor to Position-1 of Line-1
    printf(lcd_putc,"09#");          //Send, 09#, to the LCD
    fprintf(HT, "\n\r09#\n\r");      //Send, 09#, to the HyperTerminal with
                                     //a next Line & a Carriage Return

    TV=9;                            //Set the Output Voltage Target Value to 9
    delay_ms(500);
    goto SlowSt;
```

```
TV12:
    lcd_putc(0x0c);                              //Clear the LCD
    lcd_gotoxy(1,1);                             //Set the Cursor to Position-1 of Line-1
    printf(lcd_putc,"12#");                      //Send, 12#, to the LCD
    fprintf(HT, "\n\r12#\n\r");                  //Send, 12#, to the HyperTerminal with
                                                 //a next Line & a Carriage Return

    TV=12;                                       //Set the Output Voltage Target Value to 12
    delay_ms(500);
    goto SlowSt;

    SlowSt: lcd_putc(0x0c);                      //Clear the LCD
    lcd_gotoxy(1,1);                             //Set the Cursor to Position-1 of Line-1
    printf(lcd_putc,"The Reg. is in");           //Send The Fan is in, to the LCD
    lcd_gotoxy(1,2);                             //Set the Cursor to Position-1 of Line-2
    printf(lcd_putc,"Slow-Start");               //Send, Slow-Start, to the LCD
    fprintf(HT, "\n\rThe Regulator is in Slow-Start\n\r");  //Send, The Regulator is in Slow-Start, to the HyperTerminal

    DC=2;                                        //Initialize the PWM DutyCycle to 2%
    Again:
    DC=DC+2;                                     //Increase the DutyCycle by 2 Percent
    set_pwm1_duty(DC);                           //Set the PWM DutyCycle to the Specified Percentage
    delay_ms(2);                                 //Increase the PWM DutyCycle every 2 milli-second
    set_adc_channel(0);                          //Identify the Analog Input to be Sampled (Channel Input Zero)
    VS0 = read_adc();                            //Sample ADC Input Voltage-0, and place the
                                                 //10-Bit result in Variable VS0
    Vout=(VS0/SFactor)*OutVoltDivRatio;          //Convert the Sampled ADC Value VS0 to Volts
    SSV=(0.9*TV);                                //Set Slow-Start RPM Value to be 90% of the Regulated Target
    Value
    if (Vout>=SSV) goto Reg;                     //If the Output Voltage is greater than the Slow-Start Voltage Value,
                                                 //Regulate the Output Voltage
    goto Again;

    Reg:
    lcd_putc(0x0c);                              //Clear the LCD
    lcd_gotoxy(1,1);                             //Set the Cursor to Position-1 of Line-1
    printf(lcd_putc,"Vout=%lf", Vout);           //Send, Vout=, to the LCD in Long Decimal Integer Form
    fprintf(HT, "\n\rVout = %lf\n\r", Vout);     //Send, Vout =, to the HyperTerminal in Long Decimal Integer Form
                                                 //with a next Line & a Carriage Return
    lcd_gotoxy(1,2);                             //Set the Cursor to Position-1 of Line-
    printf(lcd_putc,"VS0(BCD)=%ld", VS0);        //Send, VCO(BCD)=, to the LCD in Long Decimal Integer Form
    fprintf(HT, "\n\rVS0 (BCD) = %ld\n\r", VS0); //Send, VCO(BCD) =, to the HyperTerminal in Long Decimal Integer
                                                 //Form with a next Line & a Carriage Return
    set_adc_channel(0);                          //Identify the Analog Input to be Sampled (Channel Input Zero)
    VS0 = read_adc();                            //Sample ADC Input Voltage-0, and place the
                                                 //10-Bit result in Variable VS0
    Vout=(VS0/SFactor)*OutVoltDivRatio;          //Convert the Sampled ADC Value VS0 to Volts
    if (Vout>TV) goto DDC;                       //If the Output Voltage is greater than the Target Value,
                                                 //Decrement the Ton
    if (Vout<TV) goto IDC;                       //If the Output Voltage is less than the Target Value,
                                                 //Increment the Ton
    goto DNCDC;                                  //Go to Do Not Change DutyCycle

    DDC:
    DC=DC-1;                                     //Decrease DC by 1%
    set_pwm1_duty(DC);                           //Set the PWM DutyCycle to the Specified Percentage
    goto DNCDC;                                  //Go to Do Not Change DutyCycle

    IDC:
    DC=DC+1;                                     //Increase DC by 1%
    set_pwm1_duty(DC);                           //Set the PWM DutyCycle to the Specified Percentage

    DNCDC:
    output_high(PIN_D3);                         //Scan the Keypad for an Asterisk
    delay_ms(500);
    D4 = input(PIN_D4);
    if (D4) goto Frerun;                         //*
    output_low(PIN_D3);
    goto Reg;

    Frerun:
    lcd_putc(0x0c);                              //Clear the LCD
    lcd_gotoxy(1,1);                             //Set the Cursor to Position-1 of Line-1
    printf(lcd_putc,"Vout=%lf", Vout);           //Send, Vout=, to the LCD in Long Decimal Integer Form
    fprintf(HT, "\n\rVout = %lf\n\r", Vout);     //Send, Vout =, to the HyperTerminal in Long Decimal Integer Form
                                                 //with a next Line & a Carriage Return
    lcd_gotoxy(1,2);                             //Set the Cursor to Position-1 of Line-2
    printf(lcd_putc,"VS0(BCD)=%ld", VS0);        //Send, VCO(BCD)=, to the LCD in Long Decimal Integer Form
    fprintf(HT, "\n\rVS0 (BCD) = %ld\n\r", VS0); //Send, VCO(BCD) =, to the HyperTerminal in Long Decimal Integer
                                                 //Form with a next Line & a Carriage Return
    DC=75;                                       //Set DC to 75
    set_pwm1_duty(DC);                           //Set the PWM DutyCycle to 75%
    set_adc_channel(0);                          //Identify the Analog Input to be Sampled (Channel Input Zero)
    VS0 = read_adc();                            //Sample ADC Input Voltage-0, and place the
                                                 //10-Bit result in Variable VS0
    Vout=(VS0/SFactor)*OutVoltDivRatio;          //Convert the Sampled ADC Value VS0 to Volts
    output_high(PIN_D3);                         //Scan the Keypad for a Pound-Sign
    delay_ms(500);
    D6 = input(PIN_D6);
    if (D6) goto Reg;                            //#
    output_low(PIN_D3);
    goto FreRun;
    }
```

Laboratory Experiment #2

//Lab-2B_CCS_C-code_ Keypad_Digital_Brushless_Fan_Speed_Regulator

```c
#include <18F4685.h>                          //Identify Microcontroller
#device adc=10                                //Identify ADC Bit Width
#fuses HS, NOWDT, NOPROTECT, NOLVP, MCLR      //Setup Programmer Oscillator Value
                                              //and make Master Clear Pin an
                                              //Enable Master Clear=MCLR
                                              //Disable Master Clear=NOMCLR
#use delay (clock=20000000)                   //Setup C-Code Oscillator Value
#use fast_io (d)                              //Leave the State of Port-D (all bits)
                                              //the same until changed, again

#use rs232 (baud=2400, xmit=PIN_C6, rcv=PIN_C7, stream=HT)
                                              //Set RS232 HyperTerminal
                                              //Communication Parameters
#use rs232 (baud=2400, xmit=PIN_B4, rcv=PIN_B5, invert, stream=BB)
                                              //Set RS232 Board-to-Board
                                              //Communication Parameters
#use I2C (master, SCL=PIN_B6, SDA=PIN_B7)     //Set I2C Communication Parameters
#include <lcd_flex.c>                         //Identify LCD Driver File
#include <math.h>                             //Include Math Functions

Long int Counter,DC,FS,SSV,TV;                //16-Bit Word Variables, Pulses per Second Counter,
                                              //DutyCycle, Fan Speed in RPM,
                                              //Slow-Start Value in RPM, Fan Speed,
                                              //Target Value in RPM

int PPR,Poles;                                //8-Bit Byte Variables, Pulses per Revolution,
                                              //Number of Motor Poles
short D4,D5,D6;                               //1-Bit Variable
#int_ccp2                                     //Enable the Interrupt of Capture/Compare/PWM
                                              //Pin-2 to react to Counter++
void PCount()
   {
   Counter++;                                 //Setup a Continuous Counter to Record the
   }                                          //Number of Pulse per Second Produced by the Fan
void main()
   {
   set_tris_b(0xFF);                          //Make PortB an Input Port to
                                              //Enable In-Circuit-Programming
   delay_ms(1000);                            //Pause or Delay for 1 second

   lcd_init();                                //Initialize the LCD
   lcd_putc(0x0c);                            //Clear the LCD
   lcd_gotoxy(1,1);                           //Set the Cursor to Position-1 of Line-1
   printf(lcd_putc,"Power Up");               //Send, Power Up, to the LCD

   fprintf(HT, "\n\rPower Up\n\r");           //Send, Power Up, to the HyperTerminal with
                                              //a next Line & a Carriage Return
   delay_ms(1000);                            //Pause or Delay for 1 second

   setup_port_a(AN0_to_AN3);                  //Identify all of the Analog Inputs
   setup_adc(VSS_VDD);                        //Setup the ADC Voltage Range to be
                                              //0 Volts to 5 Volts
                                              //5v-0v)/1024~=.005 Volts Per Bit

   setup_adc(ADC_CLOCK_DIV_32);               //for delay(clock=10000000),
                                              //setup_adc(ADC_CLOCK_DIV_8),
                                              //10MHz/8=1.25MHz
                                              //for delay(clock=16000000),
                                              //setup_adc(ADC_CLOCK_DIV_16),
                                              //16MHz/16=1.00MHz
                                              //for delay(clock=20000000),
                                              //setup_adc(ADC_CLOCK_DIV_16),
                                              //20MHz/16=1.25MHz
                                              //for delay(clock=24000000),
                                              //setup_adc(ADC_CLOCK_DIV_32),
                                              //24MHz/32=1.25MHz
                                              //for delay(clock=32000000),
                                              //setup_adc(ADC_CLOCK_DIV_32),
                                              //32MHz/32=1.00MHz
                                              //for delay(clock=48000000),
                                              //setup_adc(ADC_CLOCK_DIV_64),
                                              //48MHz/64=0.75MHz

   set_adc_channel(0);                        //Identify the Analog Input to be
                                              //Sampled (Channel Input Zero)

   setup_ccp1(CCP_PWM);                       //Setup Capture/Compare/PWM Pin-1 to PWM Mode

   setup_ccp2(CCP_CAPTURE_RE);                //Setup Capture/Compare/PWM Pin-2 to
                                              //Capture Mode

   setup_timer_1(T1_INTERNAL);                //Setup Timer-1, which controls the
                                              //Capture/Compare portion of
                                              //Capture/Compare/PWM Pin-2, only,
                                              //to Internal Mode
```

```
setup_timer_2(T2_DIV_BY_4, 19, 1);       //Setup Timer-2, which controls the PWM portion //of
                                         //Capture/Compare/PWM Pin-1, only.
                                         //DIV_BY_Mode, Prescale (Period), Postscale
                                         //The Following Limits are for the
                                         //PIC18F4685 Microcontroller
                                         //Mode may equal 1 or 4 or 16
                                         //Prescale (Period) may equal any integer
                                         //between 0 and 255
                                         //Postscale may equal any integer between
                                         //1 and 16 (Post = 1 to set PWM Frequency)
                                         //T2Freq=OSC/[(4)*(Mode)*(Prescale+1)]=
                                         //24MegaHz/[(4)*(4)*(PS+1)] = 75KHz
enable_interrupts(INT_CCP2);             //Enable the Internal Interrupt Input
                                         //of Capture/Compare/PWM Pin-2

enable_interrupts(GLOBAL);               //Enable Global Interrupts for the
                                         //PIC18F4685 Microcontroller

set_tris_b(0x00);                        //Make PortB an Output Port to Allow for
                                         //Normal Operation
tris_d(0xF0);                            //Make PortD, Bits-4,5,6,7 Inputs

Poles=4;
PPR=Poles/2;

kscan:

DC=0;
set_pwm1_duty(DC);                       //Set the PWM DutyCycle to 0%, (Keep Fan Off)

lcd_putc(0x0c);                          //Clear the LCD
lcd_gotoxy(1,1);                         //Set the Cursor to Position-1 of Line-1
printf(lcd_putc,"Scanning");             //Send, Scanning, to the LCD
fprintf(HT, "\n\rScanning\n\r");         //Send, Scanning, to the HyperTerminal with
                                         //a next Line & a Carriage Return
lcd_gotoxy(1,2);                         //Set the Cursor to Position-1 of Line-2
printf(lcd_putc,"The Fan is Off");       //Send, The Fan is Off, to the LCD
fprintf(HT, "\n\rThe Fan is Off\n\r");   //Send, The Fan is Off, to the HyperTerminal with
                                         //a next Line & a Carriage Return

Output_high(PIN_D0);                     //Make 0th Row High to Scan for the Following:
delay_ms(500);
D5 = input(PIN_D5);                      //Identify PortD, Bit-5 as an Input
if (D5) goto L468;                       //Check for Keypad Key "2" , Column-5
delay_ms(100);
Output_low(PIN_D0);
Output_high(PIN_D0);                     //Make 0th Row High to Scan for the Following:
delay_ms(500);
D6 = input(PIN_D6);                      //Identify PortD, Bit-4 as an Input
if (D6) goto LZ2;                        //Check for Keypad Key "3" , Column-6
delay_ms(100);
Output_low(PIN_D0);
goto Kscan;


L468:
lcd_putc(0x0c);                          //Clear the LCD
lcd_gotoxy(1,1);                         //Set the Cursor to Position-1 of Line-1
printf(lcd_putc,"2");                    //Send, 2, to the LCD
fprintf(HT, "\n\r2\n\r");                //Send, 2, to the HyperTerminal with
                                         //a next Line & a Carriage Return
Output_high(PIN_D1);                     //Make 1st Row High to Scan for the Following:
delay_ms(500);
D4 = input(PIN_D4);                      //Identify PortD, Bit-5 as an Input
if (D4) goto L24LB;                      //Check for Keypad Key "4" , Column 4
delay_ms(100);
Output_low(PIN_D1);
Output_high(PIN_D1);                     //Make 1st Row High to Scan for the Following:
delay_ms(500);
D6 = input(PIN_D6);                      //Identify PortD, Bit-6 as an Input
if (D6) goto L26LB;                      //Check for Keypad Key "6" , Column 6
delay_ms(100);
Output_low(PIN_D1);
Output_high(PIN_D2);                     //Make 2nd Row High to Scan for the Following:
delay_ms(500);
D5 = input(PIN_D5);                      //Identify PortD, Bit-6 as an Input
if (D5) goto L28LB;                      //Check for Keypad Key "8" , Column 5
delay_ms(100);
Output_low(PIN_D2);
goto L468;
```

```
LZ2:
lcd_putc(0x0c);                          //Clear the LCD
lcd_gotoxy(1,1);                         //Set the Cursor to Position-1 of Line-1
printf(lcd_putc,"3");                    //Send, 3, to the LCD
fprintf(HT, "\n\r3\n\r");                //Send, 3, to the HyperTerminal with
                                         //a next Line & a Carriage Return

Output_high(PIN_D3);                     //Make 3rd Row High to Scan for the Following:
delay_ms(500);
D5 = input(PIN_D5);                      //Identify PortD, Bit-5 as an Input
if (D5) goto L30LB;                      //Check for Keypad Key "0" , Column 5
delay_ms(100);
Output_low(PIN_D3);
Output_high(PIN_D0);                     //Make 0th Row High to Scan for the Following:
delay_ms(500);
D5 = input(PIN_D5);                      //Identify PortD, Bit-5 as an Input
if (D5) goto L32LB;                      //Check for Keypad Key "2" , Column 5
delay_ms(100);
Output_low(PIN_D0);
goto LZ2;

L24LB:
lcd_putc(0x0c);                          //Clear the LCD
lcd_gotoxy(1,1);                         //Set the Cursor to Position-1 of Line-1
printf(lcd_putc,"24");                   //Send, 24, to the LCD
fprintf(HT, "\n\r24\n\r");               //Send, 24, to the HyperTerminal with
                                         //a next Line & a Carriage Return

Output_high(PIN_D3);                     //Make 3rd Row High to Scan for the Following:
delay_ms(500);
D6 = input(PIN_D6);                      //Identify PortD, Bit-6 as an Input
if (D6) goto TV24;                       //Check for Keypad Key "#" , Column 6
delay_ms(100);
Output_low(PIN_D3);
goto L24LB;

L26LB:
lcd_putc(0x0c);                          //Clear the LCD
lcd_gotoxy(1,1);                         //Set the Cursor to Position-1 of Line-1
printf(lcd_putc,"26");                   //Send, 26, to the LCD
fprintf(HT, "\n\r26\n\r");               //Send, 26, to the HyperTerminal with
                                         //a next Line & a Carriage Return

Output_high(PIN_D3);                     //Make 3rd Row High to Scan for the Following:
delay_ms(500);
D6 = input(PIN_D6);                      //Identify PortD, Bit-6 as an Input
if (D6) goto TV26;                       //Check for Keypad Key "#" , Column 6
delay_ms(100);
Output_low(PIN_D3);
goto L26LB;

L28LB:
lcd_putc(0x0c);                          //Clear the LCD
lcd_gotoxy(1,1);                         //Set the Cursor to Position-1 of Line-1
printf(lcd_putc,"28");                   //Send, 28, to the LCD
fprintf(HT, "\n\r28\n\r");               //Send, 28, to the HyperTerminal with
                                         //a next Line & a Carriage Return

Output_high(PIN_D3);                     //Make 3rd Row High to Scan for the Following:
delay_ms(500);
D6 = input(PIN_D6);                      //Identify PortD, Bit-6 as an Input
if (D6) goto TV28;                       //Check for Keypad Key "#" , Column 6
delay_ms(100);
Output_low(PIN_D3);
goto L28LB;

L30LB:
lcd_putc(0x0c);                          //Clear the LCD
lcd_gotoxy(1,1);                         //Set the Cursor to Position-1 of Line-1
printf(lcd_putc,"30");                   //Send, 30, to the LCD
fprintf(HT, "\n\r30\n\r");               //Send, 30, to the HyperTerminal with
                                         //a next Line & a Carriage Return

Output_high(PIN_D3);                     //Make 3rd Row High to Scan for the Following:
delay_ms(500);
D6 = input(PIN_D6);                      //Identify PortD, Bit-6 as an Input
if (D6) goto TV30;                       //Check for Keypad Key "#" , Column 6
delay_ms(100);
Output_low(PIN_D3);
goto L30LB;

L32LB:
lcd_putc(0x0c);                          //Clear the LCD
lcd_gotoxy(1,1);                         //Set the Cursor to Position-1 of Line-1
printf(lcd_putc,"32");                   //Send, 32, to the LCD
fprintf(HT, "\n\r32\n\r");               //Send, 32, to the HyperTerminal with
                                         //a next Line & a Carriage Return
```

```
        Output_high(PIN_D3);          //Make 3rd Row High to Scan for the Following:
        delay_ms(500);
        D6 = input(PIN_D6);           //Identify PortD, Bit-6 as an Input
        if (D6) goto TV32;            //Check for Keypad Key "#" , Column 6
        delay_ms(100);
        Output_low(PIN_D3);
        goto L32LB;


TV24:
        lcd_putc(0x0c);               //Clear the LCD
        lcd_gotoxy(1,1);              //Set the Cursor to Position-1 of Line-1
        printf(lcd_putc,"24#");       //Send, 24#, to the LCD
        fprintf(HT, "\n\r24#\n\r");   //Send, 24#, to the HyperTerminal with
                                      //a next Line & a Carriage Return

        TV=2400;                      //Set the Output Voltage Target Value to 5
        delay_ms(500);
        goto SlowSt;

TV26:
        lcd_putc(0x0c);               //Clear the LCD
        lcd_gotoxy(1,1);              //Set the Cursor to Position-1 of Line-1
        printf(lcd_putc,"26#");       //Send, 26#, to the LCD
        fprintf(HT, "\n\r26#\n\r");   //Send, 26#, to the HyperTerminal with
                                      //a next Line & a Carriage Return

        TV=2600;                      //Set the Output Voltage Target Value to 6
        delay_ms(500);
        goto SlowSt;

TV28:
        lcd_putc(0x0c);               //Clear the LCD
        lcd_gotoxy(1,1);              //Set the Cursor to Position-1 of Line-1
        printf(lcd_putc,"28#");       //Send, 28#, to the LCD
        fprintf(HT, "\n\r28#\n\r");   //Send, 28#, to the HyperTerminal with
                                      //a next Line & a Carriage Return

        TV=2800;                      //Set the Output Voltage Target Value to 9
        delay_ms(500);
        goto SlowSt;

TV30:   l
        lcd_putc(0x0c);               //Clear the LCD
        lcd_gotoxy(1,1);              //Set the Cursor to Position-1 of Line-1
        printf(lcd_putc,"30#");       //Send, 30#, to the LCD
        fprintf(HT, "\n\r30#\n\r");   //Send, 30#, to the HyperTerminal with
                                      //a next Line & a Carriage Return

        TV=3000;                      //Set the Output Voltage Target Value to 12
        delay_ms(500);
        goto SlowSt;

TV32:
        lcd_putc(0x0c);               //Clear the LCD
        lcd_gotoxy(1,1);              //Set the Cursor to Position-1 of Line-1
        printf(lcd_putc,"32#");       //Send, 32#, to the LCD
        fprintf(HT, "\n\r32#\n\r");   //Send, 32#, to the HyperTerminal with
                                      //a next Line & a Carriage Return

        TV=3200;                      //Set the Output Voltage Target Value to 12
        delay_ms(500);
        goto SlowSt;


Slowst:
        lcd_putc(0x0c);                           //Clear the LCD
        lcd_gotoxy(1,1);                          //Set Cursor to the Position-1 of Line-1
        printf(lcd_putc,"The Fan is in");         //Send The Fan is in, to the LCD
        lcd_gotoxy(1,2);                          //Set Cursor to the Position-1 of Line-2
        printf(lcd_putc,"Slow-Start");            //Send Slow-Start, to the LCD
        fprintf(HT, "\n\rThe Fan is in Slow-Start\n\r");  //Send The Fan is in Slow-Start, to the HyperTerminal
                                                  //with a next Line & a Carriage Return

        DC=20;                        //Initialize the PWM DutyCycle to 20%
Again:  DC=DC+2;                      //Increase the DutyCycle by 2 Percent
        set_pwm1_duty(DC);            //Set the PWM DutyCycle to the Specified Percentage
        delay_ms(2);                  //Increase the PWM DutyCycle every 20 milli-second
        Counter=0;                    //Reset the Counter to Zero
        delay_ms(1000);               //Create a one Second Window for counting Fan Speed Pulses
        FS=(Counter/PPR)*60;          //Calculated Fan Speed in RPM
                                      //(Counter Pulses per Second)*
                                      //(60 Seconds per Minute)/(2 Pulses per Revolution)
        SV=(0.9*TV);                  //Set Slow-Start RPM Value to be 90% of
                                      //the Regulated Target Value
        if (FS>=SSV) goto Reg;        //If the Fan Speed is greater than the Slow-Start RPM Value,
                                      //Regulate the Fan Speed
        goto Again;
```

```c
Reg:
    while(TRUE)
        {
        Counter=0;                                          //Reset the Counter to Zero
        delay_ms(1000);                                     //Create a one Second Window for counting Fan Speed Pulses
        FS=(Counter/PPR)*60;                                //Calculated Fan Speed in RPM
                                                            //(Counter Pulses per Second)*
                                                            //(60 Seconds per Minute)/(2 Pulses per Revolution)
        lcd_putc(0x0c);                                     //Clear the LCD
        lcd_gotoxy(1,1);                                    //Set Cursor to the Position-1 of Line-1
        printf(lcd_putc,"Pulses/Sec=%ld", Counter);         //Send Pulses/Sec, to the LCD
        fprintf(HT, "\n\rPulses/Second = %ld\n\r", Counter);//Send Pulses/Second, to the HyperTerminal
                                                            //in Long Decimal Integer Form
                                                            //with a next Line & a Carriage Return
        lcd_gotoxy(1,2);                                    //Set Cursor to the Position-1 of Line-2
        printf(lcd_putc,"RPM=%ld", FS);                     //Send RPM, to the LCD in Long Decimal Integer Form
        fprintf(HT, "\n\rRPM = %ld\n\r", FS);               //Send RPM, to the HyperTerminal in Long Decimal Integer Form
                                                            //with a next Line & a Carriage Return
        Counter=0;                                          //Reset the Counter to Zero
        if (FS>TV) goto DDC;                                //If the Fan Speed is greater than the Target Value,
                                                            //Decrement the Ton
        if (FS<TV) goto IDC;                                //If the Fan Speed is less than the Target Value,
                                                            //Increment the Ton
        goto DNCDC;                                         //Go to Do Not Change DutyCycle

        DDC:     DC=DC-1;                                   //Decrease DC by 1%
        set_pwm1_duty(DC);                                  //Set the PWM DutyCycle to the Specified Percentage
        goto DNCDC;                                         //Go to Do Not Change DutyCycle

        IDC:     DC=DC+1;                                   //Increase DC by 1%
        set_pwm1_duty(DC);                                  //Set the PWM DutyCycle to the Specified Percentage

        DNCDC:
        output_high(PIN_D3);                                //Scan the Keypad for an Asterisk
        delay_ms(500);
        D4 = input(PIN_D4);
        if (D4) goto Frerun;                                //*
        delay_ms(100);
        output_low(PIN_D3);
        }
Frerun:
    while(TRUE)
        {
        Counter=0;                                          //Reset the Counter to Zero
        delay_ms(1000);
        FS=(Counter/PPR)*60;                                //Calculated Fan Speed in RPM
                                                            //(Counter Pulses per Second)*
                                                            //(60 Seconds per Minute)/(2 Pulses per Revolution)
        lcd_putc(0x0c);                                     //Clear the LCD
        lcd_gotoxy(1,1);                                    //Set Cursor to the Position-1 of Line-1
        printf(lcd_putc,"Pulses/Sec=%ld", Counter);         //Send Pulses/Sec, to the LCD in Long Decimal Integer Form
        fprintf(HT, "\n\rPulses/Sec = %ld\n\r", Counter);   //Send Pulses/Second, to the HyperTerminal in Long Decimal
                                                            //Integer Form with a next Line & a Carriage Return
        lcd_gotoxy(1,2);                                    //Set Cursor to the Position-1 of Line-2
        printf(lcd_putc,"RPM=%ld", FS);                     //Send RPM, to the LCD in Long Decimal Integer Form
        fprintf(HT, "\n\rRPM = %ld\n\r", FS);               //Send RPM, to the HyperTerminal in Long Decimal Integer Form
                                                            //with a next Line & a Carriage Return
        Counter=0;                                          //Reset the Counter to Zero

        set_pwm1_duty(75);                                  //Set the PWM DutyCycle to 75%
        output_high(PIN_D3);                                //Scan the Keypad for a Pound-Sign
        delay_ms(500);
        D6 = input(PIN_D6);
        if (D6) goto Reg;                                   //#
        delay_ms(100);
        output_low(PIN_D3);
        }
}
```

Engineering Practices for the PIC Microcontroller

By: Prof. Sal R. Riggio Jr., PhD, PE

Chapter - 5

Laboratory Experiment - 3

The Digital 4-Phase Stepper-Motor-Controller

Introduction

The purpose of this lab experiment is for the student to develop basic skills in making practical use of the Microchip **PIC18F4685** Microcontroller and become familiar with the importance, operation and hardware implementation of a Digital Stepper Motor Control Loop.

The student is expected to analyze and write **PICBasic-Pro Code** and **CCS C-code**, which enables a **4-Phase Stepper-Motor** to **rotate** as desired, thru the use of the **Keypad** input device.

It is suggested that the student **read** the **PICBasic-Pro code** and **CCS C-code**, particularly the **comments** written on each line, to learn how to write code for the PIC microcontrollers. It has been proven many times to be the most effective way to learn how generate effective code for the PIC microcontrollers. This code is shown at the end of this chapter

Portions of the Experiment Board Schematics, which are relevant to this experiment, are shown below in **Figures 5.1 & 5.2.** Please refer to **Appendix-A** for the complete set of **Experiment Board Schematics** and please refer to **Appendix-B** for the complete **Lab Board Schematic**.

Hardware Setup

1.) Connect the Square end of the USB Type-A to Type-B Cable to connector JUSB1 of the Experiment Board. Then, connect the other end of this Cable to one of the USB Inputs of the Computer.
2.) Connect the DB9 end of the DB9 Serial-to-USB Cable to connector P1 of the Experiment Board. Then, connect the USB end of this Cable to one of the other USB Inputs of the Computer. The DB9 Serial-to-USB Cable is used to communicate between the Serial HyperTerminal of the Computer and the Experiment Board.
3.) The Jumper of **JB12** is to be placed in **Positions 3-5 & 4-6.**
4.) Connect the Infrared Photo-Transistor of the Lab Board as Follows:
Connect a Wire between Pin-1 (Collector) of J1, of the IR-PT Board, and Pin-1 (FSP) of J8, of the Experiment Board. The IR-PT Board is mounted on the Lab Board. Also, Connect a Wire between Pin-2 (Emitter) of J1, of the IR-PT Board, and Pin-2 (GND) of J8, of the Experiment Board. Connect one end of a BNC to Grabber Lead Cable (Red Lead) to Pin-1 (FSP), of J8, of the Experiment Board, and the other end to Channel-1 of the Oscilloscope. Be sure to ground the Black Grabber Lead.
4.) Connect the Stepper Motor of the Lab Board as Follows:
Connect a Wire between Pin-1 (Phase-1) of J5, of the Lab Board, and Pin-2 (P1) of J4, of the Experiment Board.
Connect a Wire between Pin-2 (Phase-2) of J5, of the Lab Board, and Pin-1 (P2) of J4, of the Experiment Board.
Connect a Wire between Pin-1 (Phase-3) of J6, of the Lab Board, and Pin-2 (P3) of J3, of the Experiment Board.
Connect a Wire between Pin-2 (Phase-4) of J6, of the Lab Board, and Pin-1 (P4) of J3, of the Experiment Board.
Connect a Wire between Pin-2 (GND) of J1, of the Lab Board, and Pin-2 (GND) of J6, of the Experiment Board.

Assignment #1

Original Code:

Analyze the following;

After the **"C" Button is Pressed** on the Keypad, a **Very Slow Continuous Pulse Pattern** will appear on the Oscilloscope. This pattern can be seen by setting the **Horizontal Time-Base** to **2 Seconds per Division** and the **Vertical Scale** to **1 Volt per Division**. There are **Two Waveforms**, which are to be **Captured** by using the Oscilloscope Software Interface. Be sure that **Channel-A** of the Oscilloscope is set at **DC Coupling**.

The **First Waveform** is the **Time Measurement** between any **Two Pulses**, and the **Second Waveform** is the **Width Measurement** of any **One Pulse,** at the **50% Voltage Point**.

New Code:

Write a PICBasic-Pro Code routine and a **CCS C-code** routine, which will allow the **4-Phase Stepper Motor** to **Rotate 90 Degrees** from **North-to-East** and from **North-to-West**, on **Demand**, after the **Rotor** has been **Set** in the **North Position,** thru the use of the **Infrared Photo-Transistor Circuit**. Use **Button "1"** of the Keypad to cause the **Rotor** to **seek** and then **Stop** in the **North Position,** as a **Preset Control**, before moving the Rotor from North-to-East or from North-to-West. This Preset Control code is the same as the code, which is invoked when **"B" Button is Pressed** within the Original Code.

The Rotor must be Preset to the North Position before the following action can take place!

When the **"A" Button is Pressed** on the Keypad, the **Rotor must rotate North-to-East** and then **Stop** in the **East Position**. When the **"B" Button is Pressed** on the Keypad, the **Rotor must rotate East-to-North** and then **stop** in the **North Position,** once again.

When the **"C" Button is Pressed** on the Keypad, the **Rotor must rotate North-to-West,** and then **Stop** in the **West Position**. When the **"D" Button is Pressed** on the Keypad, the **Rotor must rotate West-to-North,** and then **stop** in the **North Position,** once again.

Figure 5.1 shows a Unipolar Stepper Motor Driver circuit which is made up of four N-Channel Power MOSFET transistors connected in the common-source inverting configuration. Each of these power MOSFET transistors is connected to one of the coils of a four phase (coil) stepper motor. When a positive pulse of voltage is delivered from Port-B, Bits-4, 5, 6 or 7, of the PIC18F4550 microcontroller, the corresponding MOSFET is turned-on into its triode region of operation. This action causes current to flow in the MOSFET and the corresponding coil of the stepper motor. This magnetic field of this energized coil causes the stepper motor to rotate some portion of 360 degrees.

Figure 5.2 shows an N-Channel MOSFET transistor connected in the common-source inverting configuration for the purpose of level-shifting an input signal of between -24 volts and +24volts to microcontroller compatible signal of between 0 volts and +5 volts. The Zener diode (DZ2) protects the gate-to-source voltage of the N-Channel MOSFET (M1) from voltages greater than 12 volts. This MOSFET has a maximum gate-to-source voltage of +/- 20 volts.

When the Collector and Emitter Output Leads of a Photo-Transistor/Light-Emitting-Diode Optical Sensor circuit are connected to input connector (J8), as shown in Figure 5.2, as pulse detection circuit is created. Each time the pointer of the stepper motor crosses the optical boundary, a negative going pulse appears a Port-C, Bit-1 of the PIC18F4685.

If the optical boundary is placed at the north pole of a compass, a designer can write code, which will indicate the current position of the stepper motor. There are many industrial applications, such as, packaging and bottle labeling process equipment, which make use of this basic mechanism.

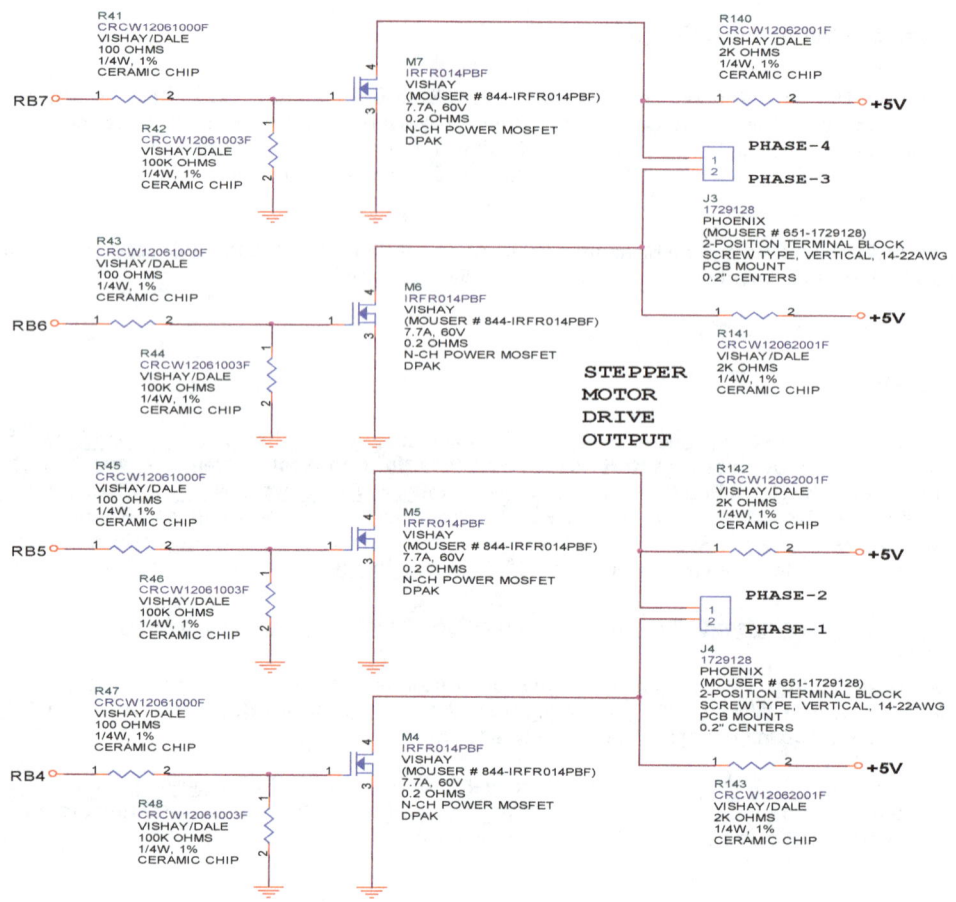

Figure 5.1: Stepper Motor Driver Circuit

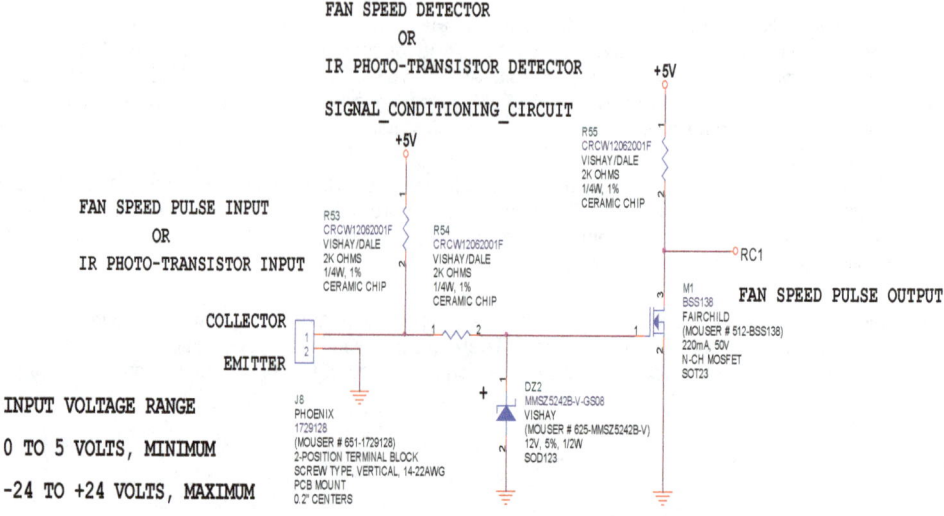

Figure 5.2: Stepper Motor Position Detector Interface Circuit

Laboratory Experiment #3

```
'Lab-3_ PICBasic-Pro_Digital_4-Phase_Stepper_Motor_Controller
INCLUDE "modedefs.bas"              'Identifies the Command File

DEFINE OSC 20                       'Set PIC-Pro Oscillator Value in Mega Hertz
DEFINE ADC_BITS 10                  'Set for 10-Bit Internal ADC

DEFINE ADC_CLOCK 5                  'DEFINE ADC_CLOCK 1, for OSC=10, MAX ADC
                                    'Clock=10MHz/8) = 1.25MHz
                                    'DEFINE ADC_CLOCK 5, for OSC=16, MAX ADC
                                    'Clock=16MHz/16) = 1.00MHz
                                    'DEFINE ADC_CLOCK 5, for OSC=20, MAX ADC
                                    'Clock=20MHz/16) = 1.25MHz
                                    'DEFINE ADC_CLOCK 2, for OSC=24, MAX ADC
                                    'Clock=24MHz/32) = 1.25MHz
                                    'DEFINE ADC_CLOCK 2, for OSC=32, MAX ADC
                                    'Clock=32MHz/32) = 1.00MHz
                                    'DEFINE ADC_CLOCK 6, for OSC=48, MAX ADC
                                    'Clock=48MHz/64) = 0.75MHz

DEFINE CCP1_BIT 2                   'Setup Hardware PWM Timer
DEFINE LCD_BITS 4                   'Setup LCD for 4-Bit Mode
DEFINE LCD_DREG PORTB               'Drive LCD from Port-B
DEFINE LCD_DBIT 0                   'Starting with Bit-0
DEFINE LCD_RSREG PORTA              'Drive LCD Register-Select from
DEFINE LCD_RSBIT 4                  'Port-A, Bit-4
DEFINE LCD_EREG PORTA               'Drive LCD Enable from
DEFINE LCD_EBIT 5                   'Port-A, Bit-5
DEFINE LCD_LINES 2                  'LCD can hold 2-Lines of Characters
DEFINE LCD_COMMANDSUS 4000          'Delay between Commands
DEFINE LCD_DATAUS 200               'Delay between Data Transfers

ADCON0 = %00001111                  'ADC Enabled, Analog Input 3 Enabled
ADCON1 = %00001011                  'ADC Voltage Range is 0 to 5 Volts &
                                    'Analog Inputs 0 Thru 3 are available
ADCON2 = %10101101                  'Right Justify to use Upper 8-Bits, only,
                                    '12*Tad, & ADC Clock=20MHz/16=1.25MHz

W0 VAR WORD                         'Set 16-Bit Word Variable,W0
W1 VAR WORD                         ' "    "      "       "    ,W1
W2 VAR WORD                         ' "    "      "       "    ,W2
W3 VAR WORD                         ' "    "      "       "    ,W3
H  VAR WORD                         ' "    "      "       "    ,H
J  VAR WORD                         ' "    "      "       "    ,J
K  VAR WORD                         ' "    "      "       "    ,K

B0 VAR BYTE                         'Set 8-Bit Byte Variable,B0
B1 VAR BYTE                         ' "    "      "       "  ,B1

TRISB=$FF                           'Make PortB an Input Port to Enable
                                    'In-Circuit-Programming
PAUSE 1000                          'Pause or Delay for 1 Seconds to Enable
                                    'In-Circuit-Programming
LCDOUT $FE,1                        'Clear the LCD
LCDOUT $FE, $80                     'Set the Cursor to Line 1, Position 0
LCDOUT "Power Up"                   'Send, Power Up, to the LCD

SERout PORTC.6,T2400,["Power Up",13,10,10]
                                    'Send "Power Up" to the HyperTerminal
                                    'with a Carriage Return & two Line Feeds,
                                    'at 2400 Bits/Second, 8-Bits, 1-Stop Bit,
                                    'No Parity, Driven True (T2400).

PAUSE 2000                          'Pause or Delay for 2 Seconds
                                    ''to Read "Power Up" in the LCD
LCDOUT $FE,1                        'Clear the LCD
TRISB=$00                           'Make PortB an Output Port to Allow for Normal Operation
TRISD=$F0                           'Make PortD, Bits-4,5,6,7 Inputs, & Bits-3
```

```
KSCAN:

        LCDOUT $FE, 1                   'Clear the LCD
        LCDOUT $FE, $80                 'Set Cursor to Line 1, Position 0
        LCDOUT "Scanning"               'Send "Scanning" to the LCD
        SERout2 PORTC.6,396,["Scanning",13,10,10]
                                        'Send "Scanning" to the HyperTerminal
                                        'with a Carriage Return & 2 Line Feeds

        HIGH PORTD.0                    'Make 0th Row High to Scan for the Following:
        PAUSE 500
        INPUT PORTD.7
        IF (PORTD.7=1) THEN ALTR        'Check for Keypad Key "A" , Column-7
        PAUSE 100
        LOW PORTD.0

        HIGH PORTD.1                    'Make 1st Row High to Scan for the Following:
        PAUSE 500
        INPUT PORTD.7
        IF (PORTD.7=1) THEN PSENSE      'Check for Keypad Key "B" , Column-7
        PAUSE 100
        LOW PORTD.1

        HIGH PORTD.2                    'Make 2nd Row High to Scan for the Following:
        PAUSE 500
        INPUT PORTD.7
        IF (PORTD.7=1) THEN CLOOP       'Check for Keypad Key "C" , Column-7
        PAUSE 100
        LOW PORTD.2

        HIGH PORTD.3                    'Make 3rd Row High to Scan for the Following:
        PAUSE 500
        INPUT PORTD.7
        IF (PORTD.7=1) THEN CCLOOP      'Check for Keypad Key "D" , Column-7
        PAUSE 100
        LOW PORTD.3

        GOTO KSCAN

        CLOOP:  LOW PORTE.2             'Turn-Off the Red LED
        LCDOUT $FE,1                    'Clear the LCD
        LCDOUT $FE, $80                 'Set Cursor to Line 1, Position 0
        LCDOUT "Clockwise"              'Send Clockwise, to the LCD
        LCDOUT $FE, $C0                 'Set Cursor to Line 2, Position 0
        LCDOUT "Rotation; Wave"         'Send Rotation; Wave, to the LCD
        SEROUT2 PORTC.6,396,["The Motor is Rotating Clockwise; Wave",13,10,10]
                                        'Send The Motor is Rotating Clockwise; Wave,
                                        'to the HyperTerminal with a Carriage Return & 2 Line Feeds

        CMLP:   PORTB=%10001111
        PAUSE 5
        PORTB=%01001111
        PAUSE 5
        PORTB=%00101111
        PAUSE 5
        PORTB=%00011111
        PAUSE 5

        GOTO CMLP

        CCLOOP: LOW PORTE.2             'Turn-Off the Red LED
        LCDOUT $FE,1                    'Clear the LCD
        LCDOUT $FE, $80                 'Set Cursor to Line 1, Position 0
        LCDOUT "Counter"                'Send Counter, to the LCD
        LCDOUT $FE, $C0                 'Set Cursor to Line 2, Position 0
        LCDOUT "Rotation; Wave"         'Send Rotation; Wave, to the LCD
        SEROUT2 PORTC.6,396,["The Motor is Rotating Counterclockwise; Wave",13,10,10]
                                        'Send The Motor is Rotating Counterclockwise; Wave,
                                        'to the HyperTerminal with a Carriage Return & 2 Line Feeds
```

```
CCMLP:   PORTB=%00011111
PAUSE 5
PORTB=%00101111
PAUSE 5
PORTB=%01001111
PAUSE 5
PORTB=%10001111
PAUSE 5

GOTO CCMLP

PSENSE: LOW PORTE.2                 'Turn-Off the Red LED
LCDOUT $FE,1                        'Clear the LCD
LCDOUT $FE, $80                     'Set Cursor to Line 1, Position 0
LCDOUT "North; Wave"                'Send, North; Wave, to the LCD
LCDOUT $FE, $C0                     'Set Cursor to Line 2, Position 0
LCDOUT "Position Search"            'Send, Position Search, to the LCD
SEROUT2 PORTC.6,396,["The Motor is Searching for the North Position; Wave",13,10,10]
                                    'Send, The Motor is Searching for the North Position; Wave,
                                    'to the HyperTerminal with a Carriage Return & 2 Line Feeds

PSLP:
PORTB=%00011111
PAUSE 5
PORTB=%00101111
PAUSE 5
PORTB=%01001111
PAUSE 5
PORTB=%10001111
PAUSE 5

INPUT PORTC.1                       'Infrared Photo-Transistor Circuit A Low on
IF PORTC.1=0 THEN PNORTH            'A Low onPortC, Bit-1 will occur when the
                                    'IR-Light-Beam is Broken
GOTO PSLP

PNORTH: H=40
ADVAN:  PORTB=%00011111
PAUSE 5
PORTB=%00101111
PAUSE 5
PORTB=%01001111
PAUSE 5
PORTB=%10001111
PAUSE 5
H=H-1
IF H!=0 THEN ADVAN

LED:
HIGH PORTE.2                        'Turn-On the Red LED
SEROUT2 PORTC.6,396,["The Motor is at the North Position",13,10,10]
                                    'Send, The Motor is at the North Position, to the
                                    'HyperTerminal with a Carriage Return & 2 Line Feeds
GOTO LED


ALTR:
LOW PORTE.2                         'Turn-Off the Red LED
LCDOUT $FE,1                        'Clear the LCD
LCDOUT $FE, $80                     'Set Cursor to Line 1, Position 0
LCDOUT "Motor is; Wave"             'Send, Motor is; Wave, to the LCD
LCDOUT $FE, $C0                     'Set Cursor to Line 2, Position 0
LCDOUT "Alternating"                'Send, Alternating, to the LCD

SEROUT2 PORTC.6,396,["The Motor is Alternating; Wave",13,10,10]
                                    'Send, The Motor is at the North Position; Wave, to the
                                    'HyperTerminal with a Carriage Return & 2 Line Feeds
ALPCN: K=12
ALPAC:   PORTB=%10001111
PAUSE 5
PORTB=%01001111
PAUSE 5
PORTB=%00101111
PAUSE 5
PORTB=%00011111
PAUSE 5
K=K-1
IF K!=0 THEN ALPAC
```

```
ALPC:   PORTB=%10001111
PAUSE 5
PORTB=%01001111
PAUSE 5
PORTB=%00101111
PAUSE 5
PORTB=%00011111
PAUSE 5

INPUT PORTC.1                     'Infrared Photo-Transistor Circuit
IF PORTC.1=0 THEN ALPCCN          'A Low onPortC, Bit-1 will occur when the
                                  'IR-Light-Beam is Broken
GOTO ALPC

ALPCCN: J=12
ALPACC: PORTB=%00011111
PAUSE 5
PORTB=%00101111
PAUSE 5
PORTB=%01001111
PAUSE 5
PORTB=%10001111
PAUSE 5
J=J-1
IF J!=0 THEN ALPACC
ALPCC:  PORTB=%00011111
PAUSE 5
PORTB=%00101111
PAUSE 5
PORTB=%01001111
PAUSE 5
PORTB=%10001111
PAUSE 5

INPUT PORTC.1                     'Infrared Photo-Transistor Circuit
IF PORTC.1=0 THEN ALPCN           'A Low onPortC, Bit-1 will occur when the
                                  'IR-Light-Beam is Broken
GOTO ALPCC

END
```

Laboratory Experiment #3

'Lab-3_ CCS_C-code_Digital_4-Phase_Stepper_Motor_Controller

```c
#include <18F4685.h>                              //Identify Microcontroller
#device adc=10                                    //Identify ADC Bit Width
#fuses HS,NOWDT,NOPROTECT,NOLVP,MCLR              //Setup Programmer Oscillator Value
                                                  //and make Master Clear Pin an
                                                  //Enable Master Clear=MCLR
                                                  //Disable Master Clear=NOMCLR
#use delay(clock=20000000)                        //Setup C-Code Oscillator Value
#use fast_io(d)                                   //Leave the State of Port-D (all bits)
                                                  //the same until changed, again

#use rs232(baud=2400, xmit=PIN_C6, rcv=PIN_C7, stream=HT)
                                                  //Set RS232 HyperTerminal
                                                  //Communication Parameters
#use rs232(baud=2400, xmit=PIN_B4, rcv=PIN_B5, invert, stream=BB)
                                                  //Set RS232 Board-to-Board
                                                  //Communication Parameters
#use I2C(master, SCL=PIN_B6, SDA=PIN_B7)          //Set I2C Communication Parameters
#include <lcd_flex.c>                             //Identify LCD Driver File
#include <math.h>                                 //Include Math Functions

int8 H,K,J;                                       //8-Bit Byte Variable
short C1,D7;                                      //1-Bit Variable

void main()
{
    set_tris_b(0xFF);                             //Make PortB an Input Port to
                                                  //Enable In-Circuit-Programming

    delay_ms(1000);                               //Pause or Delay for 1 second

    lcd_init();                                   //Initialize the LCD
    lcd_putc(0x0c);                               //Clear the LCD
    lcd_gotoxy(1,1);                              //Set the Cursor to Position-1 of Line-1
    printf(lcd_putc,"Power Up");                  //Send, Power Up, to the LCD

    fprintf(HT, "\n\rPower Up\n\r");              //Send, Power Up, to the HyperTerminal with
                                                  //a next Line & a Carriage Return
    delay_ms(1000);                               //Pause or Delay for 1 second

    setup_port_a(AN0_to_AN3);                     //Identify all of the Analog Inputs
    setup_adc(VSS_VDD);                           //Setup the ADC Voltage Range to be
                                                  //0 Volts to 5 Volts
                                                  //(5v-0v)/1024~=.005 Volts Per Bit

    setup_adc(ADC_CLOCK_DIV_32);                  //for delay(clock=10000000),
                                                  //setup_adc(ADC_CLOCK_DIV_8),
                                                  //10MHz/8=1.25MHz
                                                  //for delay(clock=16000000),
                                                  //setup_adc(ADC_CLOCK_DIV_16),
                                                  //16MHz/16=1.00MHz
                                                  //for delay(clock=20000000),
                                                  //setup_adc(ADC_CLOCK_DIV_16),
                                                  //20MHz/16=1.25MHz
                                                  //for delay(clock=24000000),
                                                  //setup_adc(ADC_CLOCK_DIV_32),
                                                  //24MHz/32=1.25MHz
                                                  //for delay(clock=32000000),
                                                  //setup_adc(ADC_CLOCK_DIV_32),
                                                  //32MHz/32=1.00MHz
                                                  //for delay(clock=48000000),
                                                  //setup_adc(ADC_CLOCK_DIV_64),
                                                  //48MHz/64=0.75MHz

    set_adc_channel(0);                           //Identify the Analog Input to be
                                                  //Sampled (Channel Input Zero)

    setup_ccp1(CCP_PWM);                          //Setup Capture/Compare/PWM Pin-1 to PWM Mode

    setup_ccp2(CCP_CAPTURE_RE);                   //Setup Capture/Compare/PWM Pin-2 to
                                                  //Capture Mode

    setup_timer_1(T1_INTERNAL);                   //Setup Timer-1, which controls the
                                                  //Capture/Compare portion of
                                                  //Capture/Compare/PWM Pin-2, only,
                                                  //to Internal Mode

    setup_timer_2(T2_DIV_BY_4, 19, 1);            //Setup Timer-2, which controls the PWM portion
                                                  //of Capture/Compare/PWM Pin-1, only.
                                                  //DIV_BY_Mode, Prescale (Period), Postscale
                                                  //The Following Limits are for the
                                                  //PIC18F4685 Microcontroller
                                                  //Mode may equal 1 or 4 or 16
                                                  //Prescale (Period) may equal any integer
                                                  //between 0 and 255
                                                  //Postscale may equal any integer between
                                                  //1 and 16 (Post = 1 to set PWM Frequency)
                                                  //T2Freq=OSC/[(4)*(Mode)*(Prescale+1)]=
                                                  //24MegaHz/[(4)*(4)*(PS+1)] = 75KHz
```

```c
        enable_interrupts(INT_CCP2);            //Enable the Internal Interrupt Input
                                                //of Capture/Compare/PWM Pin-2

        enable_interrupts(GLOBAL);              //Enable Global Interrupts for the
                                                //PIC18F4685 Microcontroller

        set_tris_b(0x00);                       //Make PortB an Output Port to Allow for
                                                //Normal Operation
        tris_d(0xF0);                           //Make PortD, Bits-4,5,6,7 Inputs
                                                //and Bits-3,2,1,0 Outputs


KSCAN:
        lcd_putc(0x0c);                         //Clear the LCD
        lcd_gotoxy(1,1)                         //Set the Cursor to Position-1 of Line-1
        printf(lcd_putc,"Scanning");            //Send, Scanning, to the LCD
        fprintf(HT, "\n\rScanning\n\r");        //Send, Scanning, to the HyperTerminal with a
                                                //next Line & a Carriage Return

        output_high(PIN_D0);                    //Make 0th Row High to Scan for the Following:
        delay_ms(500);
        D7 = input(PIN_D7);                     //Identify PortD, Bit-7 as an Input
        if (D7) goto ALTR;                      //Check for Keypad Key "A" , Column-7
        delay_ms(100);
        output_low(PIN_D0);
        output_high(PIN_D1);                    //Make 1st Row High to Scan for the Following:
        delay_ms(500);
        D7 = input(PIN_D7);                     //Identify PortD, Bit-7 as an Input
        if (D7) goto PSENSE;                    //Check for Keypad Key "B" , Column-7
        delay_ms(100);
        output_low(PIN_D1);
        output_high(PIN_D2);                    //Make 2nd Row High to Scan for the Following:
        delay_ms(500);
        D7 = input(PIN_D7);                     //Identify PortD, Bit-7 as an Input
        if (D7) goto CLOOP;                     //Check for Keypad Key "C" , Column-7
        delay_ms(100);
        output_low(PIN_D2);
        output_high(PIN_D3);                    //Make 3rd Row High to Scan for the Following:
        delay_ms(500);
        D7 = input(PIN_D7);                     //Identify PortD, Bit-7 as an Input
        if (D7) goto CCLOOP;                    //Check for Keypad Key "D" , Column-7
        delay_ms(100);
        output_low(PIN_D3);
        goto KSCAN;

CLOOP:
        output_low(PIN_E2);                     //Turn-Off the Red LED
        lcd_putc(0x0c);                         //Clear the LCD
        lcd_gotoxy(1,1);                        //Set the Cursor to Position-1 of Line-1
        printf(lcd_putc,"Clockwise");           //Send, Clockwise, to the LCD
        lcd_gotoxy(1,2);                        //Set the Cursor to Position-1 of Line-2
        printf(lcd_putc,"Rotation;  Wave");     //Send, Rotation; Wave, to the LCD
        fprintf(HT, "\n\rThe Motor is Rotating Clockwise; Wave\n\r");
                                                //Send The Motor is Rotating Clockwise; Wave,
                                                //to the HyperTerminal with a next Line & a Carrage Return

CMLP:
        output_b(0b10001111);
        delay_ms(5);
        output_b(0b01001111);
        delay_ms(5);
        output_b(0b00101111);
        delay_ms(5);
        output_b(0b00011111);
        delay_ms(5);
        goto CMLP;


CCLOOP:
        output_low(PIN_E2);                     //Turn-Off the Red LED
        lcd_putc(0x0c);                         //Clear the LCD
        lcd_gotoxy(1,1);                        //Set the Cursor to Position-1 of Line-1
        printf(lcd_putc,"Counter");             //Send, Counter, to the LCD
        lcd_gotoxy(1,2);                        //Set the Cursor to Position-1 of Line-2
        printf(lcd_putc,"Rotation;  Wave");     //Send, Rotation; Wave, to the LCD
        fprintf(HT, "\n\rThe Motor is Rotating Counterclockwise; Wave\n\r");
                                                //Send The Motor is Rotating Counterclockwise; Wave,
                                                //to the HyperTerminal with a next Line & a Carrage Return

CCMLP:
        output_b(0b00011111);
        delay_ms(5);
        output_b(0b00101111);
        delay_ms(5);
        output_b(0b01001111);
        delay_ms(5);
        output_b(0b10001111);
        delay_ms(5);
        goto CCMLP;
```

```c
PSENSE:
output_low(PIN_E2);                              //Turn-Off the Red LED
lcd_putc(0x0c);                                  //Clear the LCD
lcd_gotoxy(1,1);                                 //Set the Cursor to Position-1 of Line-1
printf(lcd_putc,"North; Wave");                  //Send, North, Wave, to the LCD
lcd_gotoxy(1,2);                                 //Set the Cursor to Position-1 of Line-2
printf(lcd_putc,"Position Search");              //Send, Position Search, to the LCD
fprintf(HT, "\n\rThe Motor is Searching for the North Position; Wave\n\r");
                                                 //Send The Motor is Searching for the North Position; Wave,
                                                 //to the HyperTerminal with a next Line & a Carrage Return
PSLP:
output_b(0b00011111);
delay_ms(5);
output_b(0b00101111);
delay_ms(5);
output_b(0b01001111);
delay_ms(5);
output_b(0b10001111);
delay_ms(5);

C1 = input(PIN_C1);                              //'Infrared Photo-Transistor Circuit'
if (!C1) goto PNORTH;                            //A Low on PortC, Bit-1 will occur when the
                                                 //IR-Light-Beam is Broken
goto PSLP;


PNORTH: H=40;
ADVAN:  output_b(0b00011111);
delay_ms(5);
output_b(0b00101111);
delay_ms(5);
output_b(0b01001111);
delay_ms(5);
output_b(0b10001111);
delay_ms(5);

H=H-1;
if (H!=0) goto ADVAN;

LED:
output_high(PIN_E2);                             //Turn-On the Red LED
fprintf(HT, "\n\rThe Motor is at the North Position\n\r");
                                                 //Send "The Motor is at the North Position,
                                                 //to the HyperTerminal with a next Line & a Carrage Return

goto LED;

ALTR:
output_low(PIN_E2);                              //Turn-Off the Red LED
lcd_putc(0x0c);                                  //Clear the LCD
printf(lcd_putc,"Motor is; Wave");               //Send, Motor is; Wave, to the LCD
lcd_gotoxy(1,2);                                 //Set the Cursor to Position-1 of Line-2
printf(lcd_putc,"Alternating");                  //Send, Alternating, to the LCD
fprintf(HT, "\n\rThe Motor is Alternating; Wave\n\r");  //Send The Motor is Alternating; Wave,
                                                 //to the HyperTerminal with a next Line & a Carrage Return

ALPCN:
K=12;
ALPAC:
output_b(0b10001111);
delay_ms(5);
output_b(0b01001111);
delay_ms(5);
output_b(0b00101111);
delay_ms(5);
output_b(0b00011111);
delay_ms(5);

K=K-1;
if (K!=0) goto ALPAC;
ALPC:
output_b(0b10001111);
delay_ms(5);
output_b(0b01001111);
delay_ms(5);
output_b(0b00101111);
delay_ms(5);
output_b(0b00011111);
delay_ms(5);

C1 = input(PIN_C1);                              //'Infrared Photo-Transistor Circuit'
if (!C1) goto ALPCCN;                            //A Low on PortC, Bit-1 will occur when the
                                                 //IR-Light-Beam is Broken
goto ALPC;
```

```
ALPCCN: J=12;
ALPACC: output_b(0b00011111);
delay_ms(5);
output_b(0b00101111);
delay_ms(5);
output_b(0b01001111);
delay_ms(5);
output_b(0b10001111);
delay_ms(5);
J=J-1;
if (J!=0) goto ALPACC;

ALPCC:
output_b(0b00011111);
delay_ms(5);
output_b(0b00101111);
delay_ms(5);
output_b(0b01001111);
delay_ms(5);
output_b(0b10001111);
delay_ms(5);

C1 = input(PIN_C1);         //'Infrared Photo-Transistor Circuit'
if (!C1) goto ALPCN;        //A Low on PortC, Bit-1 will occur when the
                            //IR-Light-Beam is Broken
goto ALPCC;

}
```

Engineering Practices for the PIC Microcontroller

By: Prof. Sal R. Riggio Jr., PhD, PE

Chapter - 6

Laboratory Experiment - 4

The Digital Electronic Thermometer

Introduction

The purpose of this lab experiment is for the student to develop basic skills in making practical use of the Microchip **PIC18F4685** Microcontroller and become familiar with the importance, operation and hardware implementation of a **Digital Electronic Thermometer**, using both the **Celsius scale** and the **Fahrenheit scale**.

The student is expected to analyze and write **PICBasic-Pro Code** and **CCS C-code**, which enables a **Temperature-Sensor** to **accurately** detect the **Air Temperature,** and **Display** the **Value** of this **Measured Air Temperature** on the **Liquid-Crystal-Display**, and on the **RS232 HyperTerminal** within a Windows based Personal Computer.

It is suggested that the student **read** the **PICBasic-Pro code** and the **CCS C-code**, particularly the **comments** written on each line, to learn how to write code for the PIC microcontrollers. It has been proven many times to be the most effective way to learn how generate effective code for the PIC microcontrollers. This code is shown at the end of this chapter.

Portions of the Experiment Board Schematics, which are relevant to this experiment, are shown in **Figure 6.1**. Please refer to **Appendix-A** for the complete set of **Experiment Board Schematics** and please refer to **Appendix-B** for the complete **Lab Board Schematic**.

Hardware Setup

1.) Connect the Square end of the USB Type-A to Type-B Cable to connector JUSB1 of the Experiment Board. Then, connect the other end of this Cable to one of the USB Inputs of the Computer.
2.) Connect the DB9 end of the DB9 Serial-to-USB Cable to connector P1 of the Experiment Board. Then, connect the USB end of this Cable to one of the other USB Inputs of the Computer. The DB9 Serial-to-USB Cable is used to communicate between the Serial HyperTerminal of the Computer and the Experiment Board.
3.) The Jumpers of **JB12** are to be placed in **Positions 3-5 & 4-6.**

Assignment #1

Analyze the **PICBasic-Pro Code** and the **CCS C-code below**, which enables the **Temperature Sensing Thermistor,** located on the experiment board, to **Accurately Detect** the **Air Temperature,** and **Display** the **Value** of the **Measured Air Temperature** on the **Liquid-Crystal-Display**, and on the **RS232 HyperTerminal**. The **Temperature** must be **Displayed** in both **Degrees Fahrenheit** and **Degrees Celsius**, **Simultaneously**.

Please refer to the **EPCOS_B57891M0102K000_NTC_Thermistor pdf Datasheet** for all device information. This includes the **Device Equation**, the **Experiment Board Equations** and **Numerical Data.**

The **PICBasic-Pro Program** and the **CCS C-code Program** must implement the **Temperature Measurements** by using the **8-Bit** ADC-Value versus Temperature Numerical Data.

Assignment #2

Write a PICBasic-Pro Code routine and a CCS C-code routine, which will cause the **Red LED** to **Flash, at a rate of speed chosen by the student,** when the **Air Temperature Exceeds 86^F (30^C)** and **Turn-Off** when **Less** than **86^F (30^C)**. Additionally, the **Student** must **Write Code** which will cause the **Yellow LED** to **Flash, at a rate of speed chosen by the student,** when the **Air Temperature** Drops **Below 32^F (0^C)** and **Turn-Off** when **Greater** than **32^F (0^C)**. The **Green LED** must be **Turned-ON** when the **Air Temperature** is between **32^F (0^C)** and **86^F (30^C)**, otherwise, the **Green LED** must **Remain Off**. The **DIP Switches** must be placed in the **Proper Position** to allow the **All LED's** to **Turn-On** and **Off**. The Temperature in **Room 310 EE-West** is approximately **22 degrees Celsius**. The Student **must** use the **Value Captured** by the ADC (W0) at each LED Switch Point **Temperature** to **Control** the On/OFF States of the LED's.

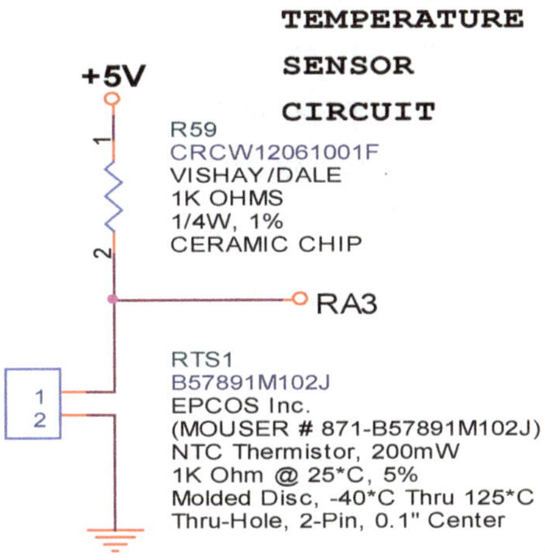

TEMPERATURE SENSOR CIRCUIT

+5V

R59
CRCW12061001F
VISHAY/DALE
1K OHMS
1/4W, 1%
CERAMIC CHIP

○ RA3

RTS1
B57891M102J
EPCOS Inc.
(MOUSER # 871-B57891M102J)
NTC Thermistor, 200mW
1K Ohm @ 25*C, 5%
Molded Disc, -40*C Thru 125*C
Thru-Hole, 2-Pin, 0.1" Center

Figure 6.1:   Negative Temperature Coefficient Thermistor Temperature Sensing Circuit

Figure 6.1 shows a voltage divider which is formed with a 1K ohm, 1%, 1/4w resistor and a Negative Temperature Coefficient Thermistor. As temperature increases, the resistance of this Thermistor decreases according to the data in Figure 6.2. This action decreases the voltage, which is applied to the input of analog-to-digital converter on Port-A, Bit-3 of the PIC18F4550 microcontroller. As the temperature decreases the voltage at Port-A, Bit-3 increases. Therefore, one can display the temperature experienced by the Thermistor on the Liquid-Crystal-Display, by writing code and creating a look-up for the data shown below.

Laboratory Experiment #4

```
'Lab-4_PICBasic-Pro_Digital_Electronic_Thermometer
INCLUDE "modedefs.bas"              'Identifies the Command File

DEFINE OSC 20                       'Set PIC-Pro Oscillator Value in Mega Hertz
DEFINE ADC_BITS 10                  'Set for 10-Bit Internal ADC
DEFINE ADC_CLOCK 5                  'DEFINE ADC_CLOCK 1, for OSC=10, MAX ADC
                                    'Clock=10MHz/8) = 1.25MHz
                                    'DEFINE ADC_CLOCK 5, for OSC=16, MAX ADC
                                    'Clock=16MHz/16) = 1.00MHz
                                    'DEFINE ADC_CLOCK 5, for OSC=20, MAX ADC
                                    'Clock=20MHz/16) = 1.25MHz
                                    'DEFINE ADC_CLOCK 2, for OSC=24, MAX ADC
                                    'Clock=24MHz/32) = 1.25MHz
                                    'DEFINE ADC_CLOCK 2, for OSC=32, MAX ADC
                                    'Clock=32MHz/32) = 1.00MHz
                                    'DEFINE ADC_CLOCK 6, for OSC=48, MAX ADC
                                    'Clock=48MHz/64) = 0.75MHz

DEFINE CCP1_BIT 2                   'Setup Hardware PWM Timer
DEFINE LCD_BITS 4                   'Setup LCD for 4-Bit Mode
DEFINE LCD_DREG PORTB               'Drive LCD from Port-B
DEFINE LCD_DBIT 0                   'Starting with Bit-0
DEFINE LCD_RSREG PORTA              'Drive LCD Register-Select from
DEFINE LCD_RSBIT 4                  'Port-A, Bit-4
DEFINE LCD_EREG PORTA               'Drive LCD Enable from
DEFINE LCD_EBIT 5                   'Port-A, Bit-5
DEFINE LCD_LINES 2                  'LCD can hold 2-Lines of Characters
DEFINE LCD_COMMANDSUS 4000          'Delay between Commands
DEFINE LCD_DATAUS 200               'Delay between Data Transfers

ADCON0 = %00001111                  'ADC Enabled, Analog Input 3 Enabled
ADCON1 = %00001011                  'ADC Voltage Range is 0 to 5 Volts &
                                    'Analog Inputs 0 Thru 3 are available
ADCON2 = %10101101                  'Right Justify to use Upper 8-Bits, only,
                                    '12*Tad, & ADC Clock=20MHz/16=1.25MHz

B0 VAR BYTE                         'Set 8-Bit Byte Variable,B0
B1 VAR BYTE                         ' "    "    "    "   ,B1

W0 VAR WORD                         'Set 16-Bit Word Variable,W0
W1 VAR WORD                         ' "    "    "    "   ,W1
W2 VAR WORD                         ' "    "    "    "   ,W2
W3 VAR WORD                         ' "    "    "    "   ,W3
TADCV VAR WORD                      ' "    "    "    "   ,TADCV
TADCV4 VAR WORD                     ' "    "    "    "   ,TADCV4
TEMPC VAR WORD                      ' "    "    "    "   ,TEMPC
TEMPCS VAR WORD                     ' "    "    "    "   ,TEMPCS
TEMPF VAR WORD                      ' "    "    "    "   ,TEMPF
TEMPFS VAR WORD                     ' "    "    "    "   ,TEMPFS

B0 VAR BYTE                         'Set 8-Bit Byte Variable,B0
B1 VAR BYTE                         ' "    "    "    "   ,B1

TRISB=$FF                           'Make PortB an Input Port to Enable
                                    'In-Circuit-Programming
PAUSE 1000                          'Pause or Delay for 1 Seconds to Enable
                                    'In-Circuit-Programming
LCDOUT $FE,1                        'Clear the LCD
LCDOUT $FE, $80                     'Set the Cursor to Line 1, Position 0
LCDOUT "Power Up"                   'Send, Power Up, to the LCD

SERout PORTC.6,T2400,["Power Up",13,10,10]
                                    'Send "Power Up" to the HyperTerminal
                                    'with a Carriage Return & two Line Feeds,
                                    'at 2400 Bits/Second, 8-Bits, 1-Stop Bit,
                                    'No Parity, Driven True (T2400).

PAUSE 2000                          'Pause or Delay for 2 Seconds
                                    "to Read "Power Up" in the LCD
LCDOUT $FE,1                        'Clear the LCD
TRISB=$00                           'Make PortB an Output Port to Allow for Normal Operation
TRISD=$F0                           'Make PortD, Bits-4,5,6,7 Inputs, & Bits-3
TRISB=$FF                           'Make PortB an Input Port to Enable 'In-Circuit-Programming'.
PAUSE                               'Pause or Delay for 1 Seconds to Enable
                                    'In-Circuit-Programming'.
```

```
LCDOUT $FE,1                              'Clear the LCD
LCDOUT $FE, $80                           'Set the Cursor to Line 1, Position 0
LCDOUT "Power Up"                         'Send, Power Up, to the LCD
SERout PORTC.6,T2400,["Power Up",13,10,10]
                                          'Send "Power Up" to the HyperTerminal
                                          'with a Carriage Return & two Line Feeds,
                                          'at 2400 Bits/Second, 8-Bits, 1-Stop Bit,
                                          'No Parity, Driven True (T2400).

PAUSE 2000                                'Pause or Delay for 2 Seconds
                                          'to Read "Power Up" in the LCD

LCDOUT $FE,1                              'Clear the LCD

TRISB=$00                                 'Make PortB an Output Port to Allow for
                                          'Normal Operation
TRISD=$F0                                 'Make PortD, Bits-4,5,6,7 Inputs, & Bits-3,2,1,0 Outputs


SAMPLE: ADCIN 3,TADCV                     'Sample ADC Input Voltage-3, and place the ADCValue
                                          '8-Bit Decimal result in the Variable TADCV
PAUSE 1

TADCV4=TADCV/4

LOOKUP2
TADCV4,[139,124,116,110,105,102,98,96,93,1,89,87,85,84,82,81,79,78,77,76,75,74,73,72,71,70,69,68,
67,66,66,65,64,64,63,62,62,61,60,60,59,58,58,57,57,56,56,55,55,54,54,53,53,52,52,51,51,50,50,49,49,
48,48,48,47,47,46,46,_

45,45,45,44,44,43,43,43,42,42,42,41,41,40,40,40,39,39,39,38,38,38,37,37,37,36,36,35,35,35,34,34,34,
33,33,33,32,32,32,31,31,31,30,30,30,30,29,29,29,28,28,28,27,27,27,26,26,26,25,25,25,24,24,24,23,23,
23,22,22,22,_

21,21,21,20,20,20,20,19,19,19,18,18,18,17,17,17,16,16,16,15,15,15,14,14,13,13,13,12,12,12,11,11,11,
10,10,10,9,9,8,8,8,7,7,7,6,6,5,5,5,4,4,3,3,2,2,2,1,1,0,0,1,1,2,2,3,3,4,4,5,5,6,6,7,7,8,8,9,10,10,11,12,12,1
3,14,14,15,16,16,17,18,19,20,21,22,23,24,25,26,27,28,29,31,32,34,35,37,39,41,43,46,48,52,55,60,66,
74,89],TEMPC

LOOKUP2 TADCV4,["+","+","+","+","+","+","+","+","+","+","+","+","+","+","+","+","+","+","+","+","+",_
"+","+","+","+","+","+","+","+","+","+","+","+","+","+","+","+","+","+","+","+","+","+","+",_
"+","+","+","+","+","+","+","+","+","+","+","+","+","+","+","+","+","+","+","+","+","+",_
"+","+","+","+","+","+","+","+","+","+","+","+","+","+","+","+","+","+","+","+","+",_
"+","+","+","+","+","+","+","+","+","+","+","+","+","+","+","+","+","+","+","+",_
"+","+","+","+","+","+","+","+","+","+","+","+","+","+","+","+","+","+","+",_
"+","+","+","+","+","+","+","+","+","+","+","+","+","+","+","+","+","+",_
"+","+","+","+","+","+","+","+","+","+","+","+","+","+","+","+","+",_
"+","+","+","+","+","+","+","+","+","+","+","+","+","+","+","+",_
"+","+","+","+","+","+","+","+","+","+","+","+","+","+","+",_
"+","+","+","+","+","+","+","+","+","+","+","+","+","+",_
"-","-","-","-","-","-","-","-","-","-","-","-","-","-","-","-","-","-","-","-","-",_
"-","-","-","-","-","-","-","-","-","-","-","-","-","-","-","-","-","-","-","-","-","-",_
"-","-","-","-","-","-","-","-","-","-","-","-","-","-","-","-","-","-","-","-","-","-","-"],TEMPCS


LOOKUP2 TADCV4,[282,256,241,230,222,215,209,204,199,195,192,188,185,182,180,177,175,172,
170,168,166,164,163,161,159,158,156,155,153,152,150,149,148,146,145,144,143,142,140,139,138,
137,136,135,134,133,132,131,130,129,_

128,128,127,126,125,124,123,122,122,121,120,119,118,118,117,116,115,115,114,113,112,112,111,
111,110,109,108,108,107,106,105,105,104,104,103,102,102,101,100,100,99,98,98,97,97,96,95,95,94,
93,93,92,92,91,_

90,90,89,89,88,87,87,86,86,85,85,84,83,83,82,82,81,80,80,79,79,78,78,77,76,76,75,75,74,74,73,72,72,
71,71,70,69,69,68,68,67,67,66,65,65,64,64,63,62,62,61,61,60,59,59,58,58,57,56,56,55,54,54,53,52,52,
51,51,50,49,_

49,48,47,47,46,45,44,44,43,42,42,41,40,39,39,38,37,36,36,35,34,33,32,32,31,30,29,28,27,26,26,25,24,
23,22,21,20,19,18,17,16,15,14,12,11,10,9,8,6,5,4,2,1,0,2,4,5,7,9,10,12,14,16,18,21,23,26,28,31,34,38,
41,45,50,55,61,68,76,87,102,128],TEMPF
```

99

```
LOOKUP2 TADCV4,["+","+","+","+","+","+","+","+","+","+","+","+","+","+","+","+","+","+","+","+",_
                "+","+","+","+","+","+","+","+","+","+","+","+","+","+","+","+","+","+","+","+",_
                "+","+","+","+","+","+","+","+","+","+","+","+","+","+","+","+","+","+","+","+",_
                "+","+","+","+","+","+","+","+","+","+","+","+","+","+","+","+","+","+","+","+",_
                "+","+","+","+","+","+","+","+","+","+","+","+","+","+","+","+","+","+","+","+",_
                "+","+","+","+","+","+","+","+","+","+","+","+","+","+","+","+","+","+","+","+",_
                "+","+","+","+","+","+","+","+","+","+","+","+","+","+","+","+","+","+","+","+",_
                "+","+","+","+","+","+","+","+","+","+","+","+","+","+","+","+","+","+","+","+",_
                "+","+","+","+","+","+","+","+","+","+","+","+","+","+","+","+","+","+","+","+",_
                "+","+","+","+","+","+","+","+","+","+","+","+","+","+","+","+","+","+","+","+",_
                "+","+","+","+","+","+","+","+","+","+","+","+","+","+","+","+","+","+","+","+",_
                "+","+","+","+","+","+","+","+","+","+","+",_
                "-","-","-","-","-","-","-","-","-","-","-","-","-","-","-","-","-","-","-","-",_
                "-","-","-"],TEMPFS


        LCDOUT $FE, 1                              'Clear the LCD
        LCDOUT $FE, $80                            'Set Cursor to Line 1, Position 0
        LCDOUT "Deg. C = ",TEMPCS,DEC TEMPC        'Send, SIGN & Deg. C = , to the LCD
        LCDOUT $FE, $C0                            'Set Cursor to Line 2, Position 0
        LCDOUT "Deg. F = ",TEMPFS,DEC TEMPF        'Send, SIGN & Deg. F = , to the LCD

        SEROUT2 PORTC.6,396,_
        [13,10,"Degrees C = ",TEMPCS,DEC TEMPC,13,10,"Degrees F = ",TEMPFS,DEC TEMPF,13,10,10]
                                                   'Send, SIGN & Degrees C = and SIGN &
                                                   'Degrees F = , to the HyperTerminal
                                                   'Carriage Return & two Line Feeds
        GOTO SAMPLE

        END
```

Laboratory Experiment #4

'Lab-4_CCS_C-code_Digital_Electronic_Thermometer

```
#include <18F4685.h>               //Identify Microcontroller
#device adc=10                     //Identify ADC Bit Width
#fuses HS,NOWDT,NOPROTECT,NOLVP,MCLR   //Setup Programmer Oscillator Value
                                   //and make Master Clear Pin an
                                   //Enable Master Clear=MCLR
                                   //Disable Master Clear=NOMCLR

#use delay(clock=20000000)         //Setup C-Code Oscillator Value
#use fast_io(d)                    //Leave the State of Port-D (all bits)
                                   //the same until changed, again

#use rs232(baud=2400, xmit=PIN_C6, rcv=PIN_C7, stream=HT)
                                   //Set RS232 HyperTerminal
                                   //Communication Parameters

#use rs232(baud=2400, xmit=PIN_B4, rcv=PIN_B5, invert, stream=BB)
                                   //Set RS232 Board-to-Board
                                   //Communication Parameters

#use I2C(master, SCL=PIN_B6, SDA=PIN_B7)
                                   //Set I2C Communication Parameters
#include <lcd_flex.c>              //Identify LCD Driver File
#include <math.h>                  //Include Math Functions

float W0,W1,TC,TF,TCAL;            //16-Bit Floating Point Integer Word Variable

void main()
    {
    set_tris_b(0xFF);              //Make PortB an Input Port to
                                   //Enable In-Circuit-Programming

    delay_ms(1000);                //Pause or Delay for 1 second

    lcd_init();                    //Initialize the LCD
    lcd_putc(0x0c);                //Clear the LCD
    lcd_gotoxy(1,1);               //Set the Cursor to Position-1 of Line-1
    printf(lcd_putc,"Power Up");   //Send, Power Up, to the LCD

    fprintf(HT, "\n\rPower Up\n\r");   //Send, Power Up, to the HyperTerminal with
                                   //a next Line & a Carriage Return
    delay_ms(1000);                //Pause or Delay for 1 second

    setup_port_a(AN0_to_AN3);      //Identify all of the Analog Inputs
    setup_adc(VSS_VDD);            //Setup the ADC Voltage Range to be
                                   //0 Volts to 5 Volts
                                   //(5v-0v)/1024~=.005 Volts Per Bit

    setup_adc(ADC_CLOCK_DIV_32);   //for delay(clock=10000000),
                                   //setup_adc(ADC_CLOCK_DIV_8),
                                   //10MHz/8=1.25MHz
                                   //for delay(clock=16000000),
                                   //setup_adc(ADC_CLOCK_DIV_16),
                                   //16MHz/16=1.00MHz
                                   //for delay(clock=20000000),
                                   //setup_adc(ADC_CLOCK_DIV_16),
                                   //20MHz/16=1.25MHz
                                   //for delay(clock=24000000),
                                   //setup_adc(ADC_CLOCK_DIV_32),
                                   //24MHz/32=1.25MHz
                                   //for delay(clock=32000000),
                                   //setup_adc(ADC_CLOCK_DIV_32),
                                   //32MHz/32=1.00MHz
                                   //for delay(clock=48000000),
                                   //setup_adc(ADC_CLOCK_DIV_64),
                                   //48MHz/64=0.75MHz

    set_adc_channel(0);            //Identify the Analog Input to be
                                   //Sampled (Channel Input Zero)

    setup_ccp1(CCP_PWM);           //Setup Capture/Compare/PWM Pin-1 to PWM Mode

    setup_ccp2(CCP_CAPTURE_RE);    //Setup Capture/Compare/PWM Pin-2 to
                                   //Capture Mode

    setup_timer_1(T1_INTERNAL);    //Setup Timer-1, which controls the
                                   //Capture/Compare portion of
                                   //Capture/Compare/PWM Pin-2, only,
                                   //to Internal Mode

    setup_timer_2(T2_DIV_BY_4, 19, 1);   //Setup Timer-2, which controls the PWM portion
                                   //of Capture/Compare/PWM Pin-1, only.
                                   //DIV_BY_Mode, Prescale (Period), Postscale
                                   //The Following Limits are for the
                                   //PIC18F4685 Microcontroller
                                   //Mode may equal 1 or 4 or 16
                                   //Prescale (Period) may equal any integer
                                   //between 0 and 255
                                   //Postscale may equal any integer between
                                   //1 and 16 (Post = 1 to set PWM Frequency)
                                   //T2Freq=OSC/[(4)*(Mode)*(Prescale+1)]=
                                   //24MegaHz/[(4)*(4)*(PS+1)] = 75KHz
```

```
enable_interrupts(INT_CCP2);              //Enable the Internal Interrupt Input
                                          //of Capture/Compare/PWM Pin-2

enable_interrupts(GLOBAL);                //Enable Global Interrupts for the
                                          //PIC18F4685 Microcontroller

set_tris_b(0x00);                         //Make PortB an Output Port to Allow for
                                          //Normal Operation
tris_d(0xF0);                             //Make PortD, Bits-4,5,6,7 Inputs
                                          //and Bits-3,2,1,0 Outputs


W0=read_adc();                            //Sample ADC Input Voltage-3, and place the 10-Bit result in
Variable (W0)
TC=(20.534*(log(3.38*((1024/W0)-1))));
TF=(TC*1.8)+32;

LCD:
lcd_putc(0x0c);                           //Clear the LCD
lcd_gotoxy(1,1);                          //Set the Cursor to Position-1 of Line-1
printf(lcd_putc,"Degrees C=%lf",TC);      //Send, Degrees C=, to the LCD in Long Integer
                                          //Floating Point Form

lcd_gotoxy(1,2);                          //Set the Cursor to Position-1 of Line-2
printf(lcd_putc,"Degrees F=%lf",TF);      //Send, Degrees F=, to the LCD in Long Integer Floating Point Form
fprintf(HT,"\n\rADC Value0=%lf",W0);      //Send, ADC Value0=, to the HyperTerminal in Long Integer
                                          //Floating Point Form
                                          //with a next Line & a Carrage Return

fprintf(HT,"\n\rDegrees C=%lf",TC);       //Send, Deg. C=, to the HyperTerminal in Long Integer
                                          //Floating Point Form
                                          //with a next Line & a Carrage Return

fprintf(HT,"\n\rDegrees F=%lf\n\r",TF);   //Send, Deg. F=, to the HyperTerminal in Long Integer
                                          //Floating Point Form
                                          //with a next Line & a Carrage Return

delay_ms(2000);                           //Pause or Delay for 2 seconds

goto gettmp;

}
```

Temp Degree-C	Temp Degree-F	Thermistor Ohms	Volts Vo (AN3)	10-Bit ADC-Value
-50	-58	38585.47	4.874	998
-49	-56	36751.38	4.868	997
-48	-54	35004.47	4.861	996
-47	-53	33340.59	4.854	994
-46	-51	31755.81	4.847	993
-45	-49	30246.36	4.840	991
-44	-47	28808.65	4.832	990
-43	-45	27439.28	4.824	988
-42	-44	26135.01	4.816	986
-41	-42	24892.73	4.807	984
-40	-40	23709.50	4.798	983
-39	-38	22582.51	4.788	981
-38	-36	21509.09	4.778	979
-37	-35	20486.70	4.767	976
-36	-33	19512.90	4.756	974
-35	-31	18585.39	4.745	972
-34	-29	17701.97	4.733	969
-33	-27	16860.54	4.720	967
-32	-26	16059.10	4.707	964
-31	-24	15295.76	4.693	961
-30	-22	14568.70	4.679	958
-29	-20	13876.21	4.664	955
-28	-18	13216.63	4.648	952
-27	-17	12588.40	4.632	949
-26	-15	11990.03	4.615	945
-25	-13	11420.11	4.597	942
-24	-11	10877.27	4.579	938
-23	-9	10360.24	4.560	934
-22	-8	9867.79	4.540	930
-21	-6	9398.74	4.519	926
-20	-4	8951.99	4.498	921
-19	-2	8526.47	4.475	917
-18	0	8121.18	4.452	912
-17	1	7735.16	4.428	907
-16	3	7367.48	4.402	902
-15	5	7017.28	4.376	896
-14	7	6683.73	4.349	891
-13	9	6366.03	4.321	885
-12	10	6063.43	4.292	879
-11	12	5775.22	4.262	873
-10	14	5500.70	4.231	866
-9	16	5239.24	4.199	860
-8	18	4990.20	4.165	853
-7	19	4753.00	4.131	846
-6	21	4527.07	4.095	839
-5	23	4311.89	4.059	831
-4	25	4106.93	4.021	823
-3	27	3911.71	3.982	816
-2	28	3725.78	3.942	807
-1	30	3548.68	3.901	799
0	32	3380.00	3.858	790
1	34	3219.34	3.815	781
2	36	3066.00	3.770	772
3	37	2920.56	3.725	763
4	39	2781.74	3.678	753
5	41	2649.51	3.630	743
6	43	2523.57	3.581	733
7	45	2403.62	3.531	723
8	46	2289.37	3.480	713
9	48	2180.55	3.428	702
10	50	2076.91	3.375	691
11	52	1978.18	3.321	680
12	54	1884.15	3.266	669
13	55	1794.59	3.211	658
14	57	1709.29	3.154	646
15	59	1628.04	3.097	634
16	61	1550.65	3.040	623
17	63	1476.94	2.981	611
18	64	1406.74	2.923	599
19	66	1339.87	2.863	586
20	68	1276.19	2.803	574
21	70	1215.52	2.743	562
22	72	1157.75	2.683	549
23	73	1102.72	2.622	537
24	75	1050.31	2.561	525
25	77	1000.38	2.500	512
26	79	952.82	2.440	500

Temp Degree-C	Temp Degree-F	Thermistor Ohms	Volts Vo (AN3)	10-Bit ADC-Value
27	81	907.53	2.379	487
28	82	864.40	2.318	475
29	84	823.31	2.258	462
30	86	784.17	2.198	450
31	88	746.90	2.138	438
32	90	711.40	2.078	426
33	91	677.58	2.020	414
34	93	645.37	1.961	402
35	95	614.70	1.903	390
36	97	585.48	1.846	378
37	99	557.65	1.790	367
38	100	531.14	1.734	355
39	102	505.90	1.680	344
40	104	481.85	1.626	333
41	106	458.95	1.573	322
42	108	437.13	1.521	311
43	109	416.35	1.470	301
44	111	396.56	1.420	291
45	113	377.71	1.371	281
46	115	359.76	1.323	271
47	117	342.66	1.276	261
48	118	326.37	1.230	252
49	120	310.86	1.186	243
50	122	296.08	1.142	234
51	124	282.01	1.100	225
52	126	268.60	1.059	217
53	127	255.83	1.019	209
54	129	243.67	0.980	201
55	131	232.09	0.942	193
56	133	221.06	0.905	185
57	135	210.55	0.870	178
58	136	200.54	0.835	171
59	138	191.01	0.802	164
60	140	181.93	0.770	158
61	142	173.28	0.738	151
62	144	165.05	0.708	145
63	145	157.20	0.679	139
64	147	149.73	0.651	133
65	149	142.61	0.624	128
66	151	135.83	0.598	122
67	153	129.38	0.573	117
68	154	123.23	0.549	112
69	156	117.37	0.525	108
70	158	111.79	0.503	103
71	160	106.48	0.481	99
72	162	101.42	0.460	94
73	163	96.60	0.440	90
74	165	92.00	0.421	86
75	167	87.63	0.403	83
76	169	83.47	0.385	79
77	171	79.00	0.368	75
78	172	75.72	0.352	72
79	174	72.12	0.336	69
80	176	68.69	0.321	66
81	178	65.43	0.307	63
82	180	62.32	0.293	60
83	181	59.35	0.280	57
84	183	56.53	0.268	55
85	185	53.85	0.255	52
86	187	51.29	0.244	50
87	189	48.85	0.233	48
88	190	46.53	0.222	46
89	192	44.32	0.212	43
90	194	42.21	0.202	41
91	196	40.20	0.193	40
92	198	38.29	0.184	38
93	199	36.47	0.176	36
94	201	34.74	0.168	34
95	203	33.09	0.160	33
96	205	31.51	0.153	31
97	207	30.02	0.146	30
98	208	28.59	0.139	28
99	210	27.23	0.133	27
100	212	25.94	0.126	26
101	214	24.70	0.121	25
102	216	23.53	0.118	24

Temp Degree-C	Temp Degree-F	Thermistor Ohms	Volts Vo (AN3)	10-Bit ADC-Value
103	217	22.41	0.101	22
104	219	21.35	0.140	21
105	221	20.33	0.100	20
106	223	19.36	0.09	19
107	225	18.44	0.09	19
108	226	17.57	0.08	18
109	228	16.73	0.08	17
110	230	15.94	0.07	16
111	232	15.18	0.05	15
112	234	14.46	0.07	15
113	235	13.77	0.06	14
114	237	13.12	0.06	13
115	239	12.49	0.06	13
116	241	11.90	0.05	12
117	243	11.33	0.05	11
118	244	10.79	0.05	11
119	246	10.28	0.05	10
120	248	9.71	0.04	10
121	250	9.33	0.04	9
122	252	8.88	0.04	9
123	253	8.46	0.04	9
124	255	8.06	0.04	8
125	257	7.68	0.03	8
126	259	7.31	0.03	7
127	261	6.96	0.03	7
128	262	6.63	0.03	7
129	264	6.32	0.03	6
130	266	6.02	0.03	6
131	268	5.73	0.02	6
132	270	5.46	0.02	6
133	271	5.20	0.02	5
134	273	4.95	0.02	5
135	275	4.72	0.02	5
136	277	4.49	0.02	5
137	279	4.28	0.02	4
138	280	4.08	0.02	4
139	282	3.88	0.02	4
140	284	3.70	0.02	4
141	286	3.52	0.02	4
142	288	3.35	0.02	3
143	289	3.19	0.02	3
144	291	3.04	0.02	3
145	293	2.90	0.01	3
146	295	2.76	0.01	3
147	297	2.63	0.01	3
148	298	2.50	0.01	3
149	300	2.39	0.01	2
150	302	2.27	0.01	2

For use with PICBasic-Pro Lookup Table

Temp Degree-C	Temp Degree-F	Thermistor Ohms	Volts Vo (AN3)	10-Bit ADC-Value
1	0.020	3.90	139	282
2	0.039	7.90	124	256
3	0.059	11.9	116	241
4	0.078	15.9	110	230
5	0.098	19.9	105	222
6	0.117	24.0	102	215
7	0.137	28.1	98	209
8	0.156	32.3	96	204
9	0.176	36.4	93	199
10	0.195	40.7	91	195
11	0.215	44.9	89	192
12	0.234	49.2	87	188
13	0.254	53.5	85	185
14	0.273	57.9	84	182
15	0.293	62.2	82	180
16	0.313	66.7	81	177
17	0.332	71.1	79	175
18	0.352	75.6	78	172
19	0.371	80.2	77	170
20	0.391	84.7	76	168
21	0.410	89.4	75	166
22	0.430	94.0	74	164
23	0.449	98.7	73	163
24	0.469	103.4	72	161
25	0.488	108.2	71	159
26	0.508	113.0	70	158
27	0.527	117.9	69	156
28	0.547	122.8	68	155
29	0.566	127.8	67	153
30	0.586	132.7	66	152
31	0.605	137.8	66	150
32	0.625	142.9	65	149
33	0.645	148.0	64	148
34	0.664	153.2	64	146
35	0.684	158.4	63	145
36	0.703	163.6	62	144
37	0.723	168.9	62	143
38	0.742	174.3	61	142
39	0.762	179.7	60	140
40	0.781	185.2	60	139
41	0.801	190.7	59	138
42	0.820	196.3	58	137
43	0.840	201.9	58	136
44	0.859	207.5	57	135
45	0.879	213.3	57	134
46	0.898	219.0	56	133
47	0.918	224.9	56	132
48	0.938	230.8	55	131
49	0.95 5	236.7	55	130
50	0.977	242.7	54	129
51	0.996	248.8	54	128
52	1.016	254.9	53	128
53	1.035	261.1	53	127
54	1.055	267.3	52	126
55	1.074	273.6	52	125
56	1.094	280.0	51	124
57	1.113	286.4	51	123
58	1.133	292.9	50	122
59	1.152	299.5	50	122
60	1.172	306.1	49	121
61	1.191	312.8	49	120
62	1.211	319.6	48	119
63	1.230	326.4	48	118
64	1.255	333.3	48	118
65	1.270	340.3	47	117
66	1.289	347.4	47	116
67	1.309	354.5	46	115

Temp Degree-C	Temp Degree-F	Thermistor Ohms	Volts Vo (AN3)	10-Bit ADC-Value

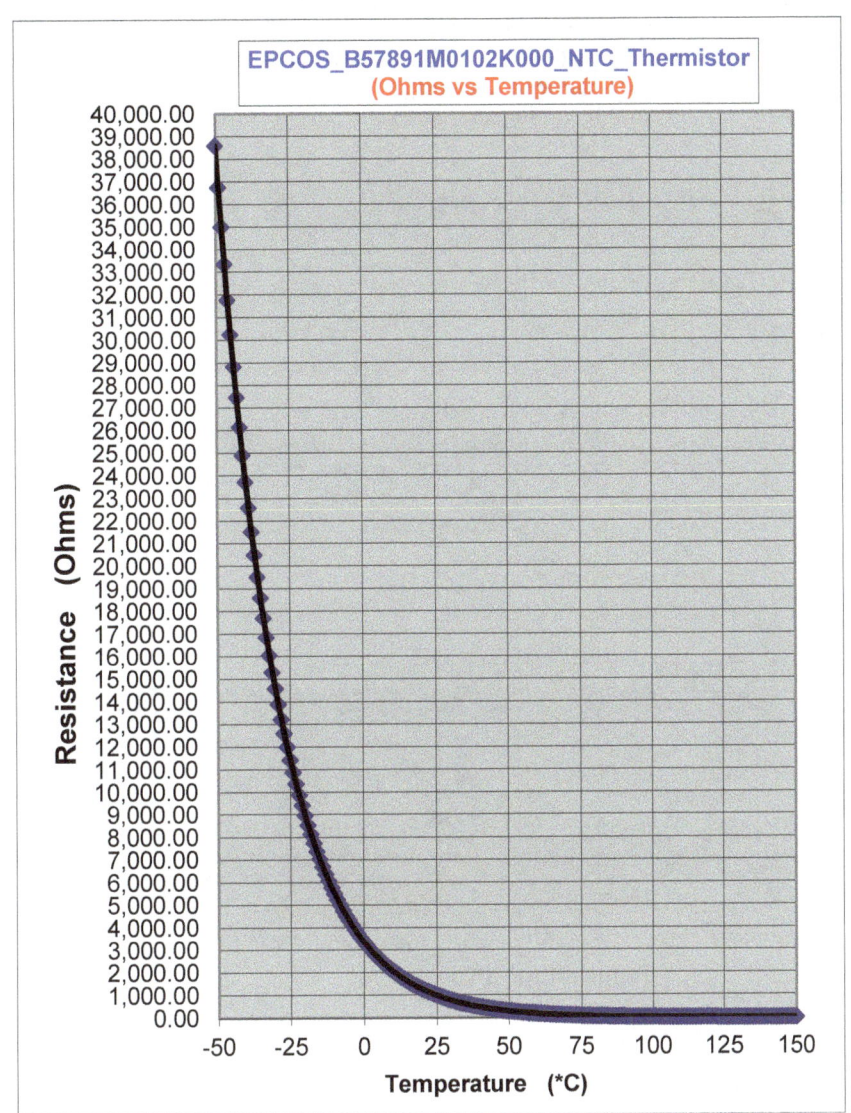

Res=3380*exp(-0.0487*Temp)

Figure 6.2: Data & Curves for Negative Temperature Coefficient Thermistor

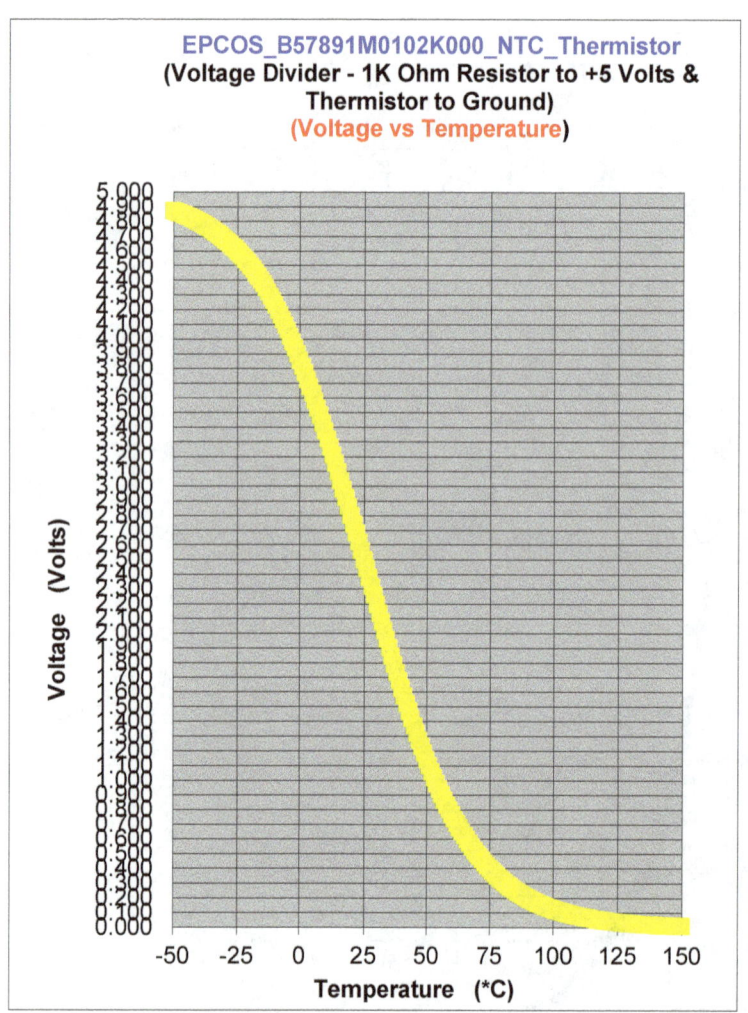

Vo(AN3)=(Res/(Res+1000))*5

Figure 6.3: Data & Curves for Negative Temperature Coefficient Thermistor

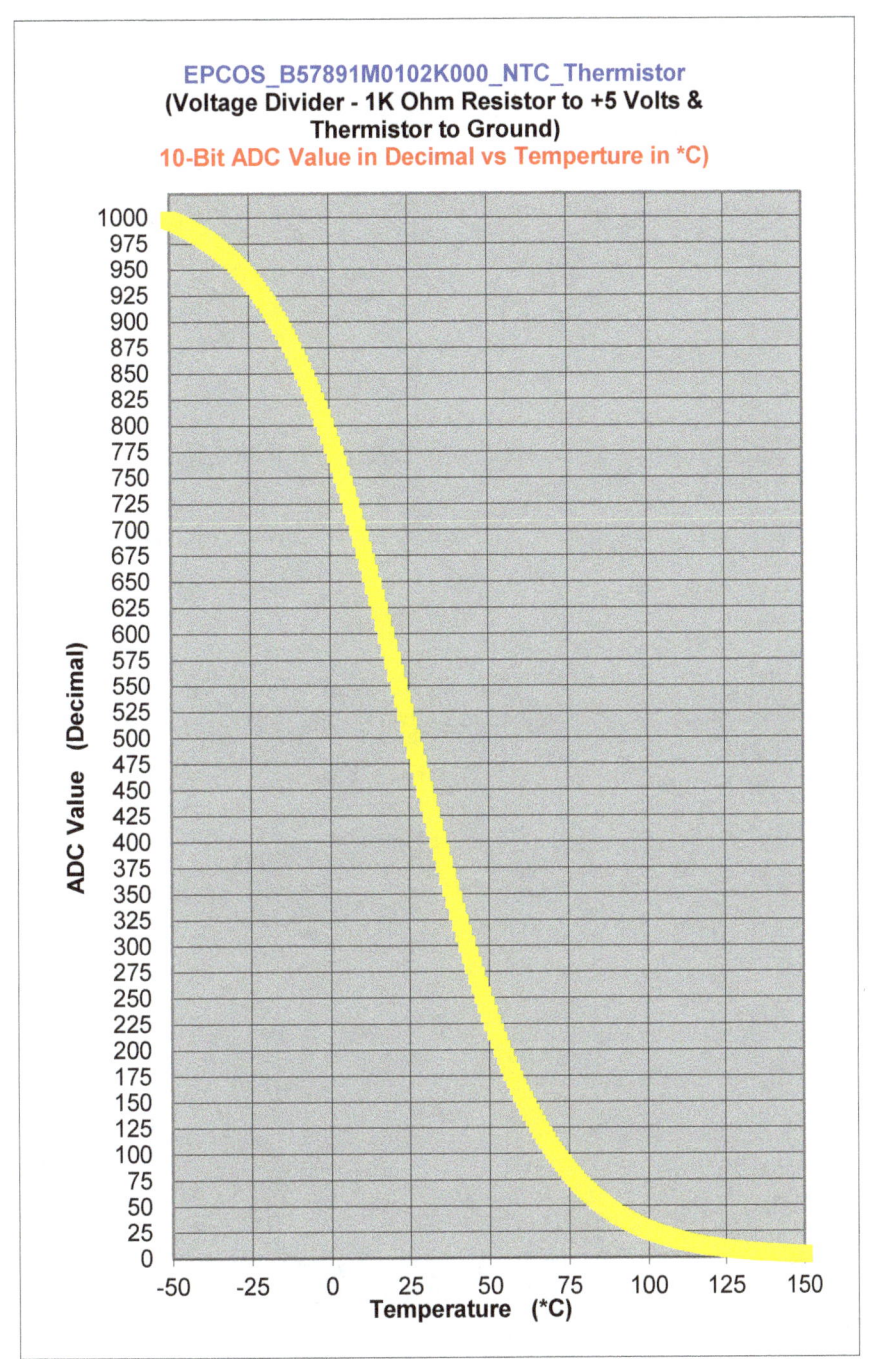

10-Bit ADCValue=(Vo(AN3)/5)*1024

Figure 6.4: Data & Curves for Negative Temperature Coefficient Thermistor

Engineering Practices for the PIC Microcontroller

By: Prof. Sal R. Riggio Jr., PhD, PE

Chapter - 7

Laboratory Experiment - 5

The I2C Real-Time Programmable Clock/Calendar

Introduction

The purpose of this lab experiment is for the student to develop basic skills in making practical use of the Microchip **PIC18F4685** Microcontroller and become familiar with the importance, operation and software and hardware implementation of an I2C Bus device, such as, a Real-Time-Clock-Calendar or an EEPROM.

The student is expected to analyze and write **PICBasic-Pro Code** and **CCS C-code**, which enables the proper operation of the **I2C_Real_Time_Programmable_Clock_Calendar.**

It is suggested that the student **read** the **PICBasic-Pro code** and the **CCS C-code**, particularly the **comments** written on each line, to learn how to write code for the PIC microcontrollers. It has been proven many times to be the most effective way to learn how generate effective code for the PIC microcontrollers. This code is shown at the end of this chapter

Portions of the Experiment Board Schematics, which are relevant to this experiment, are shown below in **Figure 7.1**. Please refer to Appendix-A for the complete set of **Experiment Board Schematics** and please refer to **Appendix-B** for the complete **Lab Board Schematic**.

Hardware Setup

1.) Connect the Square end of the USB Type-A to Type-B Cable to connector JUSB1 of the Experiment Board. Then, connect the other end of this Cable to one of the USB Inputs of the Computer.
2.) Connect the DB9 end of the DB9 Serial-to-USB Cable to connector P1 of the Experiment Board. Then, connect the USB end of this Cable to one of the other USB Inputs of the Computer. The DB9 Serial-to-USB Cable is used to communicate between the Serial HyperTerminal of the Computer and the Experiment Board.
3.) The Jumper of **JB12** is to be placed in **Positions 3-5 & 4-6.**

Assignment #1

Analyze the **PICBasic-Pro Code** and the **CCS C-code** below, which enables the **I2C Real_Time_Clock_Calandar,** located on the experiment board, to **Display** the correct **Seconds, Minutes, Hours, Day, Date, Month** and **Year**, in both **Military** and **Standard Time (AM & PM)**, on the **Liquid-Crystal-Display**, and on the **RS232 HyperTerminal**.

Assignment #2

"The Alarm Clock"

Write a **PICBasic-Pro Code** routine and a **CCS C-code** routine, which will cause the **Red LED** to **Flash, Continuously, One-Half** of a **Second On**, and then **One-Half** of a **Second Off**, until the **Alarm** is **Deactivated**. When **Connector J8, Pin-1** is **Connected** to **J8, Pin-2**, of the **Experiment Board**, with a **Jumper Wire** or an **External Switch Set** in the **Closed Position**, the **Alarm** is **Set** to **Flash** the **Red LED** at the **Specified Time**. Once **Activated**, the **Red LED** will continue to **Flash** for **3 Minutes**, or until the **Jumper Wire** is **Removed** or the **External Switch** is **Opened**. If the **Jumper Wire** is **Replaced** or the **External Switch** is **Closed**, within the **First Minute**, the **Red LED** will continue to **Flash** for the **Remainder** of the **3 Minutes** "Limited Loop" Cycle. The Student may **Set** the **Alarm Activation** to occur at anytime, by Identifying the **Hour**, the **Minute** and the **Day of the Week**. The **Clock** and **Calendar Values DO NOT** have to be **Continuously Updated** and **Displayed** on the **Liquid-Crystal-Display** and the **HyperTerminal**, while the **Alarm** is **Flashing**. A **One** to **Three Second Delay** before updating the Liquid-Crystal-Display and the HyperTerminal is acceptable.

The Student may **Set** the **Alarm Time** within the **Alarm MicroCode**.

Connector J8, of the Experiment Board, **Controls** the **Logical State** of **Port-C, Bit-1**, as shown in **Figure 7.1**.

Jumper-In or External Switch Closed means that **Port-C, Bit-1** equals a **Logical "1"**

Jumper-Out or External Switch Open means that **Port-C, Bit-1** equals a **Logical "0"**

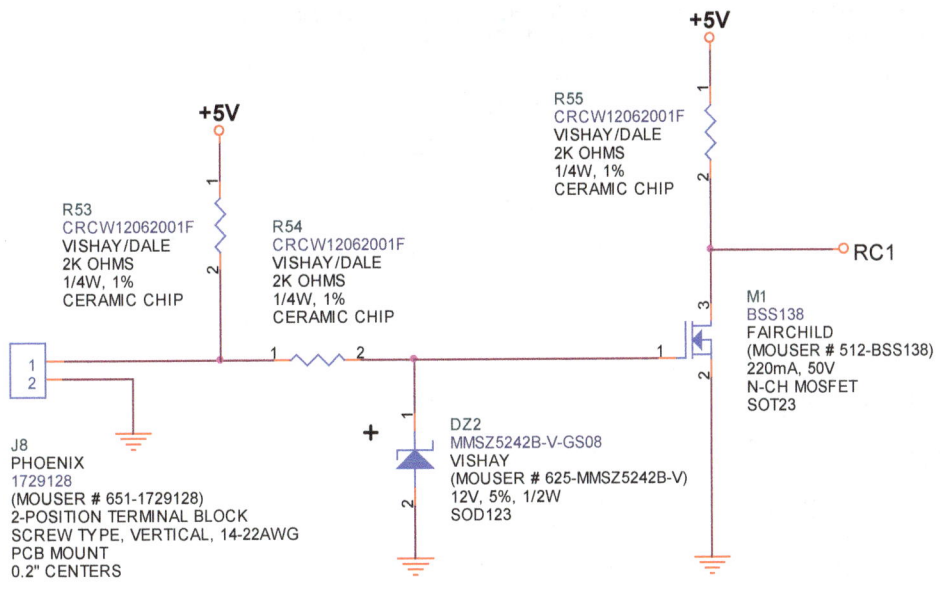

Figure 7.1: Logical Control for Port-C, Bit-1 with a Switch or Jumper connected to J8

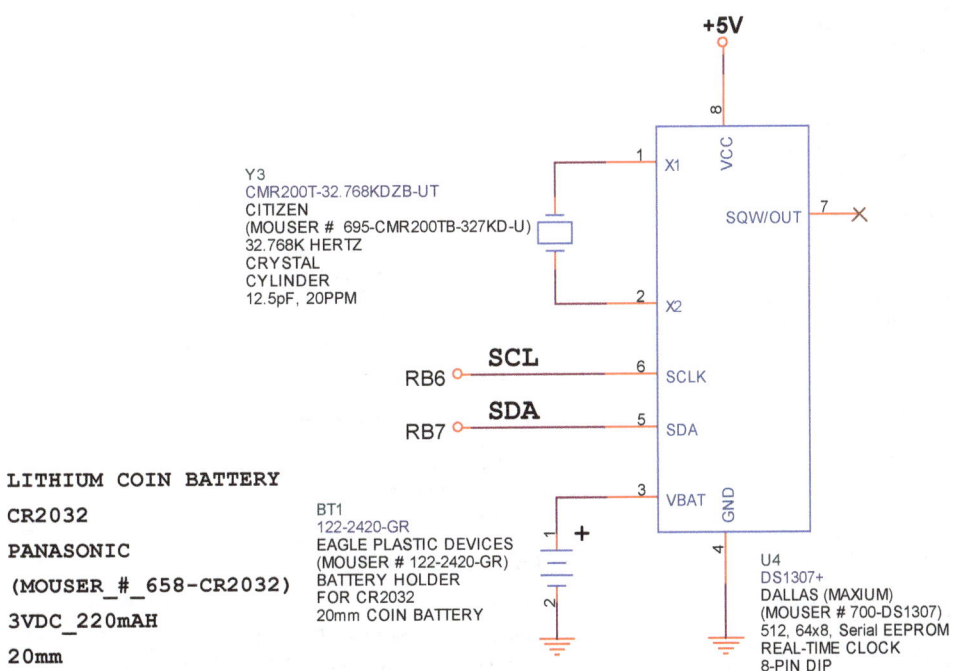

Figure 7.2: I2C Real-Time Programmable Clock-Calendar Module

Laboratory Experiment #5

```
'Lab-5A_PICBasic-Pro_I2C_Real_Time_Programmable_Clock_Calendar_Set
INCLUDE "modedefs.bas"              'Identifies the Command File

DEFINE OSC 20                       'Set PIC-Pro Oscillator Value in Mega Hertz
DEFINE ADC_BITS 10                  'Set for 10-Bit Internal ADC

DEFINE ADC_CLOCK 5                  'DEFINE ADC_CLOCK 1, for OSC=10, MAX ADC
                                    'Clock=10MHz/8) = 1.25MHz
                                    'DEFINE ADC_CLOCK 5, for OSC=16, MAX ADC
                                    'Clock=16MHz/16) = 1.00MHz
                                    'DEFINE ADC_CLOCK 5, for OSC=20, MAX ADC
                                    'Clock=20MHz/16) = 1.25MHz
                                    'DEFINE ADC_CLOCK 2, for OSC=24, MAX ADC
                                    'Clock=24MHz/32) = 1.25MHz
                                    'DEFINE ADC_CLOCK 2, for OSC=32, MAX ADC
                                    'Clock=32MHz/32) = 1.00MHz
                                    'DEFINE ADC_CLOCK 6, for OSC=48, MAX ADC
                                    'Clock=48MHz/64) = 0.75MHz

DEFINE CCP1_BIT 2                   'Setup Hardware PWM Timer
DEFINE LCD_BITS 4                   'Setup LCD for 4-Bit Mode
DEFINE LCD_DREG PORTB               'Drive LCD from Port-B
DEFINE LCD_DBIT 0                   'Starting with Bit-0
DEFINE LCD_RSREG PORTA              'Drive LCD Register-Select from
DEFINE LCD_RSBIT 4                  'Port-A, Bit-4
DEFINE LCD_EREG PORTA               'Drive LCD Enable from
DEFINE LCD_EBIT 5                   'Port-A, Bit-5
DEFINE LCD_LINES 2                  'LCD can hold 2-Lines of Characters
DEFINE LCD_COMMANDSUS 4000          'Delay between Commands
DEFINE LCD_DATAUS 200               'Delay between Data Transfers

ADCON0 = %00001111                  'ADC Enabled, Analog Input 3 Enabled
ADCON1 = %00001011                  'ADC Voltage Range is 0 to 5 Volts &
                                    'Analog Inputs 0 Thru 3 are available
ADCON2 = %10101101                  'Right Justify to use Upper 8-Bits, only,
                                    '12*Tad, & ADC Clock=20MHz/16=1.25MHz

W0 VAR WORD                         'Set 16-Bit Word Variable,W0
W1 VAR WORD                         '  "      "    "     "    ,W1
W2 VAR WORD                         '  "      "    "     "    ,W2
W3 VAR WORD                         '  "      "    "     "    ,W3

B0 VAR BYTE                         'Set 8-Bit Byte Variable,B0
B1 VAR BYTE                         '  "   "    "     "    ,B1
RTCSec VAR BYTE                     '  "   "    "     "    ,RTCSec
RTCMin VAR BYTE                     '  "   "    "     "    ,RTCMin
RTCHour VAR BYTE                    '  "   "    "     "    ,RTCHour (Military Time)
RTCDay VAR BYTE                     '  "   "    "     "    ,RTCDay
RTCDate VAR BYTE                    '  "   "    "     "    ,RTCDate
RTCMonth VAR BYTE                   '  "   "    "     "    ,RTCMonth
RTCYear VAR BYTE                    '  "   "    "     "    ,RTCYear
RTCCtrl VAR BYTE                    '  "   "    "     "    ,RTCCtrl
RTCHr VAR BYTE                      '  "   "    "     "    ,RTCHr (Civilian Time AM/PM)

TRISB=$FF                           'Make PortB an Input Port to Enable
                                    'In-Circuit-Programming
PAUSE 1000                          'Pause or Delay for 1 Seconds to Enable
                                    'In-Circuit-Programming
LCDOUT $FE,1                        'Clear the LCD
LCDOUT $FE, $80                     'Set the Cursor to Line 1, Position 0
LCDOUT "Power Up"                   'Send, Power Up, to the LCD

SERout PORTC.6,T2400,["Power Up",13,10,10]
                                    'Send "Power Up" to the HyperTerminal
                                    'with a Carriage Return & two Line Feeds,
                                    'at 2400 Bits/Second, 8-Bits, 1-Stop Bit,
                                    'No Parity, Driven True (T2400).

PAUSE 2000                          'Pause or Delay for 2 Seconds
                                    ''to Read "Power Up" in the LCD

LCDOUT $FE,1                        'Clear the LCD
TRISB=$00                           'Make PortB an Output Port to Allow for Normal Operation
TRISD=$F0                           'Make PortD, Bits-4,5,6,7 Inputs, & Bits-3
```

```
RTCSec=$00              'RTCSec   00 --> 59 Seconds
RTCMin=$50              'RTCMin   00 --> 59 Minutes
RTCHour=$12             'RTCHour  00 --> 23 Hour of the Day (Military Time)
RTCDay=$02              'RTCDay   01 --> 07 Sunday Thru Saturday
RTCDate=$24             'RTCDate  01 --> 31 Date of the Month
RTCMonth=$01            'RTCMonth 01 --> 12 January Thru December
RTCYear=$11             'RTCYear  00 --> 99 Year
RTCCtrl=$00             'RTCCtrl  Square-Wave Out Control (Not Used)
RTCHr=$01               'RTCHr  Hour of the Day (Civilian Time AM/PM)

PAUSE 10
I2CWRITE PORTB.7,PORTB.6,$D0,$00,
[RTCSec,RTCMin,RTCHour,RTCDay,RTCDate,RTCMonth,RTCYear,RTCCtrl]
                        'PORTB.7 is the Data
                        'PORTB.6 is the Clock
                        'D0 is the Hex Address of the Part
                        '00 is the Starting Address of the Memory in HEX
PAUSE 10
LCDOUT $FE,1            'Clear the LCD
LCDOUT $FE, $80         'Set Cursor to Line 1, Position 0
LCDOUT "Clock/Calendar" 'Send, Clock/Calendar, to the LCD
LCDOUT $FE, $C0         'Set Cursor to Line 2, Position 0
LCDOUT "is Set"         'Send, is Set, to the LCD

SERout2 PORTC.6,396,["Clock/Calendar is Set",13,10,10]
                        'Send, Clock/Calendar is Set, to the HyperTerminal
                        'with a Carriage Return & two Line Feeds
                        'T2400 Means: 2400 Baud, True Polarity,
                        'No Parity, 8-Bits, 1-Stop Bit
                        'See the SEROUT Command in the
                        'PICBasic-Pro Manual
End
```

Laboratory Experiment #5

```
'Lab-5B_PICBasic-Pro_I2C_Real_Time_Programmable_Clock_Calendar_Run
 INCLUDE "modedefs.bas"             'Identifies the Command File

 DEFINE OSC 20                      'Set PIC-Pro Oscillator Value in Mega Hertz
 DEFINE ADC_BITS 10                 'Set for 10-Bit Internal ADC

 DEFINE ADC_CLOCK 5                 'DEFINE ADC_CLOCK 1, for OSC=10, MAX ADC
                                    'Clock=10MHz/8) = 1.25MHz
                                    'DEFINE ADC_CLOCK 5, for OSC=16, MAX ADC
                                    'Clock=16MHz/16) = 1.00MHz
                                    'DEFINE ADC_CLOCK 5, for OSC=20, MAX ADC
                                    'Clock=20MHz/16) = 1.25MHz
                                    'DEFINE ADC_CLOCK 2, for OSC=24, MAX ADC
                                    'Clock=24MHz/32) = 1.25MHz
                                    'DEFINE ADC_CLOCK 2, for OSC=32, MAX ADC
                                    'Clock=32MHz/32) = 1.00MHz
                                    'DEFINE ADC_CLOCK 6, for OSC=48, MAX ADC
                                    'Clock=48MHz/64) = 0.75MHz

 DEFINE CCP1_BIT 2                  'Setup Hardware PWM Timer
 DEFINE LCD_BITS 4                  'Setup LCD for 4-Bit Mode
 DEFINE LCD_DREG PORTB              'Drive LCD from Port-B
 DEFINE LCD_DBIT 0                  'Starting with Bit-0
 DEFINE LCD_RSREG PORTA             'Drive LCD Register-Select from
 DEFINE LCD_RSBIT 4                 'Port-A, Bit-4
 DEFINE LCD_EREG PORTA              'Drive LCD Enable from
 DEFINE LCD_EBIT 5                  'Port-A, Bit-5
 DEFINE LCD_LINES 2                 'LCD can hold 2-Lines of Characters
 DEFINE LCD_COMMANDSUS 4000         'Delay between Commands
 DEFINE LCD_DATAUS 200              'Delay between Data Transfers

 ADCON0 = %00001111                 'ADC Enabled, Analog Input 3 Enabled
 ADCON1 = %00001011                 'ADC Voltage Range is 0 to 5 Volts &
                                    'Analog Inputs 0 Thru 3 are available
 ADCON2 = %10101101                 'Right Justify to use Upper 8-Bits, only,
                                    '12*Tad, & ADC Clock=20MHz/16=1.25MHz

 W0 VAR WORD                        'Set 16-Bit Word Variable,W0
 W1 VAR WORD                        '  "    "    "    "    "  ,W1
 W2 VAR WORD                        '  "    "    "    "    "  ,W2
 W3 VAR WORD                        '  "    "    "    "    "  ,W3

 B0 VAR BYTE                        'Set 8-Bit Byte Variable,B0
 B1 VAR BYTE                        '  "   "    "    "      ,B1
 RTCSec VAR BYTE                    '  "   "    "    "      ,RTCSec
 RTCMin VAR BYTE                    '  "   "    "    "      ,RTCMin
 RTCHour VAR BYTE                   '  "   "    "    "      ,RTCHour  (Military Time)
 RTCDay VAR BYTE                    '  "   "    "    "      ,RTCDay
 RTCDate VAR BYTE                   '  "   "    "    "      ,RTCDate
 RTCMonth VAR BYTE                  '  "   "    "    "      ,RTCMonth
 RTCYear VAR BYTE                   '  "   "    "    "      ,RTCYear
 RTCCtrl VAR BYTE                   '  "   "    "    "      ,RTCCtrl
 RTCHr VAR BYTE                     '  "   "    "    "      ,RTCHr  (Civilian Time AM/PM)


 TRISB=$FF                          'Make PortB an Input Port to Enable
                                    'In-Circuit-Programming
 PAUSE 1000                         'Pause or Delay for 1 Seconds to Enable
                                    'In-Circuit-Programming
 LCDOUT $FE,1                       'Clear the LCD
 LCDOUT $FE, $80                    'Set the Cursor to Line 1, Position 0
 LCDOUT "Power Up"                  'Send, Power Up, to the LCD

 SERout PORTC.6,T2400,["Power Up",13,10,10]
                                    'Send "Power Up" to the HyperTerminal
                                    'with a Carriage Return & two Line Feeds,
                                    'at 2400 Bits/Second, 8-Bits, 1-Stop Bit,
                                    'No Parity, Driven True (T2400).

 PAUSE 2000                         'Pause or Delay for 2 Seconds
                                    'to Read "Power Up" in the LCD
 LCDOUT $FE,1                       'Clear the LCD
 TRISB=$00                          'Make PortB an Output Port to Allow for Normal Operation
 TRISD=$F0                          'Make PortD, Bits-4,5,6,7 Inputs, & Bits-3
```

```
'RTCSec    00 --> 59 Seconds
'RTCMin    00 --> 59 Minutes
'RTCHour   00 --> 23 Hour of the Day (Military Time)
'RTCDay    01 --> 07 Sunday Thru Saturday
'RTCDate   01 --> 31 Date of the Month
'RTCMonth  01 --> 12 January Thru December
'RTCYear   00 --> 99 Year
'RTCCtrl   Square-Wave Out Control (Not Used)
'RTCHr     Hour of the Day (Civilian Time AM/PM)

LOOP:
PAUSE 10
I2CREAD PORTB.7,PORTB.6,$D0,$00,[RTCSec,RTCMin,RTCHour,RTCDay,RTCDate,
RTCMonth,RTCYear,RTCCtrl]
                                'PORTB.7 is The Data
                                'PORTB.6 is the Clock,
                                'D0 is the Hex Address of the Part
                                '00 is the Starting of the Memory in HEX
PAUSE 10

IF RTCHOUR=$00 THEN RTCHR=$12
IF RTCHOUR=$01 THEN RTCHR=$01
IF RTCHOUR=$02 THEN RTCHR=$02
IF RTCHOUR=$03 THEN RTCHR=$03
IF RTCHOUR=$04 THEN RTCHR=$04
IF RTCHOUR=$05 THEN RTCHR=$05
IF RTCHOUR=$06 THEN RTCHR=$06
IF RTCHOUR=$07 THEN RTCHR=$07
IF RTCHOUR=$08 THEN RTCHR=$08
IF RTCHOUR=$09 THEN RTCHR=$09
IF RTCHOUR=$10 THEN RTCHR=$10
IF RTCHOUR=$11 THEN RTCHR=$11
IF RTCHOUR=$12 THEN RTCHR=$12
IF RTCHOUR=$13 THEN RTCHR=$1
IF RTCHOUR=$14 THEN RTCHR=$2
IF RTCHOUR=$15 THEN RTCHR=$3
IF RTCHOUR=$16 THEN RTCHR=$4
IF RTCHOUR=$17 THEN RTCHR=$5
IF RTCHOUR=$18 THEN RTCHR=$6
IF RTCHOUR=$19 THEN RTCHR=$7
IF RTCHOUR=$20 THEN RTCHR=$8
IF RTCHOUR=$21 THEN RTCHR=$9
IF RTCHOUR=$22 THEN RTCHR=$10
IF RTCHOUR=$23 THEN RTCHR=$11

IF RTCSEC=$00 THEN LCDOUT $FE, 1      'Then Clear the LCD

LCDOUT $FE, $80                       'Set Cursor to Line 1, Position 0
LCDOUT HEX2 RTCMONTH,"/",HEX2 RTCDATE,"/",HEX2 RTCYEAR
                                'Send, Month/Date/Year, to the LCD
                                'HEX2 Displays Two Hex Digits, HEX Displays
                                'One Hex Digit

SEROUT2 PORTC.6,396,[HEX2 RTCMONTH,"/",HEX2 RTCDATE,"/",HEX2 RTCYEAR,13,10]
                                'Send, Month/Date/Year, to the HyperTerminal
                                'with a Carriage Return & one Line Feeds
                                'HEX2 Displays Two Hex Digits, HEX Displays
                                'One Hex Digit
                                '396 Means: 2400, Driven, True, No Parity,
                                '8-Bits, 1-Stop Bit
                                'See Appendix-A of the PICBasic-Pro Manual

LCDOUT $FE, $88                       'Set Cursor to Line 1, Position 9
LCDOUT HEX2 RTCHOUR,":",HEX2 RTCMIN,":",HEX2 RTCSEC
                                'Send Military, Hour/Minute/Second, to the LCD
                                'HEX2 Displays Two Hex Digits, HEX Display
                                'One Hex Digit

SEROUT2 PORTC.6,396,[HEX2 RTCHOUR,":",HEX2 RTCMIN,":",HEX2 RTCSEC,13,10]
                                'Send Military, Hour/Minute/Second,
                                'to the HyperTerminal
                                'with a Carriage Return & one Line Feeds
                                'HEX2 Displays Two Hex Digits, HEX Displays
                                'One Hex Digit
                                '396 Means: 2400, Driven, True, No Parity,
                                '8-Bits, 1-Stop Bit
                                'See Appendix-A of the PICBasic-Pro Manual
```

```
LCDOUT $FE, $C5                                    'Set Cursor to Line 2, Position 6
LCDOUT HEX2 RTCHR,":",HEX2 RTCMIN,":",HEX2 RTCSEC
                                                   'Send Civilian, Hour/Minute/Second, to the LCD
                                                   'HEX2 Displays Two Hex Digits, HEX Displays
                                                   'One Hex Digit

SEROUT2 PORTC.6,396,[HEX2 RTCHR,":",HEX2 RTCMIN,":",HEX2 RTCSEC," "]
                                                   'Send Civilian, Hour/Minute/Second,
                                                   'to the HyperTerminal
                                                   'with a Space before AM or PM
                                                   'HEX2 Displays Two Hex Digits, HEX Displays
                                                   'One Hex Digit
                                                   '396 Means: 2400, Driven, True, No Parity,
                                                   '8-Bits, 1-Stop Bit
                                                   'See Appendix-A of the PICBasic-Pro Manual

if (RTCHOUR>=$12) THEN PM                          'Compute Civilian Time

AM:
LCDOUT $FE, $CE                                    'Set Cursor to Line 2, Position 15
LCDOUT "AM"                                        'Send "AM" to the LCD
SEROUT2 PORTC.6,396,["AM",13,10]
                                                   'Send "AM" to the HyperTerminal
                                                   'with a Carriage Return & one Line Feed
                                                   '396 Means: 2400 Baud, True Polarity,
                                                   'No Parity, 8-Bits, 1-Stop Bit

GOTO DAY

PM:
LCDOUT $FE, $CE                                    'Set Cursor to Line 2, Position 15
LCDOUT "PM"                                        'Send "PM" to the LCD
SEROUT2 PORTC.6,396,["PM",13,10]
                                                   'Send "PM" to the HyperTerminal
                                                   'with a Carriage Return & one Line Feed
                                                   '396 Means: 2400 Baud, True Polarity,
                                                   'No Parity, 8-Bits, 1-Stop Bit

GOTO DAY

DAY:
IF RTCDAY=$01 THEN Sun
IF RTCDAY=$02 THEN Mon
IF RTCDAY=$03 THEN Tue
IF RTCDAY=$04 THEN Wed
IF RTCDAY=$05 THEN Thu
IF RTCDAY=$06 THEN Fri
IF RTCDAY=$07 THEN Sat

Sun:
LCDOUT $FE, $C0                                    'Set Cursor to Line 2, Position 0
LCDOUT "Sun"                                       'Send, Sun, to the LCD
SEROUT2 PORTC.6,396,["Sunday",13,10,10]
                                                   'Send, Sunday, to the HyperTerminal
                                                   'with a Carriage Return & two Line Feeds
                                                   '396 Means: 2400 Baud, True Polarity,
                                                   'No Parity, 8-Bits, 1-Stop Bit

if (RTCSec=$00) THEN pause 1000                    'Then Pause for 1 Second
GOTO LOOP

Mon:
LCDOUT $FE, $C0
LCDOUT "Mon"
SEROUT2 PORTC.6,396,["Monday",13,10,10]
if (RTCSec=$00) THEN pause 1000
GOTO LOOP

Tue:
LCDOUT $FE, $C0
LCDOUT "Tue"
SEROUT2 PORTC.6,396,["Tuesday",13,10,10]
if (RTCSec=$00) THEN pause 1000
GOTO LOOP

Wed:
LCDOUT $FE, $C0
LCDOUT "Wed"
SEROUT2 PORTC.6,396,["Wednesday",13,10,10]
if (RTCSec=$00) THEN pause 1000
GOTO LOOP
```

Thu:
LCDOUT $FE, $C0
LCDOUT "Thu"
SEROUT2 PORTC.6,396,["Thursday",13,10,10]
if (RTCSec=$00) THEN pause 1000
GOTO LOOP

Fri:
LCDOUT $FE, $C0
LCDOUT "Fri"
SEROUT2 PORTC.6,396,["Friday",13,10,10]
if (RTCSec=$00) THEN pause 1000
GOTO LOOP

Sat:
LCDOUT $FE, $C0
LCDOUT "Sat"
SEROUT2 PORTC.6,396,["Saturday",13,10,10]
if (RTCSec=$00) THEN pause 1000
GOTO LOOP

END

Laboratory Experiment #5

'Lab-5C_CCS_C-Code_I2C_Real_Time_Programmable_Clock_Calendar_Set

```c
#include <18F4685.h>                              //Identify Microcontroller
#device adc=10                                    //Identify ADC Bit Width
#fuses HS,NOWDT,NOPROTECT,NOLVP,MCLR              //Setup Programmer Oscillator Value
                                                  //and make Master Clear Pin an
                                                  //Enable Master Clear=MCLR
                                                  //Disable Master Clear=NOMCLR
#use delay(clock=20000000)                        //Setup C-Code Oscillator Value
#use fast_io(d)                                   //Leave the State of Port-D (all bits)
                                                  //the same until changed, again

#use rs232(baud=2400, xmit=PIN_C6, rcv=PIN_C7, stream=HT)
                                                  //Set RS232 HyperTerminal
                                                  //Communication Parameters
#use rs232(baud=2400, xmit=PIN_B4, rcv=PIN_B5, invert, stream=BB)
                                                  //Set RS232 Board-to-Board
                                                  //Communication Parameters
#use I2C(master, SCL=PIN_B6, SDA=PIN_B7)          //Set I2C Communication Parameters
#include <lcd_flex.c>                             //Identify LCD Driver File
#include <math.h>                                 //Include Math Functions

float W0,W1,TC,TF,TCAL; //16-Bit Floating Point Integer Word Variable

void main()
    {
    set_tris_b(0xFF);                             //Make PortB an Input Port to
                                                  //Enable In-Circuit-Programming
    delay_ms(1000);                               //Pause or Delay for 1 second

    lcd_init();                                   //Initialize the LCD
    lcd_putc(0x0c);                               //Clear the LCD
    lcd_gotoxy(1,1);                              //Set the Cursor to Position-1 of Line-1
    printf(lcd_putc,"Power Up");                  //Send, Power Up, to the LCD

    fprintf(HT, "\n\rPower Up\n\r");              //Send, Power Up, to the HyperTerminal with
                                                  //a next Line & a Carriage Return
    delay_ms(1000);                               //Pause or Delay for 1 second

    setup_port_a(AN0_to_AN3);                     //Identify all of the Analog Inputs
    setup_adc(VSS_VDD);                           //Setup the ADC Voltage Range to be
                                                  //0 Volts to 5 Volts
                                                  //(5v-0v)/1024~=.005 Volts Per Bit

    setup_adc(ADC_CLOCK_DIV_32);                  //for delay(clock=10000000),
                                                  //setup_adc(ADC_CLOCK_DIV_8),
                                                  //10MHz/8=1.25MHz
                                                  //for delay(clock=16000000),
                                                  //setup_adc(ADC_CLOCK_DIV_16),
                                                  //16MHz/16=1.00MHz
                                                  //for delay(clock=20000000),
                                                  //setup_adc(ADC_CLOCK_DIV_16),
                                                  //20MHz/16=1.25MHz
                                                  //for delay(clock=24000000),
                                                  //setup_adc(ADC_CLOCK_DIV_32),
                                                  //24MHz/32=1.25MHz
                                                  //for delay(clock=32000000),
                                                  //setup_adc(ADC_CLOCK_DIV_32),
                                                  //32MHz/32=1.00MHz
                                                  //for delay(clock=48000000),
                                                  //setup_adc(ADC_CLOCK_DIV_64),
                                                  //48MHz/64=0.75MHz

    set_adc_channel(0);                           //Identify the Analog Input to be
                                                  //Sampled (Channel Input Zero)

    setup_ccp1(CCP_PWM);                          //Setup Capture/Compare/PWM Pin-1 to PWM Mode

    setup_ccp2(CCP_CAPTURE_RE);                   //Setup Capture/Compare/PWM Pin-2 to
                                                  //Capture Mode

    setup_timer_1(T1_INTERNAL);                   //Setup Timer-1, which controls the
                                                  //Capture/Compare portion of
                                                  //Capture/Compare/PWM Pin-2, only,
                                                  //to Internal Mode

    setup_timer_2(T2_DIV_BY_4, 19, 1);            //Setup Timer-2, which controls the PWM portion
                                                  //of Capture/Compare/PWM Pin-1, only.
                                                  //DIV_BY_Mode, Prescale (Period), Postscale
                                                  //The Following Limits are for the
                                                  //PIC18F4685 Microcontroller
                                                  //Mode may equal 1 or 4 or 16
                                                  //Prescale (Period) may equal any integer
                                                  //between 0 and 255
                                                  //Postscale may equal any integer between
                                                  //1 and 16 (Post = 1 to set PWM Frequency)
                                                  //T2Freq=OSC/[(4)*(Mode)*(Prescale+1)]=
                                                  //24MegaHz/[(4)*(4)*(PS+1)] = 75KHz
```

```c
enable_interrupts(INT_CCP2);            //Enable the Internal Interrupt Input
                                        //of Capture/Compare/PWM Pin-2

enable_interrupts(GLOBAL);              //Enable Global Interrupts for the
                                        //PIC18F4685 Microcontroller

set_tris_b(0x00);                       //Make PortB an Output Port to Allow for
                                        //Normal Operation
tris_d(0xF0);                           //Make PortD, Bits-4,5,6,7 Inputs
                                        //and Bits-3,2,1,0 Outputs


I2C_START();
I2C_WRITE(0xD0);                        //I2C Device Address
I2C_WRITE(0x00);                        //Initialize I2C Device
I2C_WRITE(0x00);                        //Second
I2C_WRITE(0x47);                        //Minute
I2C_WRITE(0x22);                        //Hour
I2C_WRITE(0x02);                        //Day of the Week
I2C_WRITE(0x23);                        //Date
I2C_WRITE(0x07);                        //Month
I2C_WRITE(0x07);                        //Year
I2C_WRITE(0x80);                        //Unused Square-wave Enable
I2C_STOP();

lcd_putc(0x0c);                         //Clear the LCD
lcd_gotoxy(1,1);                        //Set the Cursor to Position-1 of Line-1
printf(lcd_putc,"Clock/Calendar");      //Send, Clock/Calendar, to the LCD
lcd_gotoxy(1,2);                        //Set the Cursor to Position-1 of Line-2
printf(lcd_putc,"is Set");              //Send, is Set, to the LCD
fprintf(HT, "\n\rThe Clock/Calendar is Set\n\r");   //Send, The Clock/Calendar is Set,"
                                        //to the HyperTerminal with a next Line & a Carriage Return

}
```

'Lab-5D_CCS_C-Code_I2C_Real_Time_Programmable_Clock_Calendar_Run

```c
#include <18F4685.h>                          //Identify Microcontroller
#device adc=10                                //Identify ADC Bit Width
#fuses HS,NOWDT,NOPROTECT,NOLVP,MCLR          //Setup Programmer Oscillator Value
                                              //and make Master Clear Pin an
                                              //Enable Master Clear=MCLR
                                              //Disable Master Clear=NOMCLR
#use delay(clock=20000000)                    //Setup C-Code Oscillator Value
#use fast_io(d)                               //Leave the State of Port-D (all bits)
                                              //the same until changed, again

#use rs232(baud=2400, xmit=PIN_C6, rcv=PIN_C7, stream=HT)
                                              //Set RS232 HyperTerminal
                                              //Communication Parameters
#use rs232(baud=2400, xmit=PIN_B4, rcv=PIN_B5, invert, stream=BB)
                                              //Set RS232 Board-to-Board
                                              //Communication Parameters
#use I2C(master, SCL=PIN_B6, SDA=PIN_B7)      //Set I2C Communication Parameters
#include <lcd_flex.c>                         //Identify LCD Driver File
#include <math.h>                             //Include Math Functions

int RTCSec,RTCMin,RTCHour,RTCDay,RTCDate,
RTCMonth,RTCYear,RTCCtrl,RTCHr;               //16-Bit Word Variable

void main()
   {
   set_tris_b(0xFF);                          //Make PortB an Input Port to
                                              //Enable In-Circuit-Programming
   delay_ms(1000);                            //Pause or Delay for 1 second

   lcd_init();                                //Initialize the LCD
   lcd_putc(0x0c);                            //Clear the LCD
   lcd_gotoxy(1,1);                           //Set the Cursor to Position-1 of Line-1
   printf(lcd_putc,"Power Up");               //Send, Power Up, to the LCD

   fprintf(HT, "\n\rPower Up\n\r");           //Send, Power Up, to the HyperTerminal with
                                              //a next Line & a Carriage Return
   delay_ms(1000);                            //Pause or Delay for 1 second

   setup_port_a(AN0_to_AN3);                  //Identify all of the Analog Inputs
   setup_adc(VSS_VDD);                        //Setup the ADC Voltage Range to be
                                              //0 Volts to 5 Volts
                                              //5v-0v)/1024~=.005 Volts Per Bit

   setup_adc(ADC_CLOCK_DIV_32);               //for delay(clock=10000000),
                                              //setup_adc(ADC_CLOCK_DIV_8),
                                              //10MHz/8=1.25MHz
                                              //for delay(clock=16000000),
                                              //setup_adc(ADC_CLOCK_DIV_16),
                                              //16MHz/16=1.00MHz
                                              //for delay(clock=20000000),
                                              //setup_adc(ADC_CLOCK_DIV_16),
                                              //20MHz/16=1.25MHz
                                              //for delay(clock=24000000),
                                              //setup_adc(ADC_CLOCK_DIV_32),
                                              //24MHz/32=1.25MHz
                                              //for delay(clock=32000000),
                                              //setup_adc(ADC_CLOCK_DIV_32),
                                              //32MHz/32=1.00MHz
                                              //for delay(clock=48000000),
                                              //setup_adc(ADC_CLOCK_DIV_64),
                                              //48MHz/64=0.75MHz

   set_adc_channel(0);                        //Identify the Analog Input to be
                                              //Sampled (Channel Input Zero)

   setup_ccp1(CCP_PWM);                       //Setup Capture/Compare/PWM Pin-1 to PWM Mode

   setup_ccp2(CCP_CAPTURE_RE);                //Setup Capture/Compare/PWM Pin-2 to
                                              //Capture Mode

   setup_timer_1(T1_INTERNAL);                //Setup Timer-1, which controls the
                                              //Capture/Compare portion of
                                              //Capture/Compare/PWM Pin-2, only,
                                              //to Internal Mode

   setup_timer_2(T2_DIV_BY_4, 19, 1);         //Setup Timer-2, which controls the PWM portion
                                              //of Capture/Compare/PWM Pin-1, only.
                                              //DIV_BY_Mode, Prescale (Period), Postscale
                                              //The Following Limits are for the
                                              //PIC18F4685 Microcontroller
                                              //Mode may equal 1 or 4 or 16
                                              //Prescale (Period) may equal any integer
                                              //between 0 and 255
                                              //Postscale may equal any integer between
                                              //1 and 16 (Post = 1 to set PWM Frequency)
                                              //T2Freq=OSC/[(4)*(Mode)*(Prescale+1)]=
                                              //24MegaHz/[(4)*(4)*(PS+1)] = 75KHz
```

```c
enable_interrupts(INT_CCP2);            //Enable the Internal Interrupt Input
                                        //of Capture/Compare/PWM Pin-2

enable_interrupts(GLOBAL);              //Enable Global Interrupts for the
                                        //PIC18F4685 Microcontroller

set_tris_b(0x00);                       //Make PortB an Output Port to Allow for
                                        //Normal Operation
tris_d(0xF0);                           //Make PortD, Bits-4,5,6,7 Inputs
                                        //and Bits-3,2,1,0 Outputs

/*
RTCSec          00 --> 59 Seconds
RTCMin          00 --> 59 Minutes
RTCHour         00 --> 23 Hour of the Day (Military Time)
RTCDay          01 --> 07 Sunday Thru Saturday
RTCDate         01 --> 31 Date of the Month
RTCMonth        01 --> 12 January Thru December
RTCYear         00 --> 99 Year
RTCCtrl         Square-Wave Out Control (Not Used)
RTCHr           Hour of the Day (AM/PM)
*/

Start:
I2C_START();
I2C_WRITE(0xD0);
I2C_WRITE(0x00);
I2C_START();
I2C_WRITE(0xD1);
RTCSec=I2C_READ();
RTCMin=I2C_READ();
RTCHour=I2C_READ();
RTCDay=I2C_READ();
RTCDate=I2C_READ();
RTCMonth=I2C_READ();
RTCYear=I2C_READ();
RTCCtrl=I2C_READ(0);
I2C_STOP();

if (RTCHour==0x00) RTCHR=0x12;
if (RTCHour==0x01) RTCHR=0x01;
if (RTCHour==0x02) RTCHR=0x02;
if (RTCHour==0x03) RTCHR=0x03;
if (RTCHour==0x04) RTCHR=0x04;
if (RTCHour==0x05) RTCHR=0x05;
if (RTCHour==0x06) RTCHR=0x06;
if (RTCHour==0x07) RTCHR=0x07;
if (RTCHour==0x08) RTCHR=0x08;
if (RTCHour==0x09) RTCHR=0x09;
if (RTCHour==0x10) RTCHR=0x10;
if (RTCHour==0x11) RTCHR=0x11;
if (RTCHour==0x12) RTCHR=0x12;
if (RTCHour==0x13) RTCHR=0x01;
if (RTCHour==0x14) RTCHR=0x02;
if (RTCHour==0x15) RTCHR=0x03;
if (RTCHour==0x16) RTCHR=0x04;
if (RTCHour==0x17) RTCHR=0x05;
if (RTCHour==0x18) RTCHR=0x06;
if (RTCHour==0x19) RTCHR=0x07;
if (RTCHour==0x20) RTCHR=0x08;
if (RTCHour==0x21) RTCHR=0x09;
if (RTCHour==0x22) RTCHR=0x10;
if (RTCHour==0x23) RTCHR=0x11;

if (RTCHour>=0x12) goto PMDay;

if (RTCDay==0x01) goto SunAM;
if (RTCDay==0x02) goto MonAM;
if (RTCDay==0x03) goto TueAM;
if (RTCDay==0x04) goto WedAM;
if (RTCDay==0x05) goto ThuAM;
if (RTCDay==0x06) goto FriAM;
if (RTCDay==0x07) goto SatAM;

PMDay:

if (RTCDay==0x01) goto SunPM;
if (RTCDay==0x02) goto MonPM;
if (RTCDay==0x03) goto TuePM;
if (RTCDay==0x04) goto WedPM;
if (RTCDay==0x05) goto ThuPM;
if (RTCDay==0x06) goto FriPM;
if (RTCDay==0x07) goto SatPM;
```

```
SunAM:
fprintf(HT, "\f\Sunday\r\n");
fprintf(HT, "\%02X/\%02X/\%02X\r\n",RTCMonth,RTCDate,RTCYear);
fprintf(HT, "\%02X:\%02X:\%02X\r\n",RTCHour,RTCMin,RTCSec);
fprintf(HT, "\%02X:\%02X:\%02X AM",RTCHr,RTCMin,RTCSec);
lcd_putc(0x0c);
lcd_gotoxy(1,1);
printf(lcd_putc,"\%02X/\%02X/\%01X",RTCMonth,RTCDate,RTCYear);
lcd_gotoxy(9,1);
printf(lcd_putc,"\%02X:\%02X:\%02X",RTCHour,RTCMin,RTCSec);
lcd_gotoxy(1,2);
printf(lcd_putc,"Sun");
lcd_gotoxy(6,2);
printf(lcd_putc,"\%02X:\%02X:\%02X AM",RTCHr,RTCMin,RTCSec);
delay_ms(1000);
goto Start;

MonAM:
fprintf(HT, "\f\Monday\r\n");
fprintf(HT, "\%02X/\%02X/\%02X\r\n",RTCMonth,RTCDate,RTCYear);
fprintf(HT, "\%02X:\%02X:\%02X\r\n",RTCHour,RTCMin,RTCSec);
fprintf(HT, "\%02X:\%02X:\%02X AM",RTCHr,RTCMin,RTCSec);
lcd_putc(0x0c);
lcd_gotoxy(1,1);
printf(lcd_putc,"\%02X/\%02X/\%01X",RTCMonth,RTCDate,RTCYear);
lcd_gotoxy(9,1);
printf(lcd_putc,"\%02X:\%02X:\%02X",RTCHour,RTCMin,RTCSec);
lcd_gotoxy(1,2);
printf(lcd_putc,"Mon");
lcd_gotoxy(6,2);
printf(lcd_putc,"\%02X:\%02X:\%02X AM",RTCHr,RTCMin,RTCSec);
delay_ms(1000);
goto Start;

TueAM:
fprintf(HT, "\f\Tuesday\r\n");
fprintf(HT, "\%02X/\%02X/\%02X\r\n",RTCMonth,RTCDate,RTCYear);
fprintf(HT, "\%02X:\%02X:\%02X\r\n",RTCHour,RTCMin,RTCSec);
fprintf(HT, "\%02X:\%02X:\%02X AM",RTCHr,RTCMin,RTCSec);
lcd_putc(0x0c);
lcd_gotoxy(1,1);
printf(lcd_putc,"\%02X/\%02X/\%01X",RTCMonth,RTCDate,RTCYear);
lcd_gotoxy(9,1);
printf(lcd_putc,"\%02X:\%02X:\%02X",RTCHour,RTCMin,RTCSec);
lcd_gotoxy(1,2);
printf(lcd_putc,"Tue");
lcd_gotoxy(6,2);
printf(lcd_putc,"\%02X:\%02X:\%02X AM",RTCHr,RTCMin,RTCSec);
delay_ms(1000);
goto Start;

WedAM:
fprintf(HT, "\f\Wednesday\r\n");
fprintf(HT, "\%02X/\%02X/\%02X\r\n",RTCMonth,RTCDate,RTCYear);
fprintf(HT, "\%02X:\%02X:\%02X\r\n",RTCHour,RTCMin,RTCSec);
fprintf(HT, "\%02X:\%02X:\%02X AM",RTCHr,RTCMin,RTCSec);
lcd_putc(0x0c);
lcd_gotoxy(1,1);
printf(lcd_putc,"\%02X/\%02X/\%01X",RTCMonth,RTCDate,RTCYear);
lcd_gotoxy(9,1);
printf(lcd_putc,"\%02X:\%02X:\%02X",RTCHour,RTCMin,RTCSec);
lcd_gotoxy(1,2);
printf(lcd_putc,"Wed");
lcd_gotoxy(6,2);
printf(lcd_putc,"\%02X:\%02X:\%02X AM",RTCHr,RTCMin,RTCSec);
delay_ms(1000);
goto Start;

ThuAM:
fprintf(HT, "\f\Thursday\r\n");
fprintf(HT, "\%02X/\%02X/\%02X\r\n",RTCMonth,RTCDate,RTCYear);
fprintf(HT, "\%02X:\%02X:\%02X\r\n",RTCHour,RTCMin,RTCSec);
fprintf(HT, "\%02X:\%02X:\%02X AM",RTCHr,RTCMin,RTCSec);
lcd_putc(0x0c);
lcd_gotoxy(1,1);
printf(lcd_putc,"\%02X/\%02X/\%01X",RTCMonth,RTCDate,RTCYear);
lcd_gotoxy(9,1);
printf(lcd_putc,"\%02X:\%02X:\%02X",RTCHour,RTCMin,RTCSec);
lcd_gotoxy(1,2);
printf(lcd_putc,"Thu");
lcd_gotoxy(6,2);
printf(lcd_putc,"\%02X:\%02X:\%02X AM",RTCHr,RTCMin,RTCSec);
delay_ms(1000);
goto Start;
```

```
FriAM:
fprintf(HT, "\f\Friday\r\n");
fprintf(HT, "\%02X/\%02X/\%02X\r\n",RTCMonth,RTCDate,RTCYear);
fprintf(HT, "\%02X:\%02X:\%02X\r\n",RTCHour,RTCMin,RTCSec);
fprintf(HT, "\%02X:\%02X:\%02X AM",RTCHr,RTCMin,RTCSec);
lcd_putc(0x0c);
lcd_gotoxy(1,1);
printf(lcd_putc,"\%02X/\%02X/\%01X",RTCMonth,RTCDate,RTCYear);
lcd_gotoxy(9,1);
printf(lcd_putc,"\%02X:\%02X:\%02X",RTCHour,RTCMin,RTCSec);
lcd_gotoxy(1,2);
printf(lcd_putc,"Fri");
lcd_gotoxy(6,2);
printf(lcd_putc,"\%02X:\%02X:\%02X AM",RTCHr,RTCMin,RTCSec);
delay_ms(1000);
goto Start;

SatAM:
fprintf(HT, "\f\Saturday\r\n");
fprintf(HT, "\%02X/\%02X/\%02X\r\n",RTCMonth,RTCDate,RTCYear);
fprintf(HT, "\%02X:\%02X:\%02X\r\n",RTCHour,RTCMin,RTCSec);
fprintf(HT, "\%02X:\%02X:\%02X AM",RTCHr,RTCMin,RTCSec);
lcd_putc(0x0c);
lcd_gotoxy(1,1);
printf(lcd_putc,"\%02X/\%02X/\%01X",RTCMonth,RTCDate,RTCYear);
lcd_gotoxy(9,1);
printf(lcd_putc,"\%02X:\%02X:\%02X",RTCHour,RTCMin,RTCSec);
lcd_gotoxy(1,2);
printf(lcd_putc,"Sat");
lcd_gotoxy(6,2);
printf(lcd_putc,"\%02X:\%02X:\%02X AM",RTCHr,RTCMin,RTCSec);
delay_ms(1000);
goto Start;

SunPM:
fprintf(HT, "\f\Sunday\r\n");
fprintf(HT, "\%02X/\%02X/\%02X\r\n",RTCMonth,RTCDate,RTCYear);
fprintf(HT, "\%02X:\%02X:\%02X\r\n",RTCHour,RTCMin,RTCSec);
fprintf(HT, "\%02X:\%02X:\%02X PM",RTCHr,RTCMin,RTCSec);
lcd_putc(0x0c);
lcd_gotoxy(1,1);
printf(lcd_putc,"\%02X/\%02X/\%01X",RTCMonth,RTCDate,RTCYear);
lcd_gotoxy(9,1);
printf(lcd_putc,"\%02X:\%02X:\%02X",RTCHour,RTCMin,RTCSec);
lcd_gotoxy(1,2);
printf(lcd_putc,"Sun");
lcd_gotoxy(6,2);
printf(lcd_putc,"\%02X:\%02X:\%02X PM",RTCHr,RTCMin,RTCSec);
delay_ms(1000);
goto Start;

MonPM:
fprintf(HT, "\f\Monday\r\n");
fprintf(HT, "\%02X/\%02X/\%02X\r\n",RTCMonth,RTCDate,RTCYear);
fprintf(HT, "\%02X:\%02X:\%02X\r\n",RTCHour,RTCMin,RTCSec);
fprintf(HT, "\%02X:\%02X:\%02X PM",RTCHr,RTCMin,RTCSec);
lcd_putc(0x0c);
lcd_gotoxy(1,1);
printf(lcd_putc,"\%02X/\%02X/\%01X",RTCMonth,RTCDate,RTCYear);
lcd_gotoxy(9,1);
printf(lcd_putc,"\%02X:\%02X:\%02X",RTCHour,RTCMin,RTCSec);
lcd_gotoxy(1,2);
printf(lcd_putc,"Mon");
lcd_gotoxy(6,2);
printf(lcd_putc,"\%02X:\%02X:\%02X PM",RTCHr,RTCMin,RTCSec);
delay_ms(1000);
goto Start;

TuePM:
fprintf(HT, "\f\Tuesday\r\n");
fprintf(HT, "\%02X/\%02X/\%02X\r\n",RTCMonth,RTCDate,RTCYear);
fprintf(HT, " \%02X:\%02X:\%02X\r\n",RTCHour,RTCMin,RTCSec);
fprintf(HT, "\%02X:\%02X:\%02X PM",RTCHr,RTCMin,RTCSec);
lcd_putc(0x0c);
lcd_gotoxy(1,1);
printf(lcd_putc,"\%02X/\%02X/\%01X",RTCMonth,RTCDate,RTCYear);
lcd_gotoxy(9,1);
printf(lcd_putc,"\%02X:\%02X:\%02X",RTCHour,RTCMin,RTCSec);
lcd_gotoxy(1,2);
printf(lcd_putc,"Tue");
lcd_gotoxy(6,2);
printf(lcd_putc,"\%02X:\%02X:\%02X PM",RTCHr,RTCMin,RTCSec);
delay_ms(1000);
goto Start;
```

```c
WedPM:
fprintf(HT, "\f\Wednesday\r\n");
fprintf(HT, "\%02X/\%02X/\%02X\r\n",RTCMonth,RTCDate,RTCYear);
fprintf(HT, "\%02X:\%02X:\%02X\r\n",RTCHour,RTCMin,RTCSec);
fprintf(HT, "\%02X:\%02X:\%02X PM",RTCHr,RTCMin,RTCSec);
lcd_putc(0x0c);
lcd_gotoxy(1,1);
printf(lcd_putc,"\%02X/\%02X/\%01X",RTCMonth,RTCDate,RTCYear);
lcd_gotoxy(9,1);
printf(lcd_putc,"\%02X:\%02X:\%02X",RTCHour,RTCMin,RTCSec);
lcd_gotoxy(1,2);
printf(lcd_putc,"Wed");
lcd_gotoxy(6,2);
printf(lcd_putc,"\%02X:\%02X:\%02X PM",RTCHr,RTCMin,RTCSec);
delay_ms(1000);
goto Start;

ThuPM:
fprintf(HT, "\f\Thursday\r\n");
fprintf(HT, "\%02X/\%02X/\%02X\r\n",RTCMonth,RTCDate,RTCYear);
fprintf(HT, "\%02X:\%02X:\%02X\r\n",RTCHour,RTCMin,RTCSec);
fprintf(HT, "\%02X:\%02X:\%02X PM",RTCHr,RTCMin,RTCSec);
lcd_putc(0x0c);
lcd_gotoxy(1,1);
printf(lcd_putc,"\%02X/\%02X/\%01X",RTCMonth,RTCDate,RTCYear);
lcd_gotoxy(9,1);
printf(lcd_putc,"\%02X:\%02X:\%02X",RTCHour,RTCMin,RTCSec);
lcd_gotoxy(1,2);
printf(lcd_putc,"Thu");
lcd_gotoxy(6,2);
printf(lcd_putc,"\%02X:\%02X:\%02X PM",RTCHr,RTCMin,RTCSec);
delay_ms(1000);
goto Start;

FriPM:
fprintf(HT, "\f\Friday\r\n");
fprintf(HT, "\%02X/\%02X/\%02X\r\n",RTCMonth,RTCDate,RTCYear);
fprintf(HT, "\%02X:\%02X:\%02X\r\n",RTCHour,RTCMin,RTCSec);
fprintf(HT, "\%02X:\%02X:\%02X PM",RTCHr,RTCMin,RTCSec);
lcd_putc(0x0c);
lcd_gotoxy(1,1);
printf(lcd_putc,"\%02X/\%02X/\%01X",RTCMonth,RTCDate,RTCYear);
lcd_gotoxy(9,1);
printf(lcd_putc,"\%02X:\%02X:\%02X",RTCHour,RTCMin,RTCSec);
lcd_gotoxy(1,2);
printf(lcd_putc,"Fri");
lcd_gotoxy(6,2);
printf(lcd_putc,"\%02X:\%02X:\%02X PM",RTCHr,RTCMin,RTCSec);
delay_ms(1000);
goto Start;

SatPM:
fprintf(HT, "\f\Saturday\r\n");
fprintf(HT, "\%02X/\%02X/\%02X\r\n",RTCMonth,RTCDate,RTCYear);
fprintf(HT, "\%02X:\%02X:\%02X\r\n",RTCHour,RTCMin,RTCSec);
fprintf(HT, "\%02X:\%02X:\%02X PM",RTCHr,RTCMin,RTCSec);
lcd_putc(0x0c);
lcd_gotoxy(1,1);
printf(lcd_putc,"\%02X/\%02X/\%01X",RTCMonth,RTCDate,RTCYear);
lcd_gotoxy(9,1);
printf(lcd_putc,"\%02X:\%02X:\%02X",RTCHour,RTCMin,RTCSec);
lcd_gotoxy(1,2);
printf(lcd_putc,"Sat");
lcd_gotoxy(6,2);
printf(lcd_putc,"\%02X:\%02X:\%02X PM",RTCHr,RTCMin,RTCSec);
delay_ms(1000);
goto Start;
}
```

Engineering Practices for the PIC Microcontroller

By: Prof. Sal R. Riggio Jr., PhD, PE

Chapter - 8

Laboratory Experiment - 6

*The Digital Frequency Counter,
The Digital Voltmeter,
The Digital Ohmmeter,
The Digital Waveform Generator,
The One-Bit Digital-to-Analog Converter*

Introduction

The purpose of this lab experiment is for the student to develop basic skills in making practical use of the Microchip PIC18F4685 Microcontroller and become familiar with the importance, operation and software and hardware implementation of a device to measure Voltage, Resistance and Frequency. Additionally, the student will be introduced to code which generates various signal waveform.

The student is expected to analyze **PICBasic-Pro Code** and **CCS C-code**, which enables the **PIC18F4685 Microcontroller,** along with some **Analog Interface Circuits,** to **measure Voltage** and **Resistance** and **Frequency** within a **Specified Range** of **Values**. These **Voltage, Resistance** and **Frequency** values are to be displayed on the **Liquid-Crystal-Display**, and on the **RS232 HyperTerminal** within a Windows based Personal Computer.

It is suggested that the student **read** the **PICBasic-Pro code** and the **CCS C-code**, particularly the **comments** written on each line, to learn how to write code for the PIC microcontrollers. It has been proven, many times, to be the most effective way to learn how generate effective code for the PIC microcontroller. This code is shown at the end of this chapter.

Portions of the Experiment Board Schematics, which are relevant to this experiment, are shown below in **Figure 8.1.** Please refer to **Appendix-A** for the complete set of **Experiment Board Schematics** and please refer to **Appendix-B** for the complete **Lab Board Schematic**.

Hardware Setup (Common)

1.) Connect the Square end of the USB Type-A to Type-B Cable to connector JUSB1 of the Experiment Board. Then, connect the other end of this Cable to one of the USB Inputs of the Computer.
2.) Connect the DB9 end of the DB9 Serial-to-USB Cable to connector P1 of the Experiment Board. Then, connect the USB end of this Cable to one of the other USB Inputs of the Computer. The DB9 Serial-to-USB Cable is used to communicate between the Serial HyperTerminal of the Computer and the Experiment Board.

Assignment #1   "Frequency Measurement"

Hardware Setup

1.) Connect Channel-A of an Oscilloscope and the output of a Signal Generator to connector (J1) of the Experiment Board. Be sure that the Red Lead of the BNC-to-Clip-Lead Cable is connected to Pin-1 of connector (J1), and that the Black Lead is connected to Pin-2 of connector (J1), which is Ground.
2.) Set the Signal Generator to a 10 Volts Peak-to-Peak Square-Wave with Zero Voltage Offset and DC-Coupling.
3.) Connect the Jumper Blocks, as follows, for all five assignments.

   a) The Jumper of **JB6** is to be placed in **Position 3-4.**
   b) The Jumper of **JB7** is to be placed in **Position 3-4.**
   c) The Jumper of **JB8** is to be placed in **Position 1-2.**
   d) The Jumpers of **JB12** are to be placed in **Positions 3-5 & 4-6.**

Program the **PIC18F4685 Microcontroller** with the given **Digital Frequency Counter** PICBasic-Pro **Code** and vary the frequency of the Signal Generator from 50 Hertz to 50K Hertz. Then, Note the Values shown on the Liquid-Crystal-Display and on the HyperTerminal Display.

Assignment #2   "DC Voltage Measurement"

Hardware Setup

1.) The Jumper of JB3 and the Jumper of JB4 must each be placed in **Position 4, Only,** and one of the **Two Jumpers** of JB5 must be placed in **Position 1-2** and the other in **Position 3-4.**
2.) Connect a **Variable DC Voltage Source** of 0 to +5 Volts DC between Connector J9, Pin-1 (+V), and Connector J9, Pin-2 (-V), of the Experiment Board.
3.) Connect this **Same** Power Supply between Connector J10, Pin-1 (+V), and Connector J10, Pin-2 (-V), of the Experiment Board.

Program the **PIC18F4685 Microcontroller** with the given **Digital Voltmeter** PICBasic-Pro Code and **CCS_C-Code**. All Digital Voltmeter values should be displayed on the Liquid-Crystal-Display and on the HyperTerminal Display.

Use a **Hand-Held Voltmeter** to determine how close the Experiment Board's **LCD Voltage Values** are to that of the **Hand-Held Voltmeter**. **Vary** the **Input Voltage** from Zero to 5 Volts DC and **Check** the results against the **Hand-Held Voltmeter**. **Calibrate** the **LCD Voltage Reading** to that of the **Hand Held Voltmeter** by Adding or Subtracting the Difference in Readings from that of the **LCD**, within the Microcontroller Code. Then, **Rerun** this updated code and **Recheck** the results.

Assignment #3 "Resistance Measurement"

Hardware Setup

The Jumper of **JB3** and the Jumper of **JB4** must each be placed in **Position 3-4**, and One of the Two Jumpers of **JB5** must be placed in **Position 1-2** and the Other in **Position 3-4**.

Program the **PIC18F4685 Microcontroller** with the given <u>Digital Ohmmeter</u> **PICBasic-Pro Code** and **CCS_C-Code**. All Digital Ohmmeter values should be displayed on the Liquid-Crystal-Display and on the HyperTerminal Display.

Select a Resistor **(Res)** between the **Values** of **1K ohms** and **20K ohms** and connect one end of it to Connector **J9, Pin-1 (+V)**, and Connect the other end to Connector **J10, Pin-1 (+V)**. Use a **Hand-Held Ohmmeter** to determine how close the Experiment Board's **LCD Voltage Values** are to that of the **Hand-Held Ohmmeter**.

The following **Equations** are used to Compute the **Value** of the Selected Resistor **(Res)**. These **Equations** are Scaled to Yield the Number of Ohms without Over-Flowing the Microcontroller's Registers.

$W0 = (VJ9-VJ10)*50$
$Res = (W0/VJ10)*20$ → in Ohms

Next, Place a <u>Piece of Wire</u> between Connector **J9, Pin-1 (+V)**, and Connector **J10, Pin-1 (+V)**, of the Experiment Board. Then, Write a **PICBasic-Pro Code Routine** which will cause the **RED LED** to **Flash Twice** and the **Relay** to **Click Twice**, just after a <u>Less</u> than **One Ohm Resistance Value** is **Displayed** on the **Liquid-Crystal-Display** and on the **HyperTerminal**. This is called a **Continuity Checker**.

Assignment #4 "Waveform Generation"

Hardware Setup

1.) Connect the Channel-A of the Oscilloscope to connector (J2) of the Experiment Board. Be sure that the <u>Red Lead</u> of the BNC-to-Clip-Lead Cable is connected to Pin-1 of connector (J2), and that the <u>Black Lead</u> is connected to Pin-2 of connector (J2), which is <u>Ground</u>.

Program the **PIC18F4685 Microcontroller** with the given <u>Digital Waveform Generator</u> **PICBasic-Pro Code** and **CCS_C-Code**. Observe all of the signal waveforms that are displayed on the Oscilloscope.

Assignment #5 "One-Bit Analog-to-Digital Converter"

Hardware Setup

1.) Connect the Channel-A of the Oscilloscope to connector (J5) of the Experiment Board. Be sure that the <u>Red Lead</u> of the BNC-to-Clip-Lead Cable is connected to Pin-1 of connector (J5), and that the <u>Black Lead</u> is connected to Pin-2 of connector (J5), which is <u>Ground</u>.
2.) Connect the Channel-B of the Oscilloscope to connector (J14) of the Experiment Board. Be sure that the <u>Red Lead</u> of the BNC-to-Clip-Lead Cable is connected to Pin-1 of connector (J14), and that the <u>Black Lead</u> is connected to Pin-2 of connector (J14), which is <u>Ground</u>.

Program the **PIC18F4685 Microcontroller** with the given <u>One-Bit Analog-to-Digital Converter</u> **PICBasic-Pro Code** and **CCS_C-Code**. Observe all of the signal waveforms that are displayed on the Oscilloscope.

Ton is the **On-Time** of the Periodic Waveform (The Time the Pulse is High).

Tp is the **Period** of the Periodic Waveform (Period = 1/Operating_Frequency).

Operating_Frequency = 30K Hertz

DutyCycle = Ton/Tp

VinDC = +5 Volts

Vout (J14, Pin-1) = DutyCycle*VinDC

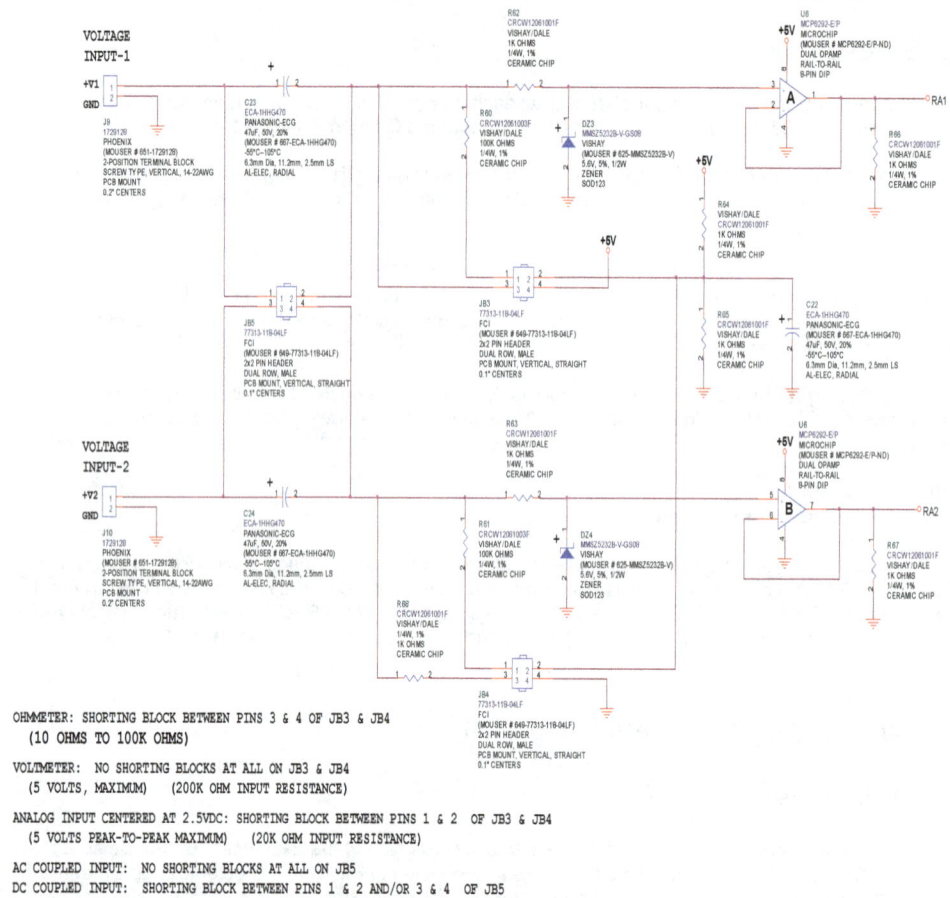

OHMMETER: SHORTING BLOCK BETWEEN PINS 3 & 4 OF JB3 & JB4
(10 OHMS TO 100K OHMS)

VOLTMETER: NO SHORTING BLOCKS AT ALL ON JB3 & JB4
(5 VOLTS, MAXIMUM)   (200K OHM INPUT RESISTANCE)

ANALOG INPUT CENTERED AT 2.5VDC: SHORTING BLOCK BETWEEN PINS 1 & 2 OF JB3 & JB4
(5 VOLTS PEAK-TO-PEAK MAXIMUM)   (20K OHM INPUT RESISTANCE)

AC COUPLED INPUT: NO SHORTING BLOCKS AT ALL ON JB5
DC COUPLED INPUT: SHORTING BLOCK BETWEEN PINS 1 & 2 AND/OR 3 & 4 OF JB5

Figure 8.1: Analog Interface Circuit for the Analog-to-Digital Converter of the Microcontroller

Figure 8.1 shows how one would pass a signal entering from an outside source to the PIC microcontrollers ADC. The in-coming signal is AC-coupled to the operational amplifier through capacitor (C23) to block any DC voltage from damaging the operational amplifier. Therefore, no shunt is placed between pins 1 & 2 of jumper block (JB5). However, the operational amplifier does need to have a shunt placed between pins 1 & 2 of jumper block (JB3) to provide a constant bias voltage of +2.5 volts to the input of the operational amplifier through resistor R60. This resistor also provides an approximate circuit input resistance of 100K ohms. Resistor (R62) and the Zener diode (DZ3) form an over-voltage protection circuit to protect the operational amplifier from being damaged due to a high transient input voltage. The 5.6 volt Zener diode conducts current only when the input voltage becomes more positive than +5.6 volts and more negative the -0.7 volts. Otherwise, the Zener diode does not have an appreciable effect on the performance of this circuit.

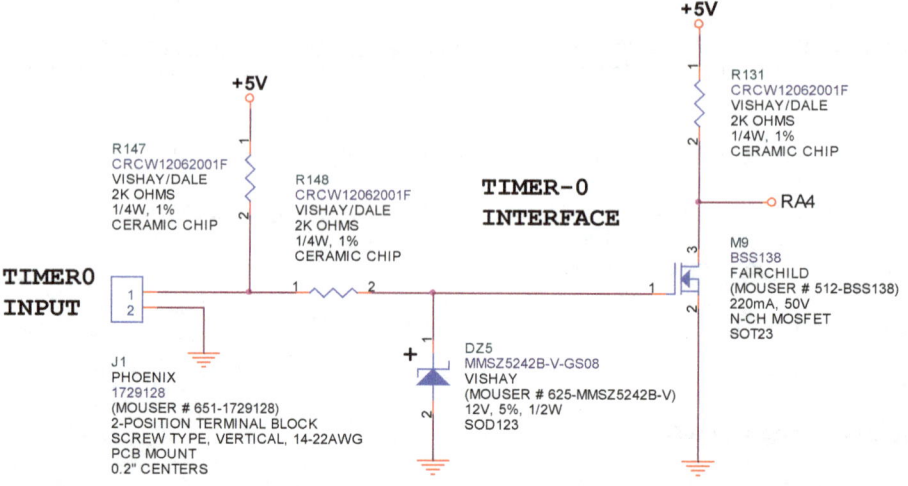

Figure 8.2: Timer-0 Interface Circuit

Laboratory Experiment #6

'Lab-6A_PICBasic-Pro_Digital_Frequency-Counter

```
INCLUDE "modedefs.bas"          'Identifies the Command File

DEFINE OSC 20                   'Set PIC-Pro Oscillator Value in Mega Hertz
DEFINE ADC_BITS 10              'Set for 10-Bit Internal ADC

DEFINE ADC_CLOCK 5              'DEFINE ADC_CLOCK 1, for OSC=10, MAX ADC
                                'Clock=10MHz/8) = 1.25MHz
                                'DEFINE ADC_CLOCK 5, for OSC=16, MAX ADC
                                'Clock=16MHz/16) = 1.00MHz
                                'DEFINE ADC_CLOCK 5, for OSC=20, MAX ADC
                                'Clock=20MHz/16) = 1.25MHz
                                'DEFINE ADC_CLOCK 2, for OSC=24, MAX ADC
                                'Clock=24MHz/32) = 1.25MHz
                                'DEFINE ADC_CLOCK 2, for OSC=32, MAX ADC
                                'Clock=32MHz/32) = 1.00MHz
                                'DEFINE ADC_CLOCK 6, for OSC=48, MAX ADC
                                'Clock=48MHz/64) = 0.75MHz

DEFINE CCP1_BIT 2               'Setup Hardware PWM Timer
DEFINE LCD_BITS 4               'Setup LCD for 4-Bit Mode
DEFINE LCD_DREG PORTB           'Drive LCD from Port-B
DEFINE LCD_DBIT 0               'Starting with Bit-0
DEFINE LCD_RSREG PORTA          'Drive LCD Register-Select from
DEFINE LCD_RSBIT 4              'Port-A, Bit-4
DEFINE LCD_EREG PORTA           'Drive LCD Enable from
DEFINE LCD_EBIT 5               'Port-A, Bit-5
DEFINE LCD_LINES 2              'LCD can hold 2-Lines of Characters
DEFINE LCD_COMMANDSUS 4000      'Delay between Commands
DEFINE LCD_DATAUS 200           'Delay between Data Transfers

ADCON0 = %00001111              'ADC Enabled, Analog Input 3 Enabled
ADCON1 = %00001011              'ADC Voltage Range is 0 to 5 Volts &
                                'Analog Inputs 0 Thru 3 are available
ADCON2 = %10101101              'Right Justify to use Upper 8-Bits, only,
                                '12*Tad, & ADC Clock=20MHz/16=1.25MHz

W0 VAR WORD                     'Set 16-Bit Word Variable,W0
W1 VAR WORD                     '  "      "       "        ,W1
W2 VAR WORD                     '  "      "       "        ,W2
W3 VAR WORD                     '  "      "       "        ,W3
W4 VAR WORD                     '  "      "       "        ,W4
W5 VAR WORD                     '  "      "       "        ,W5
W6 VAR WORD                     '  "      "       "        ,W6

B0 VAR BYTE                     'Set 8-Bit Byte Variable,B0
B1 VAR BYTE                     '  "      "      "      ,B1

TRISB=$FF                       'Make PortB an Input Port to Enable
                                'In-Circuit-Programming
PAUSE 1000                      'Pause or Delay for 1 Seconds to Enable
                                'In-Circuit-Programming
LCDOUT $FE,1                    'Clear the LCD
LCDOUT $FE, $80                 'Set the Cursor to Line 1, Position 0
LCDOUT "Power Up"               'Send, Power Up, to the LCD

SERout PORTC.6,T2400,["Power Up",13,10,10]
                                'Send "Power Up" to the HyperTerminal
                                'with a Carriage Return & two Line Feeds,
                                'at 2400 Bits/Second, 8-Bits, 1-Stop Bit,
                                'No Parity, Driven True (T2400).

PAUSE 2000                      'Pause or Delay for 2 Seconds
                                'to Read "Power Up" in the LCD
LCDOUT $FE,1                    'Clear the LCD
TRISB=$00                       'Make PortB an Output Port to Allow for Normal Operation
TRISD=$F0                       'Make PortD, Bits-4,5,6,7 Inputs, & Bits-3
```

```
KSCAN:
        LCDOUT $FE, 1                   'Clear the LCD
        LCDOUT $FE, $80                 'Set the Cursor to Line 1, Position 0
        LCDOUT "Freq Counter"           'Send, Freq Counter, to the LCD
        SERout2 PORTC.6,396,["Freq Counter",13,10,10]
                                        'Send, Freq Counter, to HyperTerminal
                                        'with a Carriage Return & Two Line Feeds
        LCDOUT $FE, $C0                 'Set the Cursor to Line 1, Position 0
        LCDOUT "Scanning"               'Send, Scanning, to the LCD
        SERout2 PORTC.6,396,["Scanning",13,10,10]
                                        'Send, Scanning, to HyperTerminal
                                        'with a Carriage Return & Two Line Feeds

        HIGH PORTD.0                    'Make 0th Row High to Scan for the Following:
        PAUSE 500
        INPUT PORTD.4
        IF (PORTD.4=1) THEN RANGE1      'Check for Keypad Key "1" , Column-4
        PAUSE 100
        INPUT PORTD.5
        IF (PORTD.5=1) THEN RANGE2      'Check for Keypad Key "2" , Column-5

        PAUSE 100
        INPUT PORTD.6
        IF (PORTD.6=1) THEN RANGE3      'Check for Keypad Key "3" , Column-6
        PAUSE 100
        LOW PORTD.0

        HIGH PORTD.3                    'Make 3th Row High to Scan for the Following:
        PAUSE 500
        INPUT PORTD.5
        IF (PORTD.5=1) THEN RANGE0      'Check for Keypad Key "0" , Column 5
        LOW PORTD.3

        GOTO KSCAN

RANGE0:
        COUNT PORTA.4,1000,W0           'Count the Number of Pulses that
                                        'occur in 1 second and place
                                        'the result in the Variable W0.
        LCDOUT $FE, 1                   'Clear the LCD
        LCDOUT $FE, $80                 'Set the Cursor to Line 1, Position 0
        LCDOUT "Freq Counter"           'Send, Freq Counter, to the LCD
        SERout2 PORTC.6,396,["Freq Counter",13,10,10]
                                        'Send, Freq Counter, to HyperTerminal
                                        'with a Carriage Return &
                                        'Two Line Feeds
        LCDOUT $FE, $C0                 'Set the Cursor to Line 2, Position 0
        LCDOUT "Freq=",DEC W0," Hz"     'Send, Freq=Value Hertz, in Decimal, to the LCD
        SEROUT2 PORTC.6,396,["Freq= ",DEC W0,"  Hz",13,10,10]
                                        'Send, Freq=Value Hertz, in Decimal,
                                        'to the HyperTerminal with a Carriage
                                        'Return and two Line Feeds
        PAUSE 100
        GOTO RANGE0

RANGE1:
        COUNT PORTA.4,100,W0            'Count the Number of Pulses that
                                        'occur in 100 milli-seconds and place
                                        'the result in the Variable W0.
        LCDOUT $FE, 1                   'Clear the LCD
        LCDOUT $FE, $80                 'Set the Cursor to Line 1, Position 0
        LCDOUT "Freq Counter"           'Send, Freq Counter, to the LCD
        SERout2 PORTC.6,396,["Freq Counter",13,10,10]
                                        'Send, Freq Counter,
                                        'to the HyperTerminal
                                        'with a Carriage Return & Two Line Feeds
        LCDOUT $FE, $C0                 'Set the Cursor to Line 2, Position 0
        LCDOUT "Freq=",DEC W0," X10-Hz" 'Send, Freq=Value Hertz,
                                        'in Decimal, to the LCD
        SEROUT2 PORTC.6,396,["Freq= ",DEC W0,"  X10-Hz",13,10,10]
                                        'Send, Freq=Value Hertz, in Decimal,
                                        'to the HyperTerminal with a Carriage
        PAUSE 10                        'Return and two Line Feeds
        GOTO RANGE1
```

```
RANGE2:
        COUNT PORTA.4,10,W0            'Count the Number of Pulses that
                                       'occur in 100 milli-seconds and place
                                       'the result in the Variable W0.
        LCDOUT $FE, 1                  'Clear the LCD
        LCDOUT $FE, $80                'Set the Cursor to Line 1, Position 0
        LCDOUT "Freq Counter"          'Send, Freq Counter, to the LCD
        SERout2 PORTC.6,396,["Freq Counter",13,10,10]
                                       'Send, Freq Counter
                                       'to the HyperTerminal
                                       'with a Carriage Return and
                                       'Two Line Feeds
        LCDOUT $FE, $C0                'Set the Cursor to Line 2, Position 0
        LCDOUT "Freq=",DEC W0," X100-Hz"
                                       'Send, Freq=Value Hertz,
                                       'in Decimal, to the LCD
        SEROUT2 PORTC.6,396,["Freq= ",DEC W0," X100-Hz",13,10,10]
                                       'Send, Freq=Value Hertz, in Decimal,
                                       'to the HyperTerminal with a Carriage
                                       'Return and two Line Feeds
        PAUSE 100
        GOTO RANGE2

RANGE3:
        COUNT PORTA.4,1,W0             'Count the Number of Pulses that
                                       'occur in 100 milli-seconds
                                       'and place the result
                                       'into the Variable W0.
        LCDOUT $FE, 1                  'Clear the LCD
        LCDOUT $FE, $80                'Set the Cursor to Line 1, Position 0
        LCDOUT "Freq Counter"          'Send, Freq Counter, to the LCD
        SERout2 PORTC.6,396,["Freq Counter",13,10,10]
                                       'Send, Freq Counter, to HyperTerminal
                                       'with a Carriage Return & Two Line Feeds
        LCDOUT $FE, $C0                'Set the Cursor to Line 2, Position 0
        LCDOUT "Freq=",DEC W0," K-Hz"  'Send, Freq=Value Hertz, in Decimal, to the LCD
        SEROUT2 PORTC.6,396,["Freq= ",DEC W0," K-Hz",13,10,10]
                                       'Send, Freq=Value Hertz, in Decimal,
                                       'to the HyperTerminal with a Carriage
                                       'Return and two Line Feeds
        PAUSE 100
        GOTO RANGE3

        END
```

Laboratory Experiment #6

```
'Lab-6B_PICBasic-Pro_Digital_Voltmeter
 INCLUDE "modedefs.bas"              'Identifies the Command File

 DEFINE OSC 20                       'Set PIC-Pro Oscillator Value in Mega Hertz
 DEFINE ADC_BITS 10                  'Set for 10-Bit Internal ADC

 DEFINE ADC_CLOCK 5                  'DEFINE ADC_CLOCK 1, for OSC=10, MAX ADC
                                     'Clock=10MHz/8) = 1.25MHz
                                     'DEFINE ADC_CLOCK 5, for OSC=16, MAX ADC
                                     'Clock=16MHz/16) = 1.00MHz
                                     'DEFINE ADC_CLOCK 5, for OSC=20, MAX ADC
                                     'Clock=20MHz/16) = 1.25MHz
                                     'DEFINE ADC_CLOCK 2, for OSC=24, MAX ADC
                                     'Clock=24MHz/32) = 1.25MHz
                                     'DEFINE ADC_CLOCK 2, for OSC=32, MAX ADC
                                     'Clock=32MHz/32) = 1.00MHz
                                     'DEFINE ADC_CLOCK 6, for OSC=48, MAX ADC
                                     'Clock=48MHz/64) = 0.75MHz

 DEFINE CCP1_BIT 2                   'Setup Hardware PWM Timer
 DEFINE LCD_BITS 4                   'Setup LCD for 4-Bit Mode
 DEFINE LCD_DREG PORTB               'Drive LCD from Port-B
 DEFINE LCD_DBIT 0                   'Starting with Bit-0
 DEFINE LCD_RSREG PORTA              'Drive LCD Register-Select from
 DEFINE LCD_RSBIT 4                  'Port-A, Bit-4
 DEFINE LCD_EREG PORTA               'Drive LCD Enable from
 DEFINE LCD_EBIT 5                   'Port-A, Bit-5
 DEFINE LCD_LINES 2                  'LCD can hold 2-Lines of Characters
 DEFINE LCD_COMMANDSUS 4000          'Delay between Commands
 DEFINE LCD_DATAUS 200               'Delay between Data Transfers

 ADCON0 = %00001111                  'ADC Enabled, Analog Input 3 Enabled
 ADCON1 = %00001011                  'ADC Voltage Range is 0 to 5 Volts &
                                     'Analog Inputs 0 Thru 3 are available
 ADCON2 = %10101101                  'Right Justify to use Upper 8-Bits, only,
                                     '12*Tad, & ADC Clock=20MHz/16=1.25MHz

 W0 VAR WORD                         'Set 16-Bit Word Variable,W0
 W1 VAR WORD                         '  "      "    "    "      ,W1
 W2 VAR WORD                         '  "      "    "    "      ,W2
 W3 VAR WORD                         '  "      "    "    "      ,W3
 W4 VAR WORD                         '  "      "    "    "      ,W4
 W5 VAR WORD                         '  "      "    "    "      ,W5
 W6 VAR WORD                         '  "      "    "    "      ,W6
 VS1 VAR WORD                        '  "      "    "    "      ,VS1
 VSWHOLE1 VAR WORD                   '  "      "    "    "      ,VSWHOLE1
 VSDECIMAL1 VAR WORD                 '  "      "    "    "      ,VSDECIMAL1
 VS2 VAR WORD                        '  "      "    "    "      ,VS2
 VSWHOLE2 VAR WORD                   '  "      "    "    "      ,VSWHOLE2
 VSDECIMAL2 VAR WORD                 '  "      "    "    "      ,VSDECIMAL2
 SFACTOR VAR WORD                    '  "      "    "    "      ,SFACTOR
 MAXVOLTRANGE VAR WORD               '  "      "    "    "      ,MAXVOLTRANGE
 LCDOC VAR WORD                      '  "      "    "    "      ,LCDOC

 B0 VAR BYTE                         'Set 8-Bit Byte Variable,B0
 B1 VAR BYTE                         '  "    "    "    "      ,B1

 TRISB=$FF                           'Make PortB an Input Port to Enable
                                     'In-Circuit-Programming
 PAUSE 1000                          'Pause or Delay for 1 Seconds to Enable
                                     'In-Circuit-Programming
 LCDOUT $FE,1                        'Clear the LCD
 LCDOUT $FE, $80                     'Set the Cursor to Line 1, Position 0
 LCDOUT "Power Up"                   'Send, Power Up, to the LCD

 SERout PORTC.6,T2400,["Power Up",13,10,10]
                                     'Send "Power Up" to the HyperTerminal
                                     'with a Carriage Return & two Line Feeds,
                                     'at 2400 Bits/Second, 8-Bits, 1-Stop Bit,
                                     'No Parity, Driven True (T2400).

 PAUSE 2000                          'Pause or Delay for 2 Seconds
                                     'to Read "Power Up" in the LCD
 LCDOUT $FE,1                        'Clear the LCD
 TRISB=$00                           'Make PortB an Output Port to Allow for Normal Operation
 TRISD=$F0                           'Make PortD, Bits-4,5,6,7 Inputs, & Bits-3
```

```
LCDOC=50                                'Initialize the LCD Output Counter
                                        'Send the Output Voltage Value to
                                        'the LCD and the HyperTerminal every
                                        'LCDOC Samples

MAXVOLTRANGE=5                          'Set the Maximum Voltage Range Value
SFACTOR=1024/MAXVOLTRANGE               'Determine the Scale Factor

SAMPLE: ADCIN 1,VS1                     'Sample ADC Input Voltage-1, and place the
                                        '10-Bit result in Variable VS1

VSWHOLE1=VS1/SFACTOR                    'Scale VS1 to Determine the WHOLE Portion of VS1
W3=VS1*10000                            'Expand VS1 to 16-Bits
W2=W3/SFACTOR                           'Scale the 16-Bit Expanded value of VS1
W1=VSWHOLE1*10000                       'Expand the WHOLE Portion of VS1 to 16-Bits
VSDECIMAL1=W2-W1                        'Determine the Decimal Portion of VS1 by Subtracting
                                        'the Expanded 16-Bit WHOLE Portion of VS1 from
                                        'the Scaled 16-Bit Expanded value of VS1

ADCIN 2,VS2                             'Sample ADC Input Voltage-2, and place the
                                        '10-Bit result in Variable VS2

VSWHOLE2=VS2/SFACTOR                    'Scale VS2 to Determine the WHOLE Portion of VS2
W6=VS2*10000                            'Expand VS2 to 16-Bits
W5=W6/SFACTOR                           'Scale the 16-Bit Expanded value of VS2
W4=VSWHOLE2*10000                       'Expand the WHOLE Portion of VS2 to 16-Bits
VSDECIMAL2=W5-W4                        'Determine the Decimal Portion of VS2 by Subtracting
                                        'the Expanded 16-Bit WHOLE Portion of VS2 from
                                        'the Scaled 16-Bit Expanded value of VS2

LCDOC=LCDOC-1                           'Decrement LCD Output Counter          So it doesn't flash as often
IF LCDOC=0 THEN LCD                     'Send the Output Voltage Value to
                                        'the LCD and the HyperTerminal every
                                        'LCDOC Samples

GOTO SAMPLE

LCD:
LCDOUT $FE, 1                           'Clear the LCD                          DEC -> 1 digit
LCDOUT $FE, $80                         'Set Cursor to Line-1, Position-0        DEC2 -> 2 digits of significance
LCDOUT "Vin1=",DEC VSWHOLE1,".",DEC2 VSDECIMAL1," Volts  "
                                        'Send, Vin1=Value Volts, in Decimal, to the LCD

LCDOUT $FE, $C0                         'Set Cursor to Line-2, Position-0
LCDOUT "Vin2=",DEC VSWHOLE2,".",DEC2 VSDECIMAL2," Volts  "
                                        'Send, Vin2=Value Volts, in Decimal, to the LCD

SEROUT2 PORTC.6,396,["Vin1= ",DEC VSWHOLE1,".",DEC2 VSDECIMAL1," Volts ",13,10,10]
                                        'Send, Vin1=Value Volts, in Decimal, to the
                                        'HyperTerminal with a Carriage
                                        'Return and two Line Feeds

SEROUT2 PORTC.6,396,["Vin2= ",DEC VSWHOLE2,".",DEC2 VSDECIMAL2," Volts ",13,10,10]
                                        'Send, Vin2=Value Volts, in Decimal, to the
                                        'HyperTerminal with a Carriage
                                        'Return and two Line Feeds

LCDOC=50                                'Initialize the LCD Output Counter
                                        'Send the Output Voltage Value to
                                        'the LCD and the HyperTerminal every
                                        'LCDOC Samples
GOTO SAMPLE

END
```

Laboratory Experiment #6

```
'Lab-6C_PICBasic-Pro_Digital_Ohmmeter
    INCLUDE "modedefs.bas"              'Identifies the Command File

    DEFINE OSC 20                       'Set PIC-Pro Oscillator Value in Mega Hertz
    DEFINE ADC_BITS 10                  'Set for 10-Bit Internal ADC

    DEFINE ADC_CLOCK 5                  'DEFINE ADC_CLOCK 1, for OSC=10, MAX ADC
                                        'Clock=10MHz/8) = 1.25MHz
                                        'DEFINE ADC_CLOCK 5, for OSC=16, MAX ADC
                                        'Clock=16MHz/16) = 1.00MHz
                                        'DEFINE ADC_CLOCK 5, for OSC=20, MAX ADC
                                        'Clock=20MHz/16) = 1.25MHz
                                        'DEFINE ADC_CLOCK 2, for OSC=24, MAX ADC
                                        'Clock=24MHz/32) = 1.25MHz
                                        'DEFINE ADC_CLOCK 2, for OSC=32, MAX ADC
                                        'Clock=32MHz/32) = 1.00MHz
                                        'DEFINE ADC_CLOCK 6, for OSC=48, MAX ADC
                                        'Clock=48MHz/64) = 0.75MHz

    DEFINE CCP1_BIT 2                   'Setup Hardware PWM Timer
    DEFINE LCD_BITS 4                   'Setup LCD for 4-Bit Mode
    DEFINE LCD_DREG PORTB               'Drive LCD from Port-B
    DEFINE LCD_DBIT 0                   'Starting with Bit-0
    DEFINE LCD_RSREG PORTA              'Drive LCD Register-Select from
    DEFINE LCD_RSBIT 4                  'Port-A, Bit-4
    DEFINE LCD_EREG PORTA               'Drive LCD Enable from
    DEFINE LCD_EBIT 5                   'Port-A, Bit-5
    DEFINE LCD_LINES 2                  'LCD can hold 2-Lines of Characters
    DEFINE LCD_COMMANDSUS 4000          'Delay between Commands
    DEFINE LCD_DATAUS 200               'Delay between Data Transfers

    ADCON0 = %00001111                  'ADC Enabled, Analog Input 3 Enabled
    ADCON1 = %00001011                  'ADC Voltage Range is 0 to 5 Volts &
                                        'Analog Inputs 0 Thru 3 are available
    ADCON2 = %10101101                  'Right Justify to use Upper 8-Bits, only,
                                        '12*Tad, & ADC Clock=20MHz/16=1.25MHz

    W0 VAR WORD                         'Set 16-Bit Word Variable,W0
    W1 VAR WORD                         ' "    "   "    "      ,W1
    W2 VAR WORD                         ' "    "   "    "      ,W2
    W3 VAR WORD                         ' "    "   "    "      ,W3
    VS1 VAR WORD                        ' "    "   "    "      ,VS1
    VS2 VAR WORD                        ' "    "   "    "      ,VS2
    RES VAR WORD                        ' "    "   "    "      ,RES
    LCDOC VAR WORD                      ' "    "   "    "      ,LODOC

    B0 VAR BYTE                         'Set 8-Bit Byte Variable,B0
    B1 VAR BYTE                         ' "    "   "    "      ,B1

    TRISB=$FF                           'Make PortB an Input Port to Enable
                                        'In-Circuit-Programming
    PAUSE 1000                          'Pause or Delay for 1 Seconds to Enable
                                        'In-Circuit-Programming
    LCDOUT $FE,1                        'Clear the LCD
    LCDOUT $FE, $80                     'Set the Cursor to Line 1, Position 0
    LCDOUT "Power Up"                   'Send, Power Up, to the LCD

    SERout PORTC.6,T2400,["Power Up",13,10,10]
                                        'Send "Power Up" to the HyperTerminal
                                        'with a Carriage Return & two Line Feeds,
                                        'at 2400 Bits/Second, 8-Bits, 1-Stop Bit,
                                        'No Parity, Driven True (T2400).

    PAUSE 2000                          'Pause or Delay for 2 Seconds
                                        "to Read "Power Up" in the LCD
    LCDOUT $FE,1                        'Clear the LCD
    TRISB=$00                           'Make PortB an Output Port to Allow for Normal Operation
    TRISD=$F0                           'Make PortD, Bits-4,5,6,7 Inputs, & Bits-3
```

```
SAMPLE: ADCIN 1,VS1              'Sample ADC Input Voltage-1, and place
                                 'the 10-Bit result in Variable VS1

       ADCIN 2,VS2               'Sample ADC Input Voltage-2, and place
                                 'the 10-Bit result in Variable VS2

       W0=(VS1-VS2)*50           'Scale the Equations to Yield the Number
       RES=(W0/VS2)*20           'of Ohms without Over-Flowing the Registers

       LCDOC=LCDOC-1             'Decrement LCD Output Counter
       IF LCDOC=0 THEN LCD       'Send the Output Voltage Value to
                                 'the LCD and the HyperTerminal every
                                 'LCDOC Samples

       GOTO SAMPLE

LCD:
       LCDOUT $FE, 1             'Clear the LCD
       LCDOUT $FE, $80           'Set Cursor to Line 1, Position 0
       LCDOUT "Resistance"       'Send, Res=Value Ohms, in Decimal to the LCD
       LCDOUT $FE, $C0           'Set Cursor to Line 1, Position 0
       LCDOUT DEC RES," = Ohms"  'Send, Res=Value Ohms, in Decimal to the LCD

       SEROUT2 PORTC.6,396,["Resistance = ",DEC RES," Ohms",13,10,10]
                                 'Send, Resistance = Value Ohms, in Decimal
                                 'to the HyperTerminal with a Carriage
                                 'Return and two Line Feeds

       LCDOC=50                  'Initialize the LCD Output Counter
                                 'Send the Output Voltage Value to
                                 'the LCD and the HyperTerminal every
                                 'LCDOC Samples
       GOTO SAMPLE

       END
```

Laboratory Experiment #6

'Lab-6D_PICBasic-Pro_Digital_Waveform_Generator

```
INCLUDE "modedefs.bas"          'Identifies the Command File

DEFINE OSC 20                   'Set PIC-Pro Oscillator Value in Mega Hertz
DEFINE ADC_BITS 10              'Set for 10-Bit Internal ADC

DEFINE ADC_CLOCK 5              'DEFINE ADC_CLOCK 1, for OSC=10, MAX ADC
                                'Clock=10MHz/8) = 1.25MHz
                                'DEFINE ADC_CLOCK 5, for OSC=16, MAX ADC
                                'Clock=16MHz/16) = 1.00MHz
                                'DEFINE ADC_CLOCK 5, for OSC=20, MAX ADC
                                'Clock=20MHz/16) = 1.25MHz
                                'DEFINE ADC_CLOCK 2, for OSC=24, MAX ADC
                                'Clock=24MHz/32) = 1.25MHz
                                'DEFINE ADC_CLOCK 2, for OSC=32, MAX ADC
                                'Clock=32MHz/32) = 1.00MHz
                                'DEFINE ADC_CLOCK 6, for OSC=48, MAX ADC
                                'Clock=48MHz/64) = 0.75MHz

DEFINE CCP1_BIT 2               'Setup Hardware PWM Timer
DEFINE LCD_BITS 4               'Setup LCD for 4-Bit Mode
DEFINE LCD_DREG PORTB           'Drive LCD from Port-B
DEFINE LCD_DBIT 0               'Starting with Bit-0
DEFINE LCD_RSREG PORTA          'Drive LCD Register-Select from
DEFINE LCD_RSBIT 4              'Port-A, Bit-4
DEFINE LCD_EREG PORTA           'Drive LCD Enable from
DEFINE LCD_EBIT 5               'Port-A, Bit-5
DEFINE LCD_LINES 2              'LCD can hold 2-Lines of Characters
DEFINE LCD_COMMANDSUS 4000      'Delay between Commands
DEFINE LCD_DATAUS 200           'Delay between Data Transfers

ADCON0 = %00001111              'ADC Enabled, Analog Input 3 Enabled
ADCON1 = %00001011              'ADC Voltage Range is 0 to 5 Volts &
                                'Analog Inputs 0 Thru 3 are available
ADCON2 = %10101101              'Right Justify to use Upper 8-Bits, only,
                                '12*Tad, & ADC Clock=20MHz/16=1.25MHz

W0 VAR WORD                     'Set 16-Bit Word Variable,W0
W1 VAR WORD                     '  "    "    "      "    ,W1
W2 VAR WORD                     '  "    "    "      "    ,W2
W3 VAR WORD                     '  "    "    "      "    ,W3
W4 VAR WORD                     '  "    "    "      "    ,W4
W5 VAR WORD                     '  "    "    "      "    ,W5
W6 VAR WORD                     '  "    "    "      "    ,W6

B0 VAR BYTE                     'Set 8-Bit Byte Variable,B0
B1 VAR BYTE                     '  "    "    "      "    ,B1

TRISB=$FF                       'Make PortB an Input Port to Enable
                                'In-Circuit-Programming
PAUSE 1000                      'Pause or Delay for 1 Seconds to Enable
                                'In-Circuit-Programming
LCDOUT $FE,1                    'Clear the LCD
LCDOUT $FE, $80                 'Set the Cursor to Line 1, Position 0
LCDOUT "Power Up"               'Send, Power Up, to the LCD

SERout PORTC.6,T2400,["Power Up",13,10,10]
                                'Send "Power Up" to the HyperTerminal
                                'with a Carriage Return & two Line Feeds,
                                'at 2400 Bits/Second, 8-Bits, 1-Stop Bit,
                                'No Parity, Driven True (T2400).

PAUSE 2000                      'Pause or Delay for 2 Seconds
                                'to Read "Power Up" in the LCD
LCDOUT $FE,1                    'Clear the LCD
TRISB=$00                       'Make PortB an Output Port to Allow for Normal Operation
TRISD=$F0                       'Make PortD, Bits-4,5,6,7 Inputs, & Bits-3

KSCAN:
LCDOUT $FE, 1                   'Clear the LCD
LCDOUT $FE, $80                 'Set the Cursor to Line 1, Position 0
LCDOUT "Scanning"               'Send, Scanning, to the LCD
SERout2 PORTC.6,396,["Scanning",13,10,10]
                                'Send, Scanning, to HyperTerminal
                                'with a Carriage Return & Two Line Feeds
```

```
        HIGH PORTD.0                    'Make 0th Row High to Scan for the Following:
        PAUSE 500
        INPUT PORTD.4
        IF PORTD.4=1 THEN FSQW          'Check for Keypad Key "1" , Column-4
        PAUSE 100
        INPUT PORTD.5
        IF PORTD.5=1 THEN FTRW          'Check for Keypad Key "2" , Column-5
        PAUSE 100
        LOW PORTD.0

        HIGH PORTD.3                    'Make 3th Row High to Scan for the Following:
        INPUT PORTD.5
        IF PORTD.5=1 THEN FSIN          'Check for Keypad Key "0" , Column 5
        PAUSE 100
        LOW PORTD.3

        GOTO KSCAN

        FSIN:
        TRISD=$00                       'Make PortD an Output Port
        LCDOUT $FE, 1                   'Clear the LCD
        LCDOUT $FE, $80                 'Set the Cursor to Line 1, Position 0
        LCDOUT "Sine-Wave"              'Send, Sine-Wave, to the LCD
        SERout2 PORTC.6,396,["Sine-Wave",13,10,10]
                                        'Send, Sine-Wave, to HyperTerminal
                                        'with a Carriage Return & Two Line Feeds
        PAUSE 500
        B0=0                            'Initialize the Sine Angle
        SINLP:  TRISD=$00               'Make PortD an Output Port

        PORTD=((((SIN B0)+127)/10)*5)+50
                                        '(+127) Removes the Negative Values
                                        'Amplitude (Peak Value) = 1/2 of 5/10
                                        'of the Maximum Output Voltage
                                        '50 Represents the DC Offset
                                        'DC Offset = 1/2 of 50/255
                                        'of the Maximum Output Voltage
                                        'The DC Offset is Measured between Zero
                                        'and the Negative Peak of the Sinewave.
                                        'To Obtain the Maximum Value, Change
                                        'Both the 5 and the 10 to a 1.

        PAUSEUS 1                       'Smallest Pause Allowed between Angle Changes
                                        'This will create, approximately, 100 Hz
        B0=B0+1                         'Increments the Sine Angle by 1 out of 255
        GOTO SINLP

        FSQW:
        TRISD=$00                       'Make PortD an Output Port
        LCDOUT $FE, 1                   'Clear the LCD
        LCDOUT $FE, $80                 'Set the Cursor to Line 1, Position 0
        LCDOUT "Square-Wave"            'Send, Square-Wave, to the LCD
        SERout2 PORTC.6,396,["Square-Wave",13,10,10]
                                        'Send, Square-Wave, to HyperTerminal
                                        'with a Carriage Return & Two Line Feeds
        PAUSE 500
        SQWLP:  TRISD=$00               'Make PortD an Output Port
        PORTD=0                         'Create a Square-Wave at PortD
                                        'Starting with Decimal 0 out of 255
                                        'of the Maximum Output Voltage.

        PAUSEUS 2                       'The Smallest PAUSE Allowed is 5 micro-seconds.
                                        '10 micro-seconds, in one 10 micro-second Step,
                                        'Plus the Approximate 1 micro-second time of
                                        "the (PORTD) statement, and the 3 micro-seconds
                                        'time of the (INPUT), (IF) and (GOTO) statements,
                                        'is chosen to create, approximately, 46KHz.

        PORTD=255
        PAUSEUS 2
        GOTO SQWLP
```

```
FTRW:
TRISD=$00                               'Make PortD an Output Port
LCDOUT $FE, 1                           'Clear the LCD
LCDOUT $FE, $80                         'Set the Cursor to Line 1, Position 0
LCDOUT "Triangle-Wave"                  'Send, Triangle-Wave, to the LCD
SERout2 PORTC.6,396,["Triangle-Wave",13,10,10]
                                        'Send, Triangle-Wave, to HyperTerminal
                                        'with a Carriage Return & Two Line Feeds
PAUSE 500
TRWLP:  TRISD=$00                       'Make PortD an Output Port
PORTD=0                                 'Create a Triangle-Wave at PortD
                                        'Starting with Decimal 0 out of 255
                                        'of the Maximum Output Voltage.

PAUSEUS 1                               'The Smallest PAUSE Allowed is 1 micro-second.
                                        '10 micro-seconds, in 1 micro-second steps,
                                        'Plus the Approximate 1 micro-second time of
                                        'the (PORTD) statement, and the 3 micro-seconds
                                        'time of the (INPUT), (IF) and (GOTO) statements,
                                        'is chosen to create, approximately, 40KHz.
PORTD=25
PAUSEUS 1
PORTD=50
PAUSEUS 1
PORTD=75
PAUSEUS 1
PORTD=100
PAUSEUS 1
PORTD=125
PAUSEUS 1
PORTD=150
PAUSEUS 1
PORTD=175
PAUSEUS 1
PORTD=200
PAUSEUS 1
PORTD=225
PAUSEUS 1
PORTD=255
PAUSEUS 1
PORTD=225
PAUSEUS 1
PORTD=200
PAUSEUS 1
PORTD=175
PAUSEUS 1
PORTD=150
PAUSEUS 1
PORTD=125
PAUSEUS 1
PORTD=100
PAUSEUS 1
PORTD=75
PAUSEUS 1
PORTD=50
PAUSEUS 1
PORTD=25

GOTO TRWLP

END
```

Laboratory Experiment #6

```
'Lab-6E_PICBasic-Pro_One-Bit_Digital-to-Analog_Converter
INCLUDE "modedefs.bas"              'Identifies the Command File

DEFINE OSC 20                       'Set PIC-Pro Oscillator Value in Mega Hertz
DEFINE ADC_BITS 10                  'Set for 10-Bit Internal ADC

DEFINE ADC_CLOCK 5                  'DEFINE ADC_CLOCK 1, for OSC=10, MAX ADC
                                    'Clock=10MHz/8) = 1.25MHz
                                    'DEFINE ADC_CLOCK 5, for OSC=16, MAX ADC
                                    'Clock=16MHz/16) = 1.00MHz
                                    'DEFINE ADC_CLOCK 5, for OSC=20, MAX ADC
                                    'Clock=20MHz/16) = 1.25MHz
                                    'DEFINE ADC_CLOCK 2, for OSC=24, MAX ADC
                                    'Clock=24MHz/32) = 1.25MHz
                                    'DEFINE ADC_CLOCK 2, for OSC=32, MAX ADC
                                    'Clock=32MHz/32) = 1.00MHz
                                    'DEFINE ADC_CLOCK 6, for OSC=48, MAX ADC
                                    'Clock=48MHz/64) = 0.75MHz

DEFINE CCP1_BIT 2                   'Setup Hardware PWM Timer
DEFINE LCD_BITS 4                   'Setup LCD for 4-Bit Mode
DEFINE LCD_DREG PORTB               'Drive LCD from Port-B
DEFINE LCD_DBIT 0                   'Starting with Bit-0
DEFINE LCD_RSREG PORTA              'Drive LCD Register-Select from
DEFINE LCD_RSBIT 4                  'Port-A, Bit-4
DEFINE LCD_EREG PORTA               'Drive LCD Enable from
DEFINE LCD_EBIT 5                   'Port-A, Bit-5
DEFINE LCD_LINES 2                  'LCD can hold 2-Lines of Characters
DEFINE LCD_COMMANDSUS 4000          'Delay between Commands
DEFINE LCD_DATAUS 200               'Delay between Data Transfers

ADCON0 = %00001111                  'ADC Enabled, Analog Input 3 Enabled
ADCON1 = %00001011                  'ADC Voltage Range is 0 to 5 Volts &
                                    'Analog Inputs 0 Thru 3 are available
ADCON2 = %10101101                  'Right Justify to use Upper 8-Bits, only,
                                    '12*Tad, & ADC Clock=20MHz/16=1.25MHz

W0 VAR WORD                         'Set 16-Bit Word Variable,W0
W1 VAR WORD                         ' "     "      "      "   ,W1
W2 VAR WORD                         ' "     "      "      "   ,W2
W3 VAR WORD                         ' "     "      "      "   ,W3
W4 VAR WORD                         ' "     "      "      "   ,W4
W5 VAR WORD                         ' "     "      "      "   ,W5
W6 VAR WORD                         ' "     "      "      "   ,W6

B0 VAR BYTE                         'Set 8-Bit Byte Variable,B0
B1 VAR BYTE                         ' "     "      "      "   ,B1
DutyCycle VAR BYTE                  ' "     "      "      "   ,DutyCycle

TRISB=$FF                           'Make PortB an Input Port to Enable
                                    'In-Circuit-Programming
PAUSE 1000                          'Pause or Delay for 1 Seconds to Enable
                                    'In-Circuit-Programming
LCDOUT $FE,1                        'Clear the LCD
LCDOUT $FE, $80                     'Set the Cursor to Line 1, Position 0
LCDOUT "Power Up"                   'Send, Power Up, to the LCD

SERout PORTC.6,T2400,["Power Up",13,10,10]
                                    'Send "Power Up" to the HyperTerminal
                                    'with a Carriage Return & two Line Feeds,
                                    'at 2400 Bits/Second, 8-Bits, 1-Stop Bit,
                                    'No Parity, Driven True (T2400).

PAUSE 2000                          'Pause or Delay for 2 Seconds
                                    'to Read "Power Up" in the LCD
LCDOUT $FE,1                        'Clear the LCD
TRISB=$00                           'Make PortB an Output Port to Allow for Normal Operation
TRISD=$F0                           'Make PortD, Bits-4,5,6,7 Inputs, & Bits-3
```

```
DutyCycle=5                          'Initialize DutyCycle to: (5/256)*100%

AGAIN:
DutyCycle=DutyCycle+5                'Increase DutyCycle to:
                                     '((Original Value + 5)/256)*100%

HPWM 1,DutyCycle,30000               'Setup Capture/Compare/PWM Pin-1 to
                                     'PWM Mode, Set Frequency to 30KHz &
                                     'Set the DutyCycle Variable equal to
                                     'any value between 0 and 255
                                     'example: DutyCycle = 128
                                     '(128/256)*100% = 50%

GOTO AGAIN

END
```

Laboratory Experiment #6

'Lab-6B_CCS_C-Code_Digital_Voltmeter

```
#include <18F4685.h>                              //Identify Microcontroller
#device adc=10                                    //Identify ADC Bit Width
#fuses HS,NOWDT,NOPROTECT,NOLVP,MCLR              //Setup Programmer Oscillator Value
                                                  //and make Master Clear Pin an
                                                  //Enable Master Clear=MCLR
                                                  //Disable Master Clear=NOMCLR

#use delay(clock=20000000)                        //Setup C-Code Oscillator Value
#use fast_io(d)                                   //Leave the State of Port-D (all bits)
                                                  //the same until changed, again
#use rs232(baud=2400, xmit=PIN_C6, rcv=PIN_C7, stream=HT)
                                                  //Set RS232 HyperTerminal
                                                  //Communication Parameters
#use rs232(baud=2400, xmit=PIN_B4, rcv=PIN_B5, invert, stream=BB)
                                                  //Set RS232 Board-to-Board
                                                  //Communication Parameters
#use I2C(master, SCL=PIN_B6, SDA=PIN_B7)          //Set I2C Communication Parameters
#include <lcd_flex.c>                             //Identify LCD Driver File
#include <math.h>                                 //Include Math Functions

float ADCValue1,ADCValue2,Vin1,Vin2;              //16-Bit Floating Point Value
unsigned long int MaxVoltRange,SFactor,LCDOC;     //16-Bit Unsigned Long Integer Value

void main()
{
    set_tris_b(0xFF);                             //Make PortB an Input Port to
                                                  //Enable In-Circuit-Programming
    delay_ms(1000);                               //Pause or Delay for 1 second

    lcd_init();                                   //Initialize the LCD
    lcd_putc(0x0c);                               //Clear the LCD
    lcd_gotoxy(1,1);                              //Set the Cursor to Position-1 of Line-1
    printf(lcd_putc,"Power Up");                  //Send, Power Up, to the LCD

    fprintf(HT, "\n\rPower Up\n\r");              //Send, Power Up, to the HyperTerminal with
                                                  //a next Line & a Carriage Return
    delay_ms(1000);                               //Pause or Delay for 1 second

    setup_port_a(AN0_to_AN3);                     //Identify all of the Analog Inputs
    setup_adc(VSS_VDD);                           //Setup the ADC Voltage Range to be
                                                  //0 Volts to 5 Volts
                                                  //5v-0v)/1024~=.005 Volts Per Bit

    setup_adc(ADC_CLOCK_DIV_32);                  //for delay(clock=10000000),
                                                  //setup_adc(ADC_CLOCK_DIV_8),
                                                  //10MHz/8=1.25MHz
                                                  //for delay(clock=16000000),
                                                  //setup_adc(ADC_CLOCK_DIV_16),
                                                  //16MHz/16=1.00MHz
                                                  //for delay(clock=20000000),
                                                  //setup_adc(ADC_CLOCK_DIV_16),
                                                  //20MHz/16=1.25MHz
                                                  //for delay(clock=24000000),
                                                  //setup_adc(ADC_CLOCK_DIV_32),
                                                  //24MHz/32=1.25MHz
                                                  //for delay(clock=32000000),
                                                  //setup_adc(ADC_CLOCK_DIV_32),
                                                  //32MHz/32=1.00MHz
                                                  //for delay(clock=48000000),
                                                  //setup_adc(ADC_CLOCK_DIV_64),
                                                  //48MHz/64=0.75MHz

    set_adc_channel(0);                           //Identify the Analog Input to be
                                                  //Sampled (Channel Input Zero)

    setup_ccp1(CCP_PWM);                          //Setup Capture/Compare/PWM Pin-1 to PWM Mode

    setup_ccp2(CCP_CAPTURE_RE);                   //Setup Capture/Compare/PWM Pin-2 to
                                                  //Capture Mode

    setup_timer_1(T1_INTERNAL);                   //Setup Timer-1, which controls the
                                                  //Capture/Compare portion of
                                                  //Capture/Compare/PWM Pin-2, only,
                                                  //to Internal Mode

    setup_timer_2(T2_DIV_BY_4, 19, 1);            //Setup Timer-2, which controls the PWM portion
                                                  //of Capture/Compare/PWM Pin-1, only.
                                                  //DIV_BY_Mode, Prescale (Period), Postscale
                                                  //The Following Limits are for the
                                                  //PIC18F4685 Microcontroller
                                                  //Mode may equal 1 or 4 or 16
                                                  //Prescale (Period) may equal any integer
                                                  //between 0 and 255
                                                  //Postscale may equal any integer between
                                                  //1 and 16 (Post = 1 to set PWM Frequency)
                                                  //T2Freq=OSC/[(4)*(Mode)*(Prescale+1)]=
                                                  //24MegaHz/[(4)*(4)*(PS+1)] = 75KHz
```

```c
enable_interrupts(INT_CCP2);                //Enable the Internal Interrupt Input
                                            //of Capture/Compare/PWM Pin-2

enable_interrupts(GLOBAL);                  //Enable Global Interrupts for the
                                            //PIC18F4685 Microcontroller

set_tris_b(0x00);                           //Make PortB an Output Port to Allow for
                                            //Normal Operation
tris_d(0xF0);                               //Make PortD, Bits-4,5,6,7 Inputs
                                            //and Bits-3,2,1,0 Outputs


LCDOC=50;                                   //Initialize the LCD Output Counter
                                            //Send the Output Voltage Value to
                                            //the LCD and the HyperTerminal every
                                            //LCDOC Samples

MaxVoltRange=5;                             //Set the Maximum ADC Voltage Range
SFactor=1024/MaxVoltRange;                  //Determine the Scale Factor


measur:
set_adc_channel(1);                         //Identify the Analog Input to be Sampled (Channel Input-1)
delay_us(100);                              //Delay to Setup Analog Multiplier
ADCValue1 = read_adc();                     //Sample ADC Input Voltage-1, and place this value into
                                            //the Variable ADCValue1

Vin1=(ADCValue1/SFactor);                   //Convert ADCValue1 to Volts

set_adc_channel(2);                         //Identify the Analog Input to be Sampled (Channel Input-2)
delay_us(100);                              //Delay to Setup Analog Multiplier
ADCValue2 = read_adc();                     //Sample ADC Input Voltage-2, and place this value into
                                            //the Variable ADCValue2

Vin2=(ADCValue2/SFactor);                   //Convert ADCValue2 to Volts

LCDOC=LCDOC-1;                              //Decrement LCD Output Counter
if (LCDOC==0) goto LCD;                     //Send the Output Voltage Value to
                                            //the LCD and the HyperTerminal every
                                            //LCDOC Samples

goto measur;

LCD:
lcd_putc(0x0c);                             //Clear the LCD
lcd_gotoxy(1,1);                            //Set the Cursor to Position-1 of Line-1
printf(lcd_putc, "Vin1=%lf", Vin1);         //Send, Vin1=, to the LCD in Floating Point Decimal Form
lcd_gotoxy(10,1);                           //Set the Cursor to Position-10 of Line-1
printf(lcd_putc, " Volts");                 //Send, Volts, to the LCD
fprintf(HT, "\n\rVin1=%lf", Vin1);          //Send, Vin1=, to the HyperTerminal in Floating Point Decimal Form
fprintf(HT, " Volts\n\r");                  //Send, Volts, to the HyperTerminal with
                                            //a next Line & a Carriage Return

lcd_gotoxy(1,2);                            //Set the Cursor to Position-1 of Line-2
printf(lcd_putc, "Vin2=%lf", Vin2);         //Send, Vin2=, to the LCD in Floating Point Decimal Form
lcd_gotoxy(10,2);                           //Set the Cursor to Position-10 of Line-2
printf(lcd_putc," Volts");                  //Send, Volts, to the LCD
fprintf(HT, "\n\rVin2=%lf", Vin2);          //Send, Vin2 = , to the HyperTerminal in
                                            //Floating Point Decimal Form

fprintf(HT, " Volts\n\r");                  //Send, Volts, to the HyperTerminal with
                                            //a next Line & a Carriage Return

LCDOC=50;                                   //Initialize the LCD Output Counter
                                            //Send the Output Voltage Value to
                                            //the LCD and the HyperTerminal every
                                            //LCDOC Samples

goto measur;

}
```

Laboratory Experiment #6

'Lab-6C_CCS_C-Code_Digital_Ohmmeter

```c
#include <18F4685.h>                              //Identify Microcontroller
#device adc=10                                    //Identify ADC Bit Width
#fuses HS,NOWDT,NOPROTECT,NOLVP,MCLR              //Setup Programmer Oscillator Value
                                                  //and make Master Clear Pin an
                                                  //Enable Master Clear=MCLR
                                                  //Disable Master Clear=NOMCLR

#use delay(clock=20000000)                        //Setup C-Code Oscillator Value
#use fast_io(d)                                   //Leave the State of Port-D (all bits)
                                                  //the same until changed, again
#use rs232(baud=2400, xmit=PIN_C6, rcv=PIN_C7, stream=HT)
                                                  //Set RS232 HyperTerminal
                                                  //Communication Parameters
#use rs232(baud=2400, xmit=PIN_B4, rcv=PIN_B5, invert, stream=BB)
                                                  //Set RS232 Board-to-Board
                                                  //Communication Parameters
#use I2C(master, SCL=PIN_B6, SDA=PIN_B7)          //Set I2C Communication Parameters
#include <lcd_flex.c>                             //Identify LCD Driver File
#include <math.h>                                 //Include Math Functions

float ADCValue1,ADCValue2,V1,V2,VD,Res;           //16-Bit Floating Point Value
unsigned long int MaxVoltRange,SFactor,LCDOC;     //16-Bit Unsigned Long Integer Value

void main()
{
set_tris_b(0xFF);                                 //Make PortB an Input Port to
                                                  //Enable In-Circuit-Programming

delay_ms(1000);                                   //Pause or Delay for 1 second

lcd_init();                                       //Initialize the LCD
lcd_putc(0x0c);                                   //Clear the LCD
lcd_gotoxy(1,1);                                  //Set the Cursor to Position-1 of Line-1
printf(lcd_putc,"Power Up");                      //Send, Power Up, to the LCD

fprintf(HT, "\n\rPower Up\n\r");                  //Send, Power Up, to the HyperTerminal with
                                                  //a next Line & a Carriage Return
delay_ms(1000);                                   //Pause or Delay for 1 second

setup_port_a(AN0_to_AN3);                         //Identify all of the Analog Inputs
setup_adc(VSS_VDD);                               //Setup the ADC Voltage Range to be
                                                  //0 Volts to 5 Volts
                                                  //5v-0v)/1024~=.005 Volts Per Bit

setup_adc(ADC_CLOCK_DIV_32);                      //for delay(clock=10000000),
                                                  //setup_adc(ADC_CLOCK_DIV_8),
                                                  //10MHz/8=1.25MHz
                                                  //for delay(clock=16000000),
                                                  //setup_adc(ADC_CLOCK_DIV_16),
                                                  //16MHz/16=1.00MHz
                                                  //for delay(clock=20000000),
                                                  //setup_adc(ADC_CLOCK_DIV_16),
                                                  //20MHz/16=1.25MHz
                                                  //for delay(clock=24000000),
                                                  //setup_adc(ADC_CLOCK_DIV_32),
                                                  //24MHz/32=1.25MHz
                                                  //for delay(clock=32000000),
                                                  //setup_adc(ADC_CLOCK_DIV_32),
                                                  //32MHz/32=1.00MHz
                                                  //for delay(clock=48000000),
                                                  //setup_adc(ADC_CLOCK_DIV_64),
                                                  //48MHz/64=0.75MHz

set_adc_channel(0);                               //Identify the Analog Input to be
                                                  //Sampled (Channel Input Zero)

setup_ccp1(CCP_PWM);                              //Setup Capture/Compare/PWM Pin-1 to PWM Mode

setup_ccp2(CCP_CAPTURE_RE);                       //Setup Capture/Compare/PWM Pin-2 to
                                                  //Capture Mode

setup_timer_1(T1_INTERNAL);                       //Setup Timer-1, which controls the
                                                  //Capture/Compare portion of
                                                  //Capture/Compare/PWM Pin-2, only,
                                                  //to Internal Mode

setup_timer_2(T2_DIV_BY_4, 19, 1);                //Setup Timer-2, which controls the PWM portion //of
                                                  //Capture/Compare/PWM Pin-1, only.
                                                  //DIV_BY_Mode, Prescale (Period), Postscale
                                                  //The Following Limits are for the
                                                  //PIC18F4685 Microcontroller
                                                  //Mode may equal 1 or 4 or 16
                                                  //Prescale (Period) may equal any integer
                                                  //between 0 and 255
                                                  //Postscale may equal any integer between
                                                  //1 and 16 (Post = 1 to set PWM Frequency)
                                                  //T2Freq=OSC/[(4)*(Mode)*(Prescale+1)]=
                                                  //24MegaHz/[(4)*(4)*(PS+1)] = 75KHz
```

```c
enable_interrupts(INT_CCP2);              //Enable the Internal Interrupt Input
                                          //of Capture/Compare/PWM Pin-2

enable_interrupts(GLOBAL);                //Enable Global Interrupts for the
                                          //PIC18F4685 Microcontroller

set_tris_b(0x00);                         //Make PortB an Output Port to Allow for
                                          //Normal Operation
tris_d(0xF0);                             //Make PortD, Bits-4,5,6,7 Inputs
                                          //and Bits-3,2,1,0 Outputs


LCDOC=50;                                 //Initialize the LCD Output Counter
                                          //Send the Output Voltage Value to
                                          //the LCD and the HyperTerminal every
                                          //LCDOC Samples


MaxVoltRange=5;                           //Set the Maximum ADC Voltage Range
SFactor=1024/MaxVoltRange;                //Determine the Scale Factor


measur:
set_adc_channel(1);                       //Identify the Analog Input to be Sampled (Channel Input-1)
delay_us(100);                            //Delay to Setup Analog Multiplier
ADCValue1 = read_adc();                   //Sample ADC Input Voltage-1, and place the 10-Bit result
                                          //in Variable ADCValue1
V1=(ADCValue1/SFactor);                   //Convert the Sampled ADC Value, ADCValue1, to V1 Volts
set_adc_channel(2);                       //Identify the Analog Input to be Sampled (Channel Input-2)
delay_us(100);                            //Delay to Setup Analog Multiplier
ADCValue2 = read_adc();                   //Sample ADC Input Voltage-2, and place the 10-Bit result
                                          //in Variable ADCValue2
V2=(ADCValue2/SFactor);                   //Convert the Sampled ADC Value, ADCValue2, to V2 Volts
VD=(V1-V2)*50;                            //Scale the Equations to Yield the Number
Res=(VD/V2)*20;                           //of Ohms without Over-Flowing the Registers

LCDOC=LCDOC-1;                            //Decrement LCD Output Counter
if (LCDOC==0) goto LCD;                   //Send the Output Voltage Value to
                                          //the LCD and the HyperTerminal every
                                          //CDOC Samples

goto measur;

LCD:
lcd_putc(0x0c);                           //Clear the LCD
lcd_gotoxy(1,1);                          //Set the Cursor to Position-1 of Line-1
printf(lcd_putc, "Resistance");           //Send, Resistance, to the LCD
lcd_gotoxy(1,2);                          //Set the Cursor to Position-1 of Line-2
printf(lcd_putc, "%lf", Res);             //Send, the Value of Res, to the LCD in
                                          //Floating Point Decimal Form
lcd_gotoxy(6,2);                          //Set the Cursor to Position-6 of Line-2
printf(lcd_putc, " Ohms");                //Send, Ohms, to the LCD
fprintf(HT, "\n\rResistance=%lf", Res);   //Send, Resistance= and the Value of Res, to the HyperTerminal in
                                          //Floating Point Decimal Form
fprintf(HT, " Ohms\n\r");                 //Send, Ohms, to the HyperTerminal with
                                          //a next Line & a Carriage Return

LCDOC=50;                                 //Initialize the LCD Output Counter
                                          //Send the Output Voltage Value to
                                          //the LCD and the HyperTerminal every
                                          //LCDOC Samples

goto measur;

}
```

Laboratory Experiment #6

'Lab-6D_CCS_C-Code_Waveform_Generator

```
#include <18F4685.h>              //Identify Microcontroller
#device adc=10                    //Identify ADC Bit Width
#fuses HS,NOWDT,NOPROTECT,NOLVP,MCLR   //Setup Programmer Oscillator Value
                                  //and make Master Clear Pin an
                                  //Enable Master Clear=MCLR
                                  //Disable Master Clear=NOMCLR
#use delay(clock=20000000)        //Setup C-Code Oscillator Value
#use fast_io(d)                   //Leave the State of Port-D (all bits)
                                  //the same until changed, again
#use rs232(baud=2400, xmit=PIN_C6, rcv=PIN_C7, stream=HT)
                                  //Set RS232 HyperTerminal
                                  //Communication Parameters
#use rs232(baud=2400, xmit=PIN_B4, rcv=PIN_B5, invert, stream=BB)
                                  //Set RS232 Board-to-Board
                                  //Communication Parameters
#use I2C(master, SCL=PIN_B6, SDA=PIN_B7)  //Set I2C Communication Parameters
#include <lcd_flex.c>             //Identify LCD Driver File
#include <math.h>                 //Include Math Functions

float B0;                         //16-Bit Floating Point Word Value
signed int B1;                    //8-Bit Signed Integer Byte Value
unsigned int DV;                  //8-Bit unsigned Integer Byte Value
unsigned int16 PV;                //16-Bit unsigned Integer Word Value
short D4,D5,D6;                   //1-Bit Value

void main()
    {
    set_tris_b(0xFF);             //Make PortB an Input Port to
                                  //Enable In-Circuit-Programming
    delay_ms(1000);               //Pause or Delay for 1 second

    lcd_init();                   //Initialize the LCD
    lcd_putc(0x0c);               //Clear the LCD
    lcd_gotoxy(1,1);              //Set the Cursor to Position-1 of Line-1
    printf(lcd_putc,"Power Up");  //Send, Power Up, to the LCD

    fprintf(HT, "\n\rPower Up\n\r");  //Send, Power Up, to the HyperTerminal with
                                  //a next Line & a Carriage Return
    delay_ms(1000);               //Pause or Delay for 1 second

    setup_port_a(AN0_to_AN3);     //Identify all of the Analog Inputs
    setup_adc(VSS_VDD);           //Setup the ADC Voltage Range to be
                                  //0 Volts to 5 Volts
                                  //5v-0v)/1024~=.005 Volts Per Bit

    setup_adc(ADC_CLOCK_DIV_32);  //for delay(clock=10000000),
                                  //setup_adc(ADC_CLOCK_DIV_8),
                                  //10MHz/8=1.25MHz
                                  //for delay(clock=16000000),
                                  //setup_adc(ADC_CLOCK_DIV_16),
                                  //16MHz/16=1.00MHz
                                  //for delay(clock=20000000),
                                  //setup_adc(ADC_CLOCK_DIV_16),
                                  //20MHz/16=1.25MHz
                                  //for delay(clock=24000000),
                                  //setup_adc(ADC_CLOCK_DIV_32),
                                  //24MHz/32=1.25MHz
                                  //for delay(clock=32000000),
                                  //setup_adc(ADC_CLOCK_DIV_32),
                                  //32MHz/32=1.00MHz
                                  //for delay(clock=48000000),
                                  //setup_adc(ADC_CLOCK_DIV_64),
                                  //48MHz/64=0.75MHz

    set_adc_channel(0);           //Identify the Analog Input to be
                                  //Sampled (Channel Input Zero)

    setup_ccp1(CCP_PWM);          //Setup Capture/Compare/PWM Pin-1 to PWM Mode

    setup_ccp2(CCP_CAPTURE_RE);   //Setup Capture/Compare/PWM Pin-2 to
                                  //Capture Mode

    setup_timer_1(T1_INTERNAL);   //Setup Timer-1, which controls the
                                  //Capture/Compare portion of
                                  //Capture/Compare/PWM Pin-2, only,
                                  //to Internal Mode

    setup_timer_2(T2_DIV_BY_4, 19, 1);  //Setup Timer-2, which controls the PWM portion //of
                                  //Capture/Compare/PWM Pin-1, only.
                                  //DIV_BY_Mode, Prescale (Period), Postscale
                                  //The Following Limits are for the
                                  //PIC18F4685 Microcontroller
                                  //Mode may equal 1 or 4 or 16
                                  //Prescale (Period) may equal any integer
                                  //between 0 and 255
                                  //Postscale may equal any integer between
                                  //1 and 16 (Post = 1 to set PWM Frequency)
                                  //T2Freq=OSC/[(4)*(Mode)*(Prescale+1)]=
                                  //24MegaHz/[(4)*(4)*(PS+1)] = 75KHz
```

```c
        enable_interrupts(INT_CCP2);              //Enable the Internal Interrupt Input
                                                  //of Capture/Compare/PWM Pin-2

        enable_interrupts(GLOBAL);                //Enable Global Interrupts for the
                                                  //PIC18F4685 Microcontroller

        set_tris_b(0x00);                         //Make PortB an Output Port to Allow for
                                                  //Normal Operation
        tris_d(0xF0);                             //Make PortD, Bits-4,5,6,7 Inputs
                                                  //and Bits-3,2,1,0 Outputs

        kscan:
        lcd_putc(0x0c);                           //Clear the LCD
        lcd_gotoxy(1,1);                          //Set the Cursor to Position 1 of Line 1
        printf(lcd_putc,"Scanning");              //Send, Scanning, to the LCD
        fprintf(HT, "\n\rScanning\n\r");          //Send, Scanning, to HyperTerminal with
                                                  //a Carriage Return & Two Line Feeds

        output_high(PIN_D0);                      //Make 0th Row High to Scan for the Following:
        delay_ms(500);
        D4 = input(PIN_D4);                       //Check for Keypad Key "1" , Column-4
        if (D4) goto FSQW;
        delay_ms(100);
        D5 = input(PIN_D5);                       //Check for Keypad Key "2" , Column-5
        if (D5) goto FTRW;
        delay_ms(100);
        output_low(PIN_D0);
        output_high(PIN_D3);                      //Make 3rd Row High to Scan for the Following:
        delay_ms(500);
        D5 = input(PIN_D5);                       //Check for Keypad Key "0" , Column-5
        if (D5) goto FSILF;
        delay_ms(100);
        D6 = input(PIN_D6);                       //Check for Keypad Key "#" , Column-6
        if (D6) goto FCWHF;
        delay_ms(100);
        output_low(PIN_D3);
        goto kscan;


        FSILF:
        set_tris_d(0x00);                         //Make PortD an Output Port
        lcd_putc(0x0c);                           //Clear the LCD
        lcd_gotoxy(1,1);                          //Set the Cursor to Position 1 of Line 1
        printf(lcd_putc, "Sine-Wave");            //Send, Sine-Wave, to the LCD
        lcd_gotoxy(1,2);                          //Set the Cursor to Position 2 of Line 1
        printf(lcd_putc,"Low Frequency");         //Send, Low Frequency, to the LCD
        fprintf(HT, "Sine-Wave; Low Frequency\n\r"); //Send, Sine-Wave; Low Frequency, to HyperTerminal with
                                                  //a Carriage Return & Two Line Feeds

        delay_ms(250);

        while(true)   for(B0=0; B0<6.283185; B0+=0.1)   //B0 is varied from Zero Radians to 2*PI Radians,
                                                  //in 0.1 Radian Steps
                    }
        B1=((100*sin(B0))+140);                   //Create a Sinewave at PortD
                                                  //100 Represents the Maximum Output Peak-to-Peak Voltage
                                                  //140 Represents the Center Value of the Sinewave, which is set at
        2.0 Volts-DC
        output_d(B1);
        delay_us(1);                              //Minimum Delay Value Allowed
                    }
        FCWHF:  set_tris_d(0x00);                 //Make PortD an Output Port
        lcd_putc(0x0c);                           //Clear the LCD
        lcd_gotoxy(1,1);                          //Set the Cursor to Position 1 of Line 1
        printf(lcd_putc, "Cosine-Wave");          //Send, Cosine-Wave, to the LCD
        lcd_gotoxy(1,2);                          //Set the Cursor to Position 1 of Line 2
        printf(lcd_putc,"High Frequency");        //Send, High Frequency, to the LCD
        fprintf(HT, "Cosine-Wave; High Frequency\n\r"); //Send, Cosine-Wave; High Frequency, to HyperTerminal with a
                                                  //Carriage Return & Two Line Feeds

        delay_ms(250);
        DV=1;                                     //Set the Decimal Delay Value (DV) (1 micro-second is
                                                  //the Minimum Value Allowed)
        PV=255;                                   //Set the Decimal Peak Value (PV) out of 255,
                                                  //of the Maximum Output Voltage

        CWHF:
        set_tris_d(0x00);                         //Make PortD an Output Port.
        output_d(PV);                             //Create a Cosine-Wave at PortD, Starting with
                                                  //the Decimal Peak Value (PV)
        delay_us(DV);                             //out of 255, of the Maximum Output Voltage.
        output_d(90*PV/100);                      //Continue Creating a Cosine-Wave at PortD with 90%
                                                  //of the Decimal Peak Value (PV)
        delay_us(DV);                             //out of 255, of the Maximum Output Voltage.
        output_d(70*PV/100);
        delay_us(DV);
        output_d(60*PV/100);
        delay_us(DV);
        output_d(40*PV/100);
        delay_us(DV);
        output_d(30*PV/100);
        delay_us(DV);
        output_d(10*PV/100);
        delay_us(DV);
```

```
            output_d(1*PV/100);
            delay_us(DV);
            output_d(10*PV/100);
            delay_us(DV);
            output_d(30*PV/100);
            delay_us(DV);
            output_d(40*PV/100);
            delay_us(DV);
            output_d(60*PV/100);
            delay_us(DV);
            output_d(70*PV/100);
            delay_us(DV);
            output_d(90*PV/100);
            delay_us(DV);

            goto CWHF;

        FSQW:
            set_tris_d(0x00);                          //Make PortD an Output Port
            lcd_putc(0x0c);                            //Clear the LCD
            lcd_gotoxy(1,1);                           //Set the Cursor to Position 1 of Line 1
            printf(lcd_putc, "Square-Wave");           //Send, Square-Wave, to the LCD
            fprintf(HT, "Square-Wave\n\r");            //Send, Square-Wave, to HyperTerminal with
                                                       //a Carriage Return & Two Line Feeds

            delay_ms(250);

        SQWLP:
            set_tris_d(0x00);                          //Make PortD an Output Port
            output_d(0);                               //Create a Square-Wave at PortD
                                                       //Starting with Decimal 0 out of 255
                                                       //of the Maximum Output Voltage
            delay_us(10);                              //The Smallest delay Allowed is 5 micro-seconds.
                                                       //10 micro-seconds, in one 10 micro-second Step,
                                                       //Plus the Approximate 0.1 micro-second time of
                                                       //the (output_d) statement,
                                                       //and the 3 micro-second time of the (input),
                                                       //(if) and (goto) statements,
                                                       //is chosen to create, approximately, 46KHz.

            output_d(255);
            delay_us(10);

            goto SQWLP;


        FTRW:
            set_tris_d(0x00);                          //Make PortD an Output Port
            lcd_putc(0x0c);                            //Clear the LCD
            lcd_gotoxy(1,1);                           //Set the Cursor to Position 1 of Line 1
            printf(lcd_putc, "Triangle-Wave");         //Send, Triangle-Wave, to the LCD
            fprintf(HT, "Triangle-Wave\n\r");          //Send, Triangle-Wave, to HyperTerminal with
                                                       //a Carriage Return & Two Line Feeds

            delay_ms(250);

        TWLP:
            set_tris_d(0x00);                          //Make PortD an Output Port
            output_d(0);                               //Create a Triangle-Wave at PortD
                                                       //Starting with Decimal 25 out of 255
                                                       //of the Maximum Output Voltage
            delay_us(1);                               //The Smallest delay Allowed is 1 micro-second.
                                                       //10 micro-seconds, in 1 micro-second steps,
                                                       //Plus the Approximate 0.1 micro-second time of
                                                       //the (output_d) statement,
                                                       //and the 3 micro-second time of the (input),
                                                       //(if) and (goto) statements,
                                                       //is chosen to create, approximately, 40KHz.

            output_d(25);
            delay_us(1);
            output_d(50);
            delay_us(1);
            output_d(75);
            delay_us(1);
            output_d(100);
            delay_us(1);
            output_d(125);
            delay_us(1);
            output_d(150);
            delay_us(1);
            output_d(175);
            delay_us(1);
            output_d(200);
            delay_us(1);
            output_d(225);
            delay_us(1);
            output_d(255);
            delay_us(1);
            output_d(225);
            delay_us(1);
            output_d(200);
            delay_us(1);
            output_d(175);
            delay_us(1);
```

```
    output_d(150);
    delay_us(1);
    output_d(125);
    delay_us(1);
    output_d(100);
    delay_us(1);
    output_d(75);
    delay_us(1);
    output_d(50);
    delay_us(1);
    output_d(25);
    delay_us(1);

    goto TWLP;
            }
```

Engineering Practices for the PIC Microcontroller

By: Prof. Sal R. Riggio Jr., PhD, PE

Chapter - 9

Laboratory Experiment - 7

*The Variable Bandwidth
Digital Low-Pass-Filter*

Introduction

The purpose of this lab experiment is for the student to develop basic skills in making practical use of the Microchip PIC18F4685 Microcontroller and become familiar with the importance, operation and hardware implementation of the techniques and methods required to create a variable bandwidth digital filter.

The student is expected to analyze the **PICBasic-Pro Code**, which allows one to **vary** the **Bandwidth** of a **Digital Low-Pass Filter**. Each **Bandwidth Setting** (Cutoff Frequency) of this Filter is to be verified with an Oscilloscope. Also, the student is to verify the **Sampling Frequency,** the **First** and **Second Null Frequencies**, and the **Response** at the **Output with** and **without** a **Reconstruction Filter**

It is suggested that the student **read** the **PICBasic-Pro**, particularly the **comments** written on each line, to learn how to write code for the PIC microcontrollers. It has been proven many times to be the most effective way to learn how generate effective code for the PIC microcontrollers. This code is shown at the end of this chapter.

Portions of the Experiment Board Schematics, which are relevant to this experiment, are shown below in **Figure 9.1**. Please refer to **Appendix-A** for the complete set of **Experiment Board Schematics** and please refer to **Appendix-B** for the complete **Lab Board Schematic**.

Hardware Setup

1.) Connect the Square end of the USB Type-A to Type-B Cable to connector JUSB1 of the Experiment Board. Then, connect the other end of this Cable to one of the USB Inputs of the Computer.
2.) Connect the DB9 end of the DB9 Serial-to-USB Cable to connector P1 of the Experiment Board. Then, connect the USB end of this Cable to one of the other USB Inputs of the Computer. The DB9 Serial-to-USB Cable is used to communicate between the Serial HyperTerminal of the Computer and the Experiment Board.
3.) The Jumpers for **JB12** are to be placed in **Positions 3-5 & 4-6.**
4.) Connect the Channel-A Input of an Oscilloscope to connector (J2) of the Experiment Board. Be sure that the Red Lead of the BNC-to-Clip-Lead Cable is connected to Pin-1 of connector (J2), and that the Black Lead is connected to Pin-2 of connector (J2), which is Ground.
5.) Connect the Channel-B Input of an Oscilloscope and the Output of a Signal Generator to connector (J9) of the Experiment Board. Be sure that the Red Lead of the BNC-to-Clip-Lead Cable is connected to Pin-1 of connector (J9), which is the +Vin-1 Pin, and that the Black Lead is connected to Pin-2 of connector (J9), which is the -Vin-1 Pin (Ground).
6. Set the output of a Signal Generator to a 2 Volts Peak-to-Peak Sine-Wave with Zero Voltage Offset and DC-Coupling. The Frequency is to be varied from 1Hz to 50KHz.
7.) Connect the Jumper Blocks as Follows:

   The Jumper of **JB3** is to be placed in **Position 1-2.**
   The Jumper of **JB4** is to be placed in **Position 1-2.**
   One of the Jumpers of **JB5** is to be placed in **Position 2, only**, and the other in **Position 4, only.**
   The Jumper of **JB6** is to be placed in **Position 3-4.**
   The Jumper of **JB8** is to be placed in **Position 1-2.**

Assignment #1

1.) Obtain Oscilloscope **Pictures** of the **Cutoff Frequencies,** for each of the **9 Moving-Average Digital Low-Pass-Filter Cutoff Frequency Settings**, which are shown below. Also, show a Picture of each **Cutoff Frequency**, **with** and **without** the **Re-Construction Filter**. The Re-Construction Filter is removed by removing the Jumper of **JB7**.

2.) **Calculate, Measure** and **Record** the **First Two Null Frequencies** of each of the **9 Moving-Average Digital Low-Pass Filter Cutoff Frequency Settings** shown below. Also, **Measure only**, each **Sampling Frequency** at Terminal **P1** of connector **(J4)**, of the experiment board.

Use the Following Table to Record the Results.

Cutoff Freq. Calculated Value	Cutoff Freq. Measured Value	1ST Zero Cr. Freq. Calculated Value	1st Zero Cr. Freq. Measured Value	2nd Zero Cr. Freq. Calculated Value	2nd Zero Cr. Freq. Measured Value	Sampling Freq. Measured Value
3475 Hz						
1630 Hz						
1200 Hz						
960 Hz						
800 Hz						
686 Hz						
428 Hz						
303 Hz						
237 Hz						

Assignment #2

Audio Test:

Connect the **Output Signal of an IPOD or a CD-Player** to connector **(J9)** of the Experiment Board. Be sure that the **Signal Lead** is **connected** to **Pin-1** of connector **(J9)**, which is the **+Vin-1 Pin,** and that the **Ground Lead** is connected to **Pin-2** of **connector (J9)**, which is the **-Vin-1 Pin** (Ground).

Connect a **Speaker or a set of Headphones** to **connector (J2)** of the Experiment Board. **Pin-1** of **connector (J2)** is the **Output Signal** and **Pin-2** is **Ground**.

Connect the Channel-B Input of an Oscilloscope to **Pin-1** of connector **(J9)**. Connect the Channel-A Input of this Oscilloscope to **Pin-1** of connector **(J2)**. Be sure to **ground** the **Oscilloscope Probe Ground Clips to Pin-2 of each connector (J2 & J9)**.

Turn on the **Audio Source** and **Listen** to the sound and **Observe** the **Sound Waveform** on the oscilloscope at the input and output of the filter, as the **Bandwidth** is **varied** throughout each of the **nine Moving-Average Digital Low-Pass-Filter Cutoff Frequencies.**

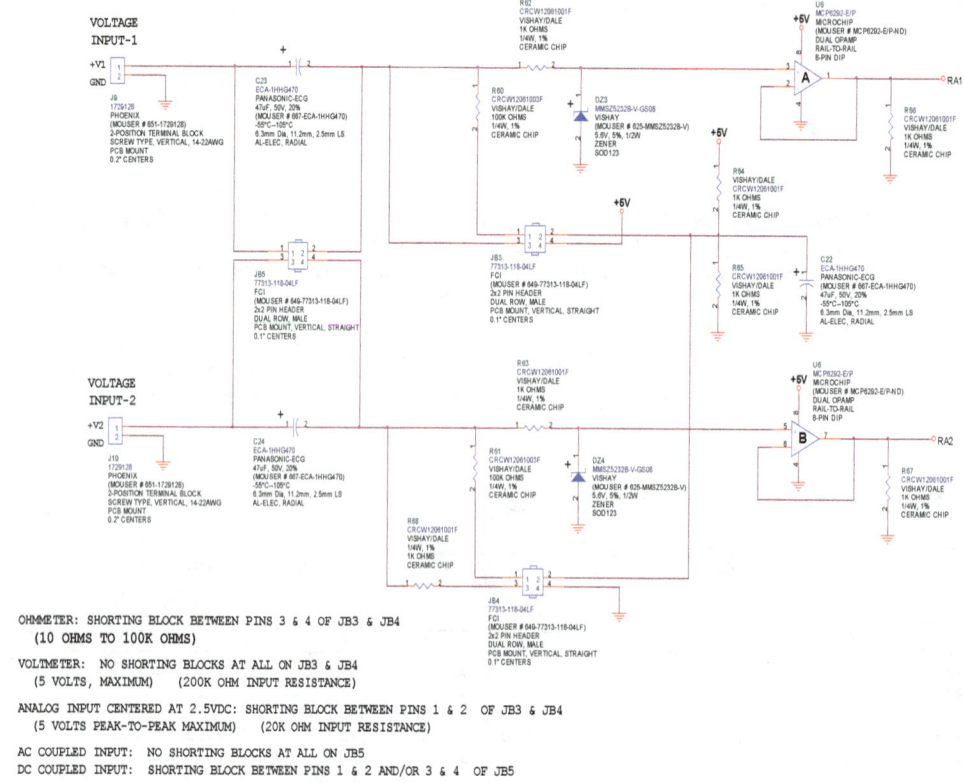

Figure 9.1: Analog-to-Digital (ADC) Converter Interface Circuit for the PIC microcontroller

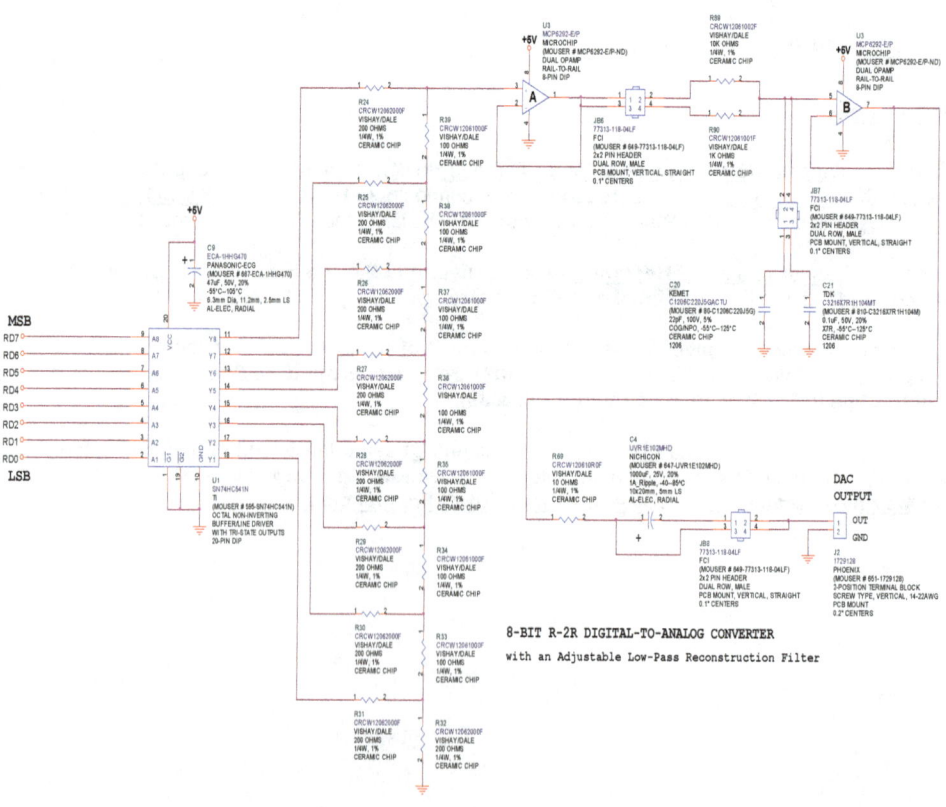

Figure 9.2: R-2R Ladder Digital-to-Analog Converter

Laboratory Experiment #7

'Lab-7_PICBasic-Pro_Variable_Bandwidth_Digital_Low_Pass_Filter

```
'M --> Averaging Terms, No Filter at M=1
'The Following is an Approximation of the Filter Response
'First Null Frequency = (Sampling Frequency)/(M)
'The Subsequent Null Frequencies are Integer Harmonics of the First Null Frequency
'Cutoff Frequency = (First Null Frequency)/2

        INCLUDE "modedefs.bas"      'Identifies the Command File

        DEFINE OSC 20               'Set PIC-Pro Oscillator Value in Mega Hertz
        DEFINE ADC_BITS 10          'Set for 10-Bit Internal ADC

        DEFINE ADC_CLOCK 5          'DEFINE ADC_CLOCK 1, for OSC=10, MAX ADC
                                    'Clock=10MHz/8) = 1.25MHz
                                    'DEFINE ADC_CLOCK 5, for OSC=16, MAX ADC
                                    'Clock=16MHz/16) = 1.00MHz
                                    'DEFINE ADC_CLOCK 5, for OSC=20, MAX ADC
                                    'Clock=20MHz/16) = 1.25MHz
                                    'DEFINE ADC_CLOCK 2, for OSC=24, MAX ADC
                                    'Clock=24MHz/32) = 1.25MHz
                                    'DEFINE ADC_CLOCK 2, for OSC=32, MAX ADC
                                    'Clock=32MHz/32) = 1.00MHz
                                    'DEFINE ADC_CLOCK 6, for OSC=48, MAX ADC
                                    'Clock=48MHz/64) = 0.75MHz

        DEFINE CCP1_BIT 2           'Setup Hardware PWM Timer
        DEFINE LCD_BITS 4           'Setup LCD for 4-Bit Mode
        DEFINE LCD_DREG PORTB       'Drive LCD from Port-B
        DEFINE LCD_DBIT 0           'Starting with Bit-0
        DEFINE LCD_RSREG PORTA      'Drive LCD Register-Select from
        DEFINE LCD_RSBIT 4          'Port-A, Bit-4
        DEFINE LCD_EREG PORTA       'Drive LCD Enable from
        DEFINE LCD_EBIT 5           'Port-A, Bit-5
        DEFINE LCD_LINES 2          'LCD can hold 2-Lines of Characters
        DEFINE LCD_COMMANDSUS 4000  'Delay between Commands
        DEFINE LCD_DATAUS 200       'Delay between Data Transfers

        ADCON0 = %00001111          'ADC Enabled, Analog Input 3 Enabled
        ADCON1 = %00001011          'ADC Voltage Range is 0 to 5 Volts &
                                    'Analog Inputs 0 Thru 3 are available
        ADCON2 = %10101101          'Right Justify to use Upper 8-Bits,
                                    'only, 12*Tad, & ADC Clock=20MHz/16=1.25MHz

        W0 VAR WORD                 'Set 16-Bit Word Variable,W0
        W1 VAR WORD                 '  "      "    "     "    ,W1
        W2 VAR WORD                 '  "      "    "     "    ,W2
        W3 VAR WORD                 '  "      "    "     "    ,W3

        B0 VAR BYTE                 'Set 8-Bit Byte Variable,B0
        B1 VAR BYTE                 '  "   "   "     "    ,B1
        DAC VAR BYTE                '  "   "   "     "    ,DAC
        S0 VAR BYTE                 '  "   "   "     "    ,S0
        S1 VAR BYTE                 '  "   "   "     "    ,S1
        S2 VAR BYTE                 '  "   "   "     "    ,S2
        S3 VAR BYTE                 '  "   "   "     "    ,S3
        S4 VAR BYTE                 '  "   "   "     "    ,S4
        S5 VAR BYTE                 '  "   "   "     "    ,S5
        S6 VAR BYTE                 '  "   "   "     "    ,S6
        S7 VAR BYTE                 '  "   "   "     "    ,S7
        S8 VAR BYTE                 '  "   "   "     "    ,S8
        S9 VAR BYTE                 '  "   "   "     "    ,S9
        S10 VAR BYTE                '  "   "   "     "    ,S10
        S11 VAR BYTE                '  "   "   "     "    ,S11
        S12 VAR BYTE                '  "   "   "     "    ,S12
        S13 VAR BYTE                '  "   "   "     "    ,S13
        S14 VAR BYTE                '  "   "   "     "    ,S14
        S15 VAR BYTE                '  "   "   "     "    ,S15
        S16 VAR BYTE                '  "   "   "     "    ,S16
        S17 VAR BYTE                '  "   "   "     "    ,S17
        S18 VAR BYTE                '  "   "   "     "    ,S18
        S19 VAR BYTE                '  "   "   "     "    ,S19

        TRISB=$FF                   'Make PortB an Input Port to Enable 'In-Circuit-Programming
        PAUSE 1000                  'Pause or Delay for 1 Seconds to Enable In-Circuit-Programming
```

```
LCDOUT $FE,1                        'Clear the LCD
LCDOUT $FE, $80                     'Set the Cursor to Line 1, Position 0
LCDOUT "Power Up"                   'Send, Power Up, to the LCD

SERout PORTC.6,T2400,["Power Up",13,10,10]
                                    'Send "Power Up" to the HyperTerminal
                                    'with a Carriage Return & two Line Feeds,
                                    'at 2400 Bits/Second, 8-Bits, 1-Stop Bit,
                                    'No Parity, Driven True (T2400).

PAUSE 2000                          'Pause or Delay for 2 Seconds
                                    'to Read "Power Up" in the LCD
LCDOUT $FE,1                        'Clear the LCD
TRISB=$00                           'Make PortB an Output Port to Allow for Normal Operation
TRISD=$F0                           'Make PortD, Bits-4,5,6,7 Inputs, & Bits-3

KSCAN:
LCDOUT $FE,1                        'Clear the LCD
LCDOUT $FE, $80                     'Set the Cursor to Line 1, Position 0
LCDOUT "Scanning"                   'Send, Scanning, to the LCD
SEROUT2 PORTC.6,396,["Scanning",13,10,10]
                                    'Send, Scanning, to the HyperTerminal
                                    'with a Carriage Return & one Line Feed

INPUT PORTD.4
INPUT PORTD.5
INPUT PORTD.6
INPUT PORTD.7

HIGH PORTD.0
PAUSE 500
IF (PORTD.4=1) THEN LPF1
PAUSE 100
IF (PORTD.5=1) THEN LPF2
PAUSE 100
IF (PORTD.6=1) THEN LPF3
PAUSE 100
IF (PORTD.7=1) THEN KSCAN
LOW PORTD.0

HIGH PORTD.1
PAUSE 500
IF (PORTD.4=1) THEN LPF4
PAUSE 100
IF (PORTD.5=1) THEN LPF5
PAUSE 100
IF (PORTD.6=1) THEN LPF6
PAUSE 100
IF (PORTD.7=1) THEN KSCAN
LOW PORTD.1

HIGH PORTD.2
PAUSE 500
IF (PORTD.4=1) THEN LPF7
PAUSE 100
IF (PORTD.5=1) THEN LPF8
PAUSE 100
IF (PORTD.6=1) THEN LPF9
PAUSE 100
IF (PORTD.7=1) THEN KSCAN
LOW PORTD.2

HIGH PORTD.3
PAUSE 500
IF (PORTD.4=1) THEN KSCAN
PAUSE 100
IF (PORTD.5=1) THEN LPF0
PAUSE 100
IF (PORTD.6=1) THEN KSCAN
PAUSE 100
IF (PORTD.7=1) THEN KSCAN
LOW PORTD.3

GOTO KSCAN
```

```
LPF0:
TRISD=0
OUTPUT PORTD

LCDOUT $FE, 1
LCDOUT $FE, $80
LCDOUT "M=1, No Filter"
SEROUT2 PORTC.6,396,["M=1, No Filter",13,10,10]
                        'Send, No Filter, to the HyperTerminal
                        'with a Carriage Return & one Line Feed
DAC=0
S0=0

LOOP0:
HIGH PORTC.2
ADCIN 1,S0
LOW PORTC.2

DAC=S0
PORTD=DAC

GOTO LOOP0


LPF1:
TRISD=0
OUTPUT PORTD

LCDOUT $FE, 1
LCDOUT $FE, $80
LCDOUT "M=2, SF=14,200Hz"
LCDOUT $FE, $C0
LCDOUT "Cutoff=3,550Hz"
SEROUT2 PORTC.6,396,["M=2 Term, SF=14,200Hz, Cutoff=3,550Hz",13,10,10]
                        'Send, M=2 Term, SF=13900Hz, Cutoff=3475Hz,
                        'to the HyperTerminal
                        'with a Carriage Return & one Line Feed

W0=0
DAC=0
S0=0
S1=0
S2=0

LOOP1:
HIGH PORTC.2
ADCIN 1,S0
LOW PORTC.2
S1=S0
W0=S1+S2
DAC=W0/2
PORTD=DAC

S2=S1


GOTO LOOP1

LPF2:
TRISD=0
OUTPUT PORTD

LCDOUT $FE, 1
LCDOUT $FE, $80
LCDOUT "M=3, SF=14,200Hz"
LCDOUT $FE, $C0
LCDOUT "Cutoff=2,367Hz"
SEROUT2 PORTC.6,396,["M=3 Term, SF=14,200Hz, Cutoff=2,367Hz",13,10,10]
                        'Send, 3-Term, SF=    , Cutoff=3333Hz,
                        'to the HyperTerminal
                        'with a Carriage Return & one Line Feed
```

```
W0=0
DAC=0
S0=0
S1=0
S2=0
S3=0

LOOP2:  HIGH PORTC.2
ADCIN 1,S0
LOW PORTC.2
S1=S0

W0=S1+S2+S3
DAC=W0/3
PORTD=DAC

S3=S2
S2=S1

GOTO LOOP2


LPF3:
TRISD=0
OUTPUT PORTD

LCDOUT $FE, 1
LCDOUT $FE, $80
LCDOUT "M=4, SF=14,200Hz"
LCDOUT $FE, $C0
LCDOUT "Cutoff = 1,775Hz"
SEROUT2 PORTC.6,396,["M=4 Term, SF=14,200Hz, Cutoff=1,775Hz",13,10,10]
                        'Send, 4-Term, SF=   , Cutoff=2500Hz,
                        'to the HyperTerminal
                        'with a Carriage Return & one Line Feed

W0=0
DAC=0
S0=0
S1=0
S2=0
S3=0
S4=0

LOOP3:
HIGH PORTC.2
ADCIN 1,S0
LOW PORTC.2
S1=S0

W0=S1+S2+S3+S4
DAC=W0/4
PORTD=DAC

S4=S3
S3=S2
S2=S1

GOTO LOOP3

LPF4:
TRISD=0
OUTPUT PORTD

LCDOUT $FE, 1
LCDOUT $FE, $80
LCDOUT "M=5, SF=14,200Hz"
LCDOUT $FE, $C0
LCDOUT "Cutoff=1,420Hz"


SEROUT2 PORTC.6,396,["M=5 Term, SF=14,200Hz, Cutoff=1,420Hz",13,10,10]
                        'Send, 5-Term, SF=   , Cutoff=2000Hz,
                        'to the HyperTerminal
                        'with a Carriage Return & one Line Feed
```

```
W0=0
DAC=0
S0=0
S1=0
S2=0
S3=0
S4=0
S5=0

LOOP4:
HIGH PORTC.2
ADCIN 1,S0
LOW PORTC.2
S1=S0

W0=S1+S2+S3+S4+S5
DAC=W0/5
PORTD=DAC

S5=S4
S4=S3
S3=S2
S2=S1
GOTO LOOP4


LPF5:
TRISD=0
OUTPUT PORTD

LCDOUT $FE, 1
LCDOUT $FE, $80
LCDOUT "M=6, SF=14,200Hz"
LCDOUT $FE, $C0
LCDOUT "Cutoff=1,184Hz"
SEROUT2 PORTC.6,396,["M=6 Term, SF=14,200Hz, Cutoff=1,184Hz",13,10,10]
                       'Send, 6-Term, SF=   , Cutoff=1667Hz,
                       ' to the HyperTerminal
                       'with a Carriage Return & one Line Feed

W0=0
DAC=0
S0=0
S1=0
S2=0
S3=0
S4=0
S5=0
S6=0

LOOP5:
HIGH PORTC.2
ADCIN 1,S0
LOW PORTC.2
S1=S0

W0=S1+S2+S3+S4+S5+S6
DAC=W0/6
PORTD=DAC

S6=S5
S5=S4
S4=S3
S3=S2
S2=S1
GOTO LOOP5
```

```
LPF6:
TRISD=0
OUTPUT PORTD

LCDOUT $FE, 1
LCDOUT $FE, $80
LCDOUT "M=7, SF=14,200Hz"
LCDOUT $FE, $C0
LCDOUT "Cutoff=1,014Hz"
SEROUT2 PORTC.6,396,["M=7 Term, SF=14,200Hz, Cutoff=1,014Hz",13,10,10]
                           'Send, 7-Term, SF=   , Cutoff=1428Hz,
                           'to the HyperTerminal
                           'with a Carriage Return & one Line Feed

W0=0
DAC=0
S0=0
S1=0
S2=0
S3=0
S4=0
S5=0
S6=0
S7=0

LOOP6:
HIGH PORTC.2
ADCIN 1,S0
LOW PORTC.2
S1=S0

W0=S1+S2+S3+S4+S5+S6+S7
DAC=W0/7
PORTD=DAC

S7=S6
S6=S5
S5=S4
S4=S3
S3=S2
S2=S1
GOTO LOOP6

LPF7:
TRISD=0
OUTPUT PORTD

LCDOUT $FE, 1
LCDOUT $FE, $80
LCDOUT "M=11, SF=14,200Hz"
LCDOUT $FE, $C0
LCDOUT "Cutoff=646Hz"
SEROUT2 PORTC.6,396,["M=11 Term, SF=14,200Hz, Cutoff=646Hz",13,10,10]
                           'Send, 11-Term, SF=   , Cutoff=909Hz,
                           'to the HyperTerminal
                           'with a Carriage Return & one Line Feed
W0=0
DAC=0
S0=0
S1=0
S2=0
S3=0
S4=0
S5=0
S6=0
S7=0
S8=0
S9=0
S10=0
S11=0
```

```
LOOP7:
HIGH PORTC.2
ADCIN 1,S0
LOW PORTC.2

S1=S0

W0=S1+S2+S3+S4+S5+S6+S7+S8+S9+S10+S11
DAC=W0/11
PORTD=DAC

S11=S10
S10=S9
S9=S8
S8=S7
S7=S6
S6=S5
S5=S4
S4=S3
S3=S2
S2=S1
GOTO LOOP7

LPF8:
TRISD=0
OUTPUT PORTD

LCDOUT $FE, 1
LCDOUT $FE, $80
LCDOUT "M=15, SF=14,200Hz"
LCDOUT $FE, $C0
LCDOUT "Cutoff=474Hz"
SEROUT2 PORTC.6,396,["M=15 Term, SF=14,200Hz, Cutoff=474Hz",13,10,10]
                        'Send, 15-Term, SF=    , Cutoff=667Hz, to the HyperTerminal
                        'with a Carriage Return & one Line Feed
W0=0
DAC=0
S0=0
S1=0
S2=0
S3=0
S4=0
S5=0
S6=0
S7=0
S8=0
S9=0
S10=0
S11=0
S12=0
S13=0
S14=0
S15=0

LOOP8:
HIGH PORTC.2
ADCIN 1,S0
LOW PORTC.2
S1=S0

W0=S1+S2+S3+S4+S5+S6+S7+S8+S9+S10+S11+S12+S13+S14+S15
DAC=W0/15
PORTD=DAC

S15=S14
S14=S13
S13=S12
S12=S11
S11=S10
S10=S9
S9=S8
S8=S7
S7=S6
S6=S5
S5=S4
S4=S3
S3=S2
S2=S1
 GOTO LOOP8
```

```
LPF9:
TRISD=0
OUTPUT PORTD

LCDOUT $FE, 1
LCDOUT $FE, $80
LCDOUT "M=19, SF=14,200Hz"
LCDOUT $FE, $C0
LCDOUT "Cutoff=374Hz"


SEROUT2 PORTC.6,396,["M=19 Term, SF=14,200Hz, Cutoff=374Hz",13,10,10]
                        'Send, 19-Term, SF=    , Cutoff=526Hz, to the HyperTerminal
                        'with a Carriage Return & one Line Feed
W0=0
DAC=0
S0=0

S1=0
S2=0
S3=0
S4=0
S5=0
S6=0
S7=0
S8=0
S9=0
S10=0
S11=0
S12=0
S13=0
S14=0
S15=0
S16=0
S17=0
S18=0
S19=0

LOOP9:
HIGH PORTC.2
ADCIN 1,S0
LOW PORTC.2
S1=S0

W0=S1+S2+S3+S4+S5+S6+S7+S8+S9+S10+S11+S12+S13+S14+S15+S16+S17+S18+S19
DAC=W0/19
PORTD=DAC

S19=S18
S18=S17
S17=S16
S16=S15
S15=S14
S14=S13
S13=S12
S12=S11
S11=S10
S10=S9
S9=S8
S8=S7
S7=S6
S6=S5
S5=S4
S4=S3
S3=S2
S2=S1
GOTO LOOP9

END
```

Engineering Practices for the PIC Microcontroller

By: Prof. Sal R. Riggio Jr., PhD, PE

Chapter - 10

Laboratory Experiment - 8

Frequency, Phase and Amplitude Modulation

Introduction

The purpose of this lab experiment is for the student to develop basic skills in making practical use of the Microchip PIC18F4685 Microcontroller and become familiar with the importance, operation and hardware implementation of the techniques and methods required to generate Frequency, Phase & Amplitude modulated waves.

The student is expected to analyze the **PICBasic-Pro Code** and the **CCS C-code**, below, which allows one to create a **20K Hertz, 50% Duty-Cycle Frequency** that can be both **Frequency Modulated (FM)** and **Phase Modulated (PM)** with a **Single Tone Message** of 100 Hertz. Additionally, the student is expected to write **PICBasic-Pro code** that will allow one to create an **Amplitude Modulated (AM) Wave** and a **Double-Sideband Suppressed Carrier (DSBSC) Wave**. The **Carrier Frequency** for each of these two waves is to be **900K Hertz** and the **Message** is to be a single tone of **600 Hertz**. The 900K Hertz carrier frequency is generated by the **Variable High Frequency Oscillator**, which is present on the experiment board, and is applied to **Port-C, Bit-3** of the PIC18F4685 microcontroller. The 600 Hertz message frequency is generated by the **Variable Low Frequency Oscillator**, which is also present on the experiment board, and is applied to **Port-C, Bit-4** of the PIC18F4685 microcontroller.

It is suggested that the student **read** the **PICBasic-Pro code** and the **CCS C-code**, particularly the **comments** written on each line, to learn how to write code for the PIC microcontrollers. It has been proven many times to be the most effective way to learn how generate effective code for the PIC microcontrollers. This code is shown at the end of this chapter

Portions of the Experiment Board Schematics, which are relevant to this experiment, are shown below in **Figures 10.1, 10.2, 10.3 & 10.4**. Please refer to **Appendix-A** for the complete set of **Experiment Board Schematics** and please refer to **Appendix-B** for the complete **Lab Board Schematic**.

Hardware Setup

1.) Connect the Square end of the USB Type-A to Type-B Cable to connector JUSB1 of the Experiment Board. Then, connect the other end of this Cable to one of the USB Inputs of the Computer.
2.) Connect the DB9 end of the DB9 Serial-to-USB Cable to connector P1 of the Experiment Board. Then, connect the USB end of this Cable to one of the other USB Inputs of the Computer. The DB9 Serial-to-USB Cable is used to communicate between the Serial HyperTerminal of the Computer and the Experiment Board.
3.) The Jumpers for **JB12** are to be placed in **Positions 3-5 & 4-6.**
4.) Connect Channel-A of an Oscilloscope, to connector (J5) of the Experiment Board. Be sure that the Probe Tip is connected to Pin-1 of connector (J5), which is the FM/PM/AM/DSBSC Modulated Wave Output Pin and that the Ground Clip of the Probe is connected to Pin-2 of connector (J5).
5.) Connect Channel-B of an Oscilloscope to the Junction of the 10K Ohm, ¼ watt Resistor and the 0.1 micro-Farad Capacitor. Be sure that the Probe Tip is connected to the R/C Junction, which is the De-Modulated FM/PM/AM/DSBSC Output Signal and that the Ground Clip of the Probe is connected to Pin-2 of connector (J5).
6.) Connect the Jumper Blocks as Follows:
   The Jumper of **JB3** is to be placed in **Position 1-2.**
   The Jumper of **JB4** is to be placed in **Position 1-2.**
   One of the Jumpers of **JB5** is to be placed in **Position 2, only**, and the other in **Position 4, only**.
   The Jumper of **JB6** is to be placed in **Position 1-2.**
   The Jumper of **JB7** is to be placed in **Position 2, only.**
   The Jumper of **JB8** is to be placed in **Position 1-2.**
7.) Setup a Signal Generator to produce a **100 Hertz** Sine-Wave of **No greater** than **4 Volts** Peak-to-Peak for **FM-Mode** Operation and **No greater** than **0.4 Volts** Peak-to-Peak for **PM-Mode (PWM)** Operation.
8.) Assemble the Circuit shown below.

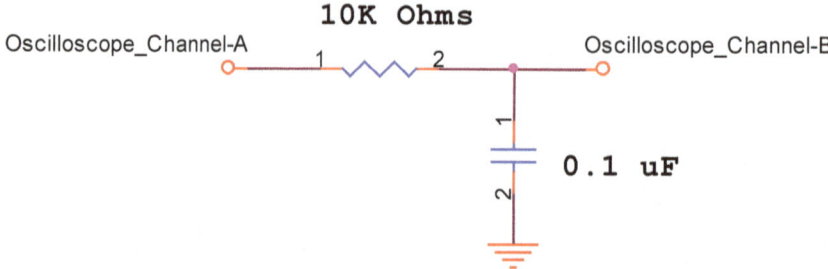

Figure 10.1:   **Low Pass Filter Circuit**

<u>Figure: 10.1</u> shows a Low-Pass-Filter circuit, which is used as a Signal Demodulator, to separate the message frequency from the carrier frequency for all four modes of modulation.

Pressing **Keypad Button #1** will produce <u>**Frequency Modulation (FM) Mode**</u>.
Pressing **Keypad Button #2** will produce <u>**Phase Modulation (PM) Mode**</u>.
Pressing **Keypad Button #3** will produce <u>**Amplitude Modulation (AM) Mode**</u>.
Pressing **Keypad Button #A** will produce <u>**Double Sideband Suppressed Carrier Modulation (DSBSC) Mode**</u>.

In <u>**FM-Mode**</u>, when the **Input Modulating Sine-Wave Amplitude** is set to <u>**Zero**</u>, the **Output Frequency** is equal to **20 KHz** and the **Duty-Cycle** is equal to **50%**. In **PM-Mode (PWM)**, when the **Input Modulating Sine-Wave Amplitude** is set to <u>**Zero**</u>, the **Output Frequency** is equal to **20 KHz** and the **Duty-Cycle** is equal to **25%**.

In order to view the **changing Frequency** & **changing Duty-Cycle**, set the **Frequency** of the **Input Modulating Sine-Wave** to **0.2 Hertz** and then vary it to **100 Hertz** while viewing the **Modulated Wave** at **Pin-1** of **J5** of the experiment board with the MyDaq device.

In order to view the **changing Frequency** & **changing Amplitude**, set the **Frequency** of the **Input Modulating Sine-Wave** to **0.2 Hertz** and then vary it to **100 Hertz** while viewing the <u>**De-Modulated Wave**</u> at the **Junction** of the **10K Ohm**, ¼ watt **Resistor** and the **0.1 micro-Farad** Capacitor with the <u>MyDaq device</u>. The Mydaq device can act as an oscilloscope or a spectrum analyzer.

1.) Using <u>**Frequency Modulation Mode**</u>, **Vary** the **Amplitude** of the **Input Sinewave** and **Measure** the <u>**Maximum Deviation Frequency**</u> from **20K Hertz** with **Beta-FM = 0.05**, at **Channel-1** of the Oscilloscope. **Capture a Picture** of this Signal and that of the **Demodulated Signal** at **Channel-2** of the O-scope, which is the output of the **Low-Pass-Filter**. A Phase-Lock-Loop or a Quadrature Detector is needed to demodulate a Frequency Modulated Wave. This circuit is much too complicated to build during one Lab Session.

2.) Therefore, a simple **Low-Pass-Filter** is used **De-modulate** the **Frequency Modulated Wave**. The Low-Pass-Filter does not perform as well as the Phase-Lock-Loop or a Quadrature Detector, but it does produce an acceptable result.

3.) Using **Phase Modulation Mode (Pulse-Width-Modulation)**, **Vary** the **Amplitude** of the **Input Sine-Wave** and **Measure** the **Maximum Deviation Duty-Cycle** from **50%** with **Beta-PM = 0.5**, at Channel-1 of the MyDaq device. **Capture** and **Store** a picture of this Signal and that of the **De-modulated Signal** at Channel-2 of the MyDaq device, which is the output of the Low-Pass-Filter.

4.) Using **Amplitude Modulation** and then **Double-Sideband Suppressed Carrier Modulation**, set the **Frequency** of the **Low Variable Frequency Oscillator** to **600 Hertz** by adjusting Trimpot (**RP3**), of the experiment board, and observe the **Waveform** at Channel-1 of the MyDaq with both the oscilloscope and spectrum analyzer features of the MyDaq device. **Capture** and **Store** a picture of this Waveform and that of the **De-modulated Signal** at Channel-2 of the MyDaq device, which is the output of the Low-Pass-Filter. The **900K Hertz Carrier Frequency** is adjusted with Trimpot (**RP2**) of the experiment board.

Assignment Background

Frequency Modulation

Beta-FM = The **Deviation Frequency**/The **Carrier Frequency**

The <u>**Deviation Frequency**</u> equals the **Carrier Frequency** minus the **Maximum Forced Frequency** due to the *<u>Positive Amplitude</u>* of the **Input Modulation Baseband Information Signal (Message)**.

In this Lab Experiment, the **Carrier Frequency** equals <u>**20K Hertz**</u>.

Phase Modulation

Beta-PM = The **Deviation Duty-Cycle**/The **Carrier Duty-Cycle**

In this Lab Experiment, the **Carrier Duty-Cycle** equals **50%**.

The <u>**Deviation Duty-Cycle**</u> equals the **Carrier Duty-Cycle** minus the **Maximum Forced Duty-Cycle** due to the *<u>Positive Amplitude</u>* of the **Input Modulation Baseband Information Signal (Message)**.

General

When using **Frequency Modulation**, the **Speed** at which the **Carrier Frequency Changes** represents the **Frequency** of the **Input Modulation Baseband Information Signal (Message)**.

When using **Phase Modulation**, the **Speed** at which the **Carrier Duty-Cycle Changes** represents the **Frequency** of the **Input Modulation Baseband Information Signal (Message)**.

The selection of <u>Beta-FM</u> and <u>Beta-PM</u> are chosen at the discretion of the system designer to prevent over-modulation.

Amplitude Modulation

An Amplitude Modulated Wave is generated when the carrier frequency is combined with the message frequency through a logical <u>AND gate</u>.

Double-Sideband Modulation Suppressed Carrier

A Double-Sideband Modulation Suppressed Carrier Modulated Wave is generated when the carrier frequency is combined with the message frequency through a logical <u>XOR gate</u>.

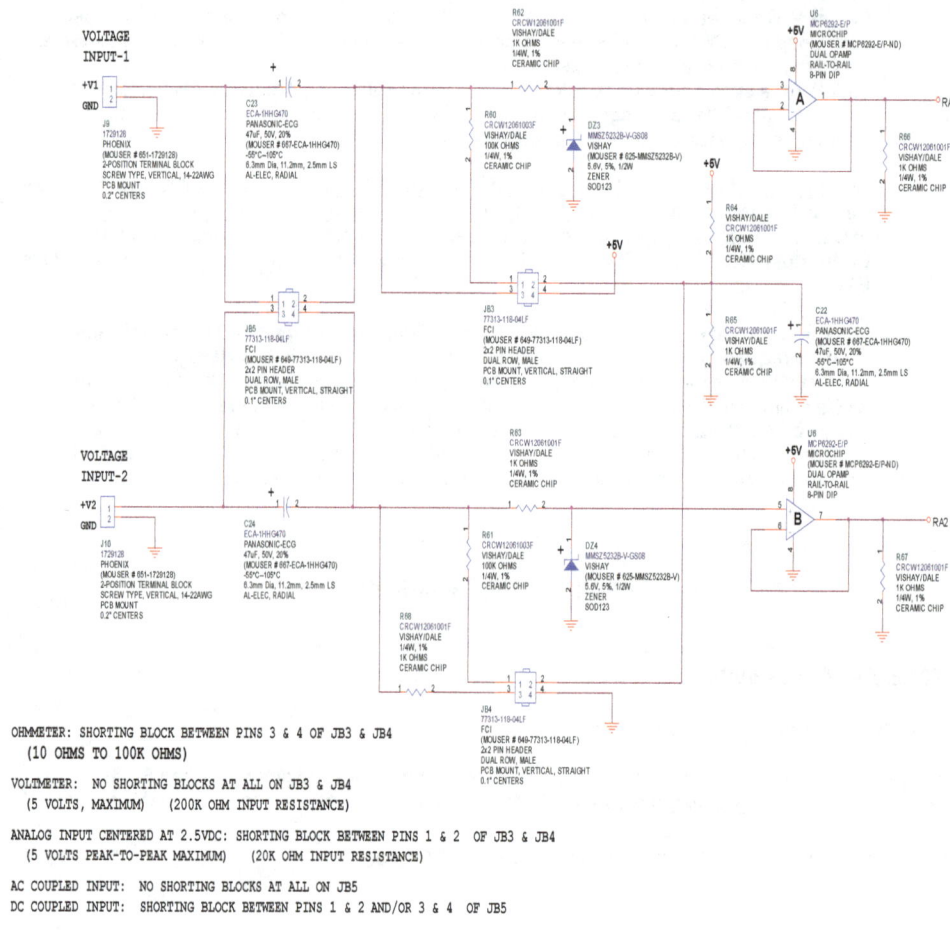

OHMMETER: SHORTING BLOCK BETWEEN PINS 3 & 4 OF JB3 & JB4
(10 OHMS TO 100K OHMS)

VOLTMETER: NO SHORTING BLOCKS AT ALL ON JB3 & JB4
(5 VOLTS, MAXIMUM)   (200K OHM INPUT RESISTANCE)

ANALOG INPUT CENTERED AT 2.5VDC: SHORTING BLOCK BETWEEN PINS 1 & 2 OF JB3 & JB4
(5 VOLTS PEAK-TO-PEAK MAXIMUM)   (20K OHM INPUT RESISTANCE)

AC COUPLED INPUT: NO SHORTING BLOCKS AT ALL ON JB5
DC COUPLED INPUT: SHORTING BLOCK BETWEEN PINS 1 & 2 AND/OR 3 & 4 OF JB5

Figure 10.2: Analog-to-Digital (ADC) Converter Interface Circuit for the PIC microcontroller

<u>Figure: 10.2</u> shows how one would pass a signal entering from an outside source to the PIC microcontrollers ADC. The in-coming signal is AC-coupled to the operational amplifier through capacitor (C23) to block any DC voltage from damaging the operational amplifier. Therefore, no shunt is placed between pins 1 & 2 of jumper block (JB5). However, the operational amplifier does need to have a shunt placed between pins 1 & 2 of jumper block (JB3) to provide a constant bias voltage of +2.5 volts to the input of the operational amplifier through resistor R60. This resistor also provides an approximate circuit input resistance of 100K ohms. Resistor (R62) and the Zener diode (DZ3) form an over-voltage protection circuit to protect the operational amplifier from being damaged due to a high transient input voltage. The 5.6 volt Zener diode conducts current only when the input voltage becomes more positive than +5.6 volts and more negative the -0.7 volts. Otherwise, the Zener diode does not have an appreciable effect on the performance of this circuit.

Figures: 10.3 & 10.4 shows one how to construct a bi-stable oscillator with a rail-to-rail operational amplifier. Rail-to-Rail means that the output voltage of the operational amplifier may me equal to the most positive supply voltage or the most negative supply voltage depending upon the value of the current output state (High or Low). The output signal from each of these oscillator circuits is a square-wave of approximately 50% duty-cycle. Oscillation occurs due to the generation of two alternating threshold voltages and a trimpot/resistor/capacitor time-delay connection at the negative input of the operational amplifier. The High Threshold Voltage is equal to 2/3 of the +5v power supply voltage and the Low Threshold Voltage is equal to 1/3 of the +5v power supply voltage. This is made possible by the three 2K ohm resistors connected at the positive input of the operational amplifier. The value of the frequency of oscillation is adjusted with the trimpot, along with the fixed value of the series resistor and the fixed value of the capacitor. The frequency of oscillation can be calculated from the following equation.

Fosc = 1/((ln2)(C(Rtrim+Rseries)))

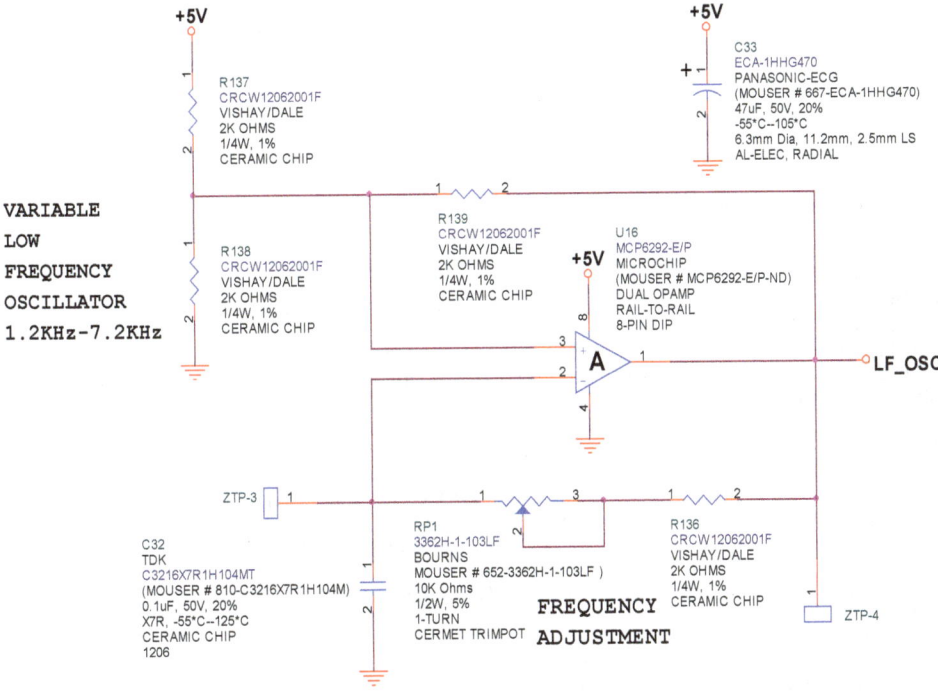

Figure 10.3:   Variable Low Frequency Oscillator Circuit

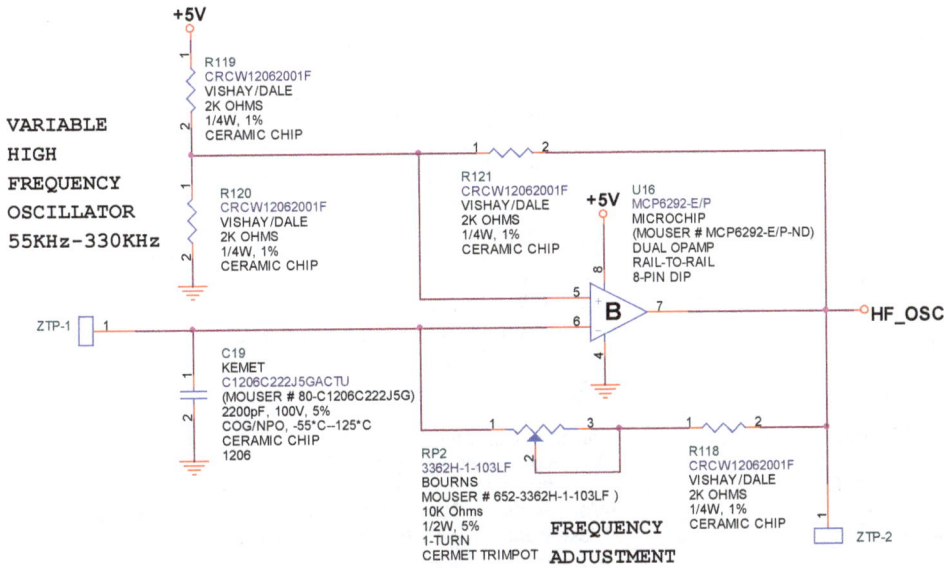

Figure 10.4:   Variable High Frequency Oscillator Circuit

Laboratory Experiment #8

```
'Lab-8_PICBasic_Pro_Frequency, Phase and Amplitude Modulation
INCLUDE "modedefs.bas"          'Identifies the Command File

DEFINE OSC 20                   'Set PIC-Pro Oscillator Value in Mega Hertz
DEFINE ADC_BITS 10              'Set for 10-Bit Internal ADC

DEFINE ADC_CLOCK 5              'DEFINE ADC_CLOCK 1, for OSC=10, MAX ADC
                                'Clock=10MHz/8) = 1.25MHz
                                'DEFINE ADC_CLOCK 5, for OSC=16, MAX ADC
                                'Clock=16MHz/16) =1.00MHz
                                'DEFINE ADC_CLOCK 5, for OSC=20, MAX ADC
                                'Clock=20MHz/16) =1.25MHz
                                'DEFINE ADC_CLOCK 2, for OSC=24, MAX ADC
                                'Clock=24MHz/32) =1.25MHz
                                'DEFINE ADC_CLOCK 2, for OSC=32, MAX ADC
                                'Clock=32MHz/32) =1.00MHz
                                'DEFINE ADC_CLOCK 6, for OSC=48, MAX ADC
                                'Clock=48MHz/64) = 0.75MHz

DEFINE CCP1_BIT 2               'Setup Hardware PWM Timer
DEFINE LCD_BITS 4               'Setup LCD for 4-Bit Mode
DEFINE LCD_DREG PORTB           'Drive LCD from Port-B
DEFINE LCD_DBIT 0               'Starting with Bit-0
DEFINE LCD_RSREG PORTA          'Drive LCD Register-Select from
DEFINE LCD_RSBIT 4              'Port-A, Bit-4
DEFINE LCD_EREG PORTA           'Drive LCD Enable from
DEFINE LCD_EBIT 5               'Port-A, Bit-5
DEFINE LCD_LINES 2              'LCD can hold 2-Lines of Characters
DEFINE LCD_COMMANDSUS 4000      'Delay between Commands
DEFINE LCD_DATAUS 200           'Delay between Data Transfers

ADCON0 = %00001111              'ADC Enabled, Analog Input 3 Enabled
ADCON1 = %00001011              'ADC Voltage Range is 0 to 5 Volts &
                                'Analog Inputs 0 Thru 3 are available
ADCON2 = %10101101              'Right Justify to use Upper 8-Bits, only,
                                '12*Tad, & ADC Clock=20MHz/16=1.25MHz

W0 VAR WORD                     'Set 16-Bit Word Variable,W0
W1 VAR WORD                     ' "    "    "    "     "   ,W1
W2 VAR WORD                     ' "    "    "    "     "   ,W2
W3 VAR WORD                     ' "    "    "    "     "   ,W3
CenFR VAR WORD                  ' "    "    "    "     "   ,CenFR
OutFR VAR WORD                  ' "    "    "    "     "   ,OutFR
SamFR VAR WORD                  ' "    "    "    "     "   ,SamFR
NoInS VAR WORD                  ' "    "    "    "     "   ,NoInS V
SamDC VAR WORD                  ' "    "    "    "     "   ,SamDC

B0 VAR BYTE                     'Set 8-Bit Byte Variable,B0
B1 VAR BYTE                     ' "    "    "    "    "  ,B1
CenDC VAR BYTE                  ' "    "    "    "    "  ,CenDC
OutDC VAR BYTE                  ' "    "    "    "    "  ,OutDC

TRISB=$FF                       'Make PortB an Input Port to Enable
                                'In-Circuit-Programming
PAUSE 1000                      'Pause or Delay for 1 Seconds to Enable
                                'In-Circuit-Programming
LCDOUT $FE,1                    'Clear the LCD
LCDOUT $FE, $80                 'Set the Cursor to Line 1, Position 0
LCDOUT "Power Up"               'Send, Power Up, to the LCD

SERout PORTC.6,T2400,["Power Up",13,10,10]
                                'Send "Power Up" to the HyperTerminal
                                'with a Carriage Return & two Line Feeds,
                                'at 2400 Bits/Second, 8-Bits, 1-Stop Bit,
                                'No Parity, Driven True (T2400).

PAUSE 2000                      'Pause or Delay for 2 Seconds
                                "to Read "Power Up" in the LCD
LCDOUT $FE,1                    'Clear the LCD
TRISB=$00                       'Make PortB an Output Port to Allow for Normal Operation
TRISD=$F0                       'Make PortD, Bits-4,5,6,7 Inputs, & Bits-3
```

```
KSCAN:
    LCDOUT $FE, 1              'Clear the LCD
    LCDOUT $FE, $80            'Set Cursor to Line 1, Position 0
    LCDOUT "The MOD. is Off"   'Send, The Reg. is Off, to the LCD
    LCDOUT $FE, $C0            'Set Cursor to Line 2, Position 0
    LCDOUT "Enter Mode"        'Send, Enter Voltage, to the LCD

    SEROUT2 PORTC.6,396,["The Modulator is Off",13,10]
                               'Send, The Regulator is Off, to the HyperTerminal
                               'with a Carriage Return & 2 Line Feeds

    SEROUT2 PORTC.6,396,["Enter Mode",13,10,10]
                               'Send, Enter Mode, to the HyperTerminal
                               'with a Carriage Return & 2 Line Feeds

    HIGH PORTD.0               'Make 0th Row High to Scan for the Following:
    PAUSE 500
    Input PORTD.4
    IF (PORTD.4=1) THEN FM     'Check for Keypad Key "1", Column-4
    PAUSE 100
    Input PORTD.5
    IF (PORTD.5=1) THEN PM     'Check for Keypad Key "2", Column-5
    PAUSE 100
    LOW PORTD.6
    IF (PORTD.4=1) THEN AM     'Check for Keypad Key "3", Column-4
    PAUSE 100
    Input PORTD.7
    IF (PORTD.5=1) THEN DS     'Check for Keypad Key "A", Column-5
    PAUSE 100
    LOW PORTD.0

    GOTO KSCAN


FM:   [Freq. Mod.]
    LCDOUT $FE, 1              'Clear the LCD
    LCDOUT $FE, $80            'Set Cursor to Line 1, Position 0
    LCDOUT "Freq. Modulation"  'Send, Freq. Modulation, to the LCD
    SERout2 PORTC.6,396,["Frequency Modulation",13,10,10]
                               'Send, Frequency Modulation,
                               'to the HyperTerminal with a Carriage Return
                               '& 2 Line Feeds

    FMLP:  ADCIN 1,SamFR       'Sample ADC Input Voltage-1, and place the
                               'Result in Variable SamFR
                               '(0 - 1023 Decimal in 4-Bit Steps)

    CenFR=20000                'Center Frequency = 20,000 Hertz  [Carrier]
    NoInS=512                  'No Input Signal
                               '(+2.5 Volt ADC Center Value = 512 Decimal Value)

    OutFR=CenFR+SamFR-NoInS    'Output Frequency
    HPWM 1,127,OutFR           'Hardware Output Frequency at a Fixed 50% Duty-Cycle

    GOTO FMLP


PM:   [Phase Mod]
    LCDOUT $FE, 1              'Clear the LCD
    LCDOUT $FE, $80            'Set Cursor to Line 1, Position 0
    LCDOUT "Phase Modulation"  'Send Freq. Modulation, to the LCD
    SERout2 PORTC.6,396,["Phase Modulation",13,10,10]
                               'Send Phase Modulation,
                               'to the HyperTerminal with a Carrage Return & 2 Line Feeds

    PMLP:
    ADCIN 1,SamDC  [duty cycle]    'Sample ADC Input Voltage-1, and place the
                                   'Result in Variable SamDC (0 - 1023 Decimal in 4-Bit Steps)

    CenDC=127                  'Center Duty-Cycle = 50%
    NoInS=512                  'No Input Signal
                               '(+2.5 Volt ADC Center Value = 512 Decimal Value)

    OutDC=CenDC+(SamDC/4)-(NoIns/4)   [div by 4 to convert to 8 bit]
                               'Output Duty-Cycle
    HPWM 1,OutDC,20000         'Hardware Output Duty-Cycle at a Fixed 20,000 Hertz

    GOTO PMLP
```

[amplitude doesn't matter]

```
AM:   Amplitude
    LCDOUT $FE, 1                       'Clear the LCD
    LCDOUT $FE, $80                     'Set Cursor to Line 1, Position 0
    LCDOUT "Amp. Modulation"            'Send, Amp. Modulation, to the LCD
    SERout2 PORTC.6,396,["Amplitude Modulation",13,10,10]
                                        'Send, Amplitude Modulation,
                                        'to the HyperTerminal
                                        'with a Carriage Return & 2 Line Feeds

AMLP:
    PORTC.2 = (PORTC.1 & PORTC.5)       'Produces an AM Wave with the Carrier Frequency
                                        'at Port-C, Bit-1 and the Message Frequency
                                        'at Port-C, Bit-5 (Logical AND)
    GOTO AMLP


DS:
    LCDOUT $FE, 1                       'Clear the LCD
    LCDOUT $FE, $80                     'Set Cursor to Line 1, Position 0
    LCDOUT "DSBSC Modulation"           'Send, DSBSC Modulation, to the LCD
    SERout2 PORTC.6,396,["DSBSC Modulation",13,10,10]
                                        'Send, DSBSC Modulation,
                                        'to the HyperTerminal with a Carriage Return
                                        '& 2 Line Feeds

DSLP:
    PORTC.2 = (PORTC.1 ^ PORTC.5)       'Produces a DSBSC Wave with the Carrier Frequency
                                        'at Port-C,'Bit-1 and the Message Frequency
                                        'at Port-C, Bit-5 (Logical XOR)
    GOTO DSLP

    END
```

Laboratory Experiment #8

'Lab-8_CCS_C-Code_Frequency, Phase and Amplitude Modulation

```c
#include <18F4685.h>                              //Identify Microcontroller
#device adc=10                                    //Identify ADC Bit Width
#fuses HS,NOWDT,NOPROTECT,NOLVP,MCLR              //Setup Programmer Oscillator Value
                                                  //and make Master Clear Pin an
                                                  //Enable Master Clear=MCLR
                                                  //Disable Master Clear=NOMCLR
#use delay(clock=20000000)                        //Setup C-Code Oscillator Value
#use fast_io(d)                                   //Leave the State of Port-D (all bits)
                                                  //the same until changed, again
#use rs232(baud=2400, xmit=PIN_C6, rcv=PIN_C7, stream=HT)
                                                  //Set RS232 HyperTerminal
                                                  //Communication Parameters
#use rs232(baud=2400, xmit=PIN_B4, rcv=PIN_B5, invert, stream=BB)
                                                  //Set RS232 Board-to-Board
                                                  //Communication Parameters
#use I2C(master, SCL=PIN_B6, SDA=PIN_B7)          //Set I2C Communication Parameters
#include <lcd_flex.c>                             //Identify LCD Driver File
#include <math.h>                                 //Include Math Functions

int8 VS,PS,DC;                                    //8-Bit Byte Variable
short D4,D5;                                      //1-Bit Variable

void main()
{
    set_tris_b(0xFF);                             //Make PortB an Input Port to
                                                  //Enable In-Circuit-Programming
    delay_ms(1000);                               //Pause or Delay for 1 second

    lcd_init();                                   //Initialize the LCD
    lcd_putc(0x0c);                               //Clear the LCD
    lcd_gotoxy(1,1);                              //Set the Cursor to Position-1 of Line-1
    printf(lcd_putc,"Power Up");                  //Send, Power Up, to the LCD

    fprintf(HT, "\n\rPower Up\n\r");              //Send, Power Up, to the HyperTerminal with
                                                  //a next Line & a Carriage Return
    delay_ms(1000);                               //Pause or Delay for 1 second

    setup_port_a(AN0_to_AN3);                     //Identify all of the Analog Inputs
    setup_adc(VSS_VDD);                           //Setup the ADC Voltage Range to be
                                                  //0 Volts to 5 Volts
                                                  //(5v-0v)/1024~=.005 Volts Per Bit

    setup_adc(ADC_CLOCK_DIV_32);                  //for delay(clock=10000000),
                                                  //setup_adc(ADC_CLOCK_DIV_8),
                                                  //10MHz/8=1.25MHz
                                                  //for delay(clock=16000000),
                                                  //setup_adc(ADC_CLOCK_DIV_16),
                                                  //16MHz/16=1.00MHz
                                                  //for delay(clock=20000000),
                                                  //setup_adc(ADC_CLOCK_DIV_16),
                                                  //20MHz/16=1.25MHz
                                                  //for delay(clock=24000000),
                                                  //setup_adc(ADC_CLOCK_DIV_32),
                                                  //24MHz/32=1.25MHz
                                                  //for delay(clock=32000000),
                                                  //setup_adc(ADC_CLOCK_DIV_32),
                                                  //32MHz/32=1.00MHz
                                                  //for delay(clock=48000000),
                                                  //setup_adc(ADC_CLOCK_DIV_64),
                                                  //48MHz/64=0.75MHz

    set_adc_channel(0);                           //Identify the Analog Input to be
                                                  //Sampled (Channel Input Zero)

    setup_ccp1(CCP_PWM);                          //Setup Capture/Compare/PWM Pin-1 to PWM Mode

    setup_ccp2(CCP_CAPTURE_RE);                   //Setup Capture/Compare/PWM Pin-2 to
                                                  //Capture Mode

    setup_timer_1(T1_INTERNAL);                   //Setup Timer-1, which controls the
                                                  //Capture/Compare portion of
                                                  //Capture/Compare/PWM Pin-2, only,
                                                  //to Internal Mode

    setup_timer_2(T2_DIV_BY_4, 19, 1);            //Setup Timer-2, which controls the PWM portion
                                                  //of Capture/Compare/PWM Pin-1, only.
                                                  //DIV_BY_Mode, Prescale (Period), Postscale
                                                  //The Following Limits are for the
                                                  //PIC18F4685 Microcontroller
                                                  //Mode may equal 1 or 4 or 16
                                                  //Prescale (Period) may equal any integer
                                                  //between 0 and 255
                                                  //Postscale may equal any integer between
                                                  //1 and 16 (Post = 1 to set PWM Frequency)
                                                  //T2Freq=OSC/[(4)*(Mode)*(Prescale+1)]=
                                                  //24MegaHz/[(4)*(4)*(PS+1)] = 75KHz
```

```c
    enable_interrupts(INT_CCP2);                //Enable the Internal Interrupt Input
                                                //of Capture/Compare/PWM Pin-2

    enable_interrupts(GLOBAL);                  //Enable Global Interrupts for the
                                                //PIC18F4685 Microcontroller

    set_tris_b(0x00);                           //Make PortB an Output Port to Allow for
                                                //Normal Operation
    tris_d(0xF0);                               //Make PortD, Bits-4,5,6,7 Inputs
                                                //and Bits-3,2,1,0 Outputs

kscan:
    lcd_putc(0x0c);                             //Clear the LCD
    lcd_gotoxy(1,1);                            //Set the Cursor to Position 1 of Line 1
    printf(lcd_putc,"The MOD. is Off");         //Send, The MOD. is Off, to the LCD
    lcd_gotoxy(1,2);                            //Set the Cursor to Position 1 of Line 2
    printf(lcd_putc,"Enter Mode");              //Send, Enter Mode, to the LCD
    output_high(PIN_D0);                        //Make 0th Row High to Scan for the Following:
    delay_ms(500);
    D4 = input(PIN_D4);                         //Check for Keypad Key "1" , Column-4
    if (D4) goto FM;
    delay_ms(100);
    output_low(PIN_D0);
    output_high(PIN_D0);                        //Make 0th Row High to Scan for the Following:
    delay_ms(500);
    D5 = input(PIN_D5);                         //Check for Keypad Key "2" , Column-5
    if (D5) goto PM;
    delay_ms(100);
    output_low(PIN_D0);
    goto KSCAN;


FM:
    lcd_putc(0x0c);                             //Clear the LCD
    lcd_gotoxy(1,1);                            //Set the Cursor to Position-1 of Line-1
    printf(lcd_putc,"Freq. Modulation");        //Send, Freq. Modulation, to the LCD
    printf("\n\rFrequency Modulation\n\r");     //Send, Frequency Modulation, to the HyperTerminal with
                                                //a next Line & a Carriage Return

fmlp:
    set_adc_channel(1);                         //Identify the Analog Input to be Sampled (Channel Input-1)
    VS = read_adc();                            //Sample ADC Input Voltage-1, and place this value into
                                                //the 8-Bit Variable VS

    set_pwm1_duty(50);                          //Set the PWM DutyCycle to 50% (Constant)
    PS=1+VS;
    setup_timer_2(T2_DIV_BY_4, PS, 1);          //Setup Timer-2, which controls the PWM portion of
                                                //Capture/Compare/PWM Pin-1, only.
                                                //DIV_BY_Mode, Prescale (Period), Postscale
                                                //The Following Limits are for the PIC18F4550 Microcontroller
                                                //Mode may equal 1 or 4 or 16
                                                //Prescale (Period) may equal any integer between 0 and 255
                                                //Postscale may equal any integer between 1 and 16
                                                //(Post = 1 to set PWM Frequency)
                                                //T2Freq=OSC/[(4)*(Mode)*
                                                //(Prescale+1)]= 24MegaHz/[(4)*(4)*(PS+1)] = 5.86KHz To 750KHz
                                                //(Frequency=11.7KHz at VS=127, which occurs at +2.5VDC at
                                                //the Input of AN1)

    goto fmlp;


PM:
    lcd_putc(0x0c);                             //Clear the LCD
    lcd_gotoxy(1,1);                            //Set the Cursor to Position-1 of Line-1
    printf(lcd_putc,"Phase Modulation");        //Send, Phase Modulation, to the LCD
    printf("\n\rPhase Modulation\n\r");         //Send, Phase Modulation, to the HyperTerminal with
                                                //a next Line & a Carriage Return

pmlp:
    set_adc_channel(1);                         //Identify the Analog Input to be Sampled (Channel Input-1)
    VS = read_adc();                            //Sample ADC Input Voltage-1, and place this value into
                                                //the 8-Bit Variable VS

    setup_timer_2(T2_DIV_BY_4, 127, 1);         //Setup Timer-2, which controls the PWM portion of
                                                //Capture/Compare/PWM Pin-1, only.
                                                //DIV_BY_Mode, Prescale (Period), Postscale
                                                //The Following Limits are for the PIC18F4685 Microcontroller
                                                //Mode may equal 1 or 4 or 16
                                                //Prescale (Period) may equal any integer between 0 and 255
                                                //Postscale may equal any integer between 1 and 16
                                                //(Post = 1 to set PWM Frequency)
                                                //T2Freq=OSC/[(4)*(Mode)*
                                                //Prescale+1)]= 24MegaHz/[(4)*(4)*(127+1)] = 11.7KHz (Constant)

    DC=10+((VS/255)*80);                        //10% To 90% DutyCycle, (DutyCycle=50% at VS=127,
                                                //which occurs at +2.5VDC at the Input of AN1)
    set_pwm1_duty(DC);                          //Set the PWM DutyCycle to DC Percent

    goto pmlp;

}
```

Engineering Practices for the PIC Microcontroller

By: Prof. Sal R. Riggio Jr., PhD, PE

Chapter - 11

Oscillators
And
Important Microcontroller Interface Circuits

A Microcontroller is furnished with two different types of Oscillators, along with a Phase-Lock-Loop, a number of Frequency Dividers and a number of Multiplexers. All of which are used to change frequency values and to route various signals of different frequencies as needed. The first type of oscillator, as well as the most accurate oscillator, is called the **Pierce/Colpitts Crystal Oscillator**. This is because it is derived from the combination of the Pierce series resonant Crystal oscillator and the Colpitts parallel resonant LC oscillator. The second type of oscillator is called the **RC-Oscillator**. This oscillator is a **1st Order** Single Time-Constant oscillator circuit. The Pierce/Colpitts oscillator circuit is a **2nd Order** oscillator circuit. It is also worth noting that the Colpitts parallel resonant LC Oscillator can be changed to a Crystal Oscillator by replacing the Inductor (L1) with a crystal. The following material is dedicated to these three circuits.

Additionally, this chapter presents material regarding some special purpose interface circuits that are used as signal conditioning circuits between the microcontroller and the outside world. These circuits are separated into two categories. The first category is called the **Analog-to-Bistable Comparator and Hysteresis Comparator Noise Rejecting Circuits.** The second category is called **Level-Shifting Circuits.** These circuits are used to convert CMOS Logic Voltage Levels (Logic "0" =0v & Logic "1" =5v) to **RS232** Logic Voltage Levels (Logic "0" = (-3v to -15v) and Logic "1" = (+3v to +15v)). Converting **Unipolar** Logic Signal Levels (CMOS) to **Bipolar** Logic Signal Levels (RS232), along with providing for **Isolated Circuit Paths**, yields a much greater Signal-to-Ratio within any electronic communication system. This is especially important for long distance wired electronic communication systems. Each of these circuits is presented in both **Inverting** and **Non-inverting** circuit form. The **Level-shifting Circuits** are also presented in both **Isolated** and **Non-Isolated** circuit form. An additional circuit, which is called the **Isolated Low Power Converter**, is presented because it is a necessary component of any Signal Isolation System. All of these circuits are presented in schematic form with the appropriate equations and transfer curves.

Let us begin with a discussion of Oscillators. In general, **all 2nd order oscillators** must meet the following Basic Criteria, except for point number 5, in order to produce and maintain a Single Output Frequency of Oscillation.

BASIC Criteria for Oscillation

1.) There must be a sufficient amount of BJT, MOSFET or CMOS Inverter Open-Feedback-Loop Voltage-Gain (AVOL), such that, the **"Separated" Open-Feedback-Loop Voltage-Gain (AVSOL)** is equal to **One**, after accounting for circuit losses. The **Closed-Feedback-Loop Voltage-Gain (AVCL)** is determined by the of BJT, MOSFET or CMOS Inverter active device.

2.) The **Phase-Shift** thru the **"Separated" Open-Feedback-Loop Voltage-Gain (AVSOL)** must be equal to **Zero Degrees**. This means that the Phase Difference between Vin of the Test Signal Generator and the Feedback Voltage (Vfb) must be **Zero**. This measurement can only be accomplished by separating the loop at the input of the CMOS Inverter, or the BJT Base, or the MOSFET Gate and inserting a signal generator (Vin), at this point, and comparing the phase of signal generator (Vin) with the phase of the signal across capacitor (C2), which is the feedback voltage (Vfb).

3.) A **Frequency Selective Circuit** must be present within the closed-feedback-loop circuit in order to force a specific **Value** for the **Frequency of Oscillation**.

4.) The frequency selective circuit must be at least **2nd Order** to produce a **Sinusoidal** Waveform at the output of the oscillator. This means that, at least **Two Energy Storing Components** must be present in this circuit. Either 1-Inductor and 1-Capacitor, or 2-Capacitors or 2-Inductors are required to set the specific **Value** of the frequency of oscillation.

5.) If the frequency selective circuit is **1st Order**, then a Piecewise Linear Waveform, such as, a **Pulse, Square, Triangle, Saw-tooth** or **Exponential** waveform will be produced at the output of the oscillator. Only **One** Energy Storing Component need be present in this circuit (1-Inductor or 1-Capacitor) to set the specific **Value** of the frequency of oscillation.

6.) The definition of an oscillator is that, it is a circuit, which **generates** an **output** signal **without** any **input** signal. At first, this sounds like the impossible perpetual motion machine. However, it is not, because of the presence of the input DC energy from the power supply connected to the oscillator circuit. In addition, how does any oscillator start to oscillate and produce an output signal waveform? All **2nd Order** oscillator circuits start oscillating from the **Noise Generated** when the power switch is switched to the power-on position. All **1st Order** circuits start oscillating from the **stored energy** or **lack of stored energy** within the single energy-storing component within the oscillator circuit.

Since the Pierce/Crystal Oscillator is a derivative of the Colpitts LC Sinusoidal Oscillator, we must first understand the operation of the **Colpitts LC Sinusoidal Oscillator** in order to properly analyze and understand the Pierce/Colpitts Crystal Oscillator. The following circuit models and equations explain the operation these oscillator circuits.

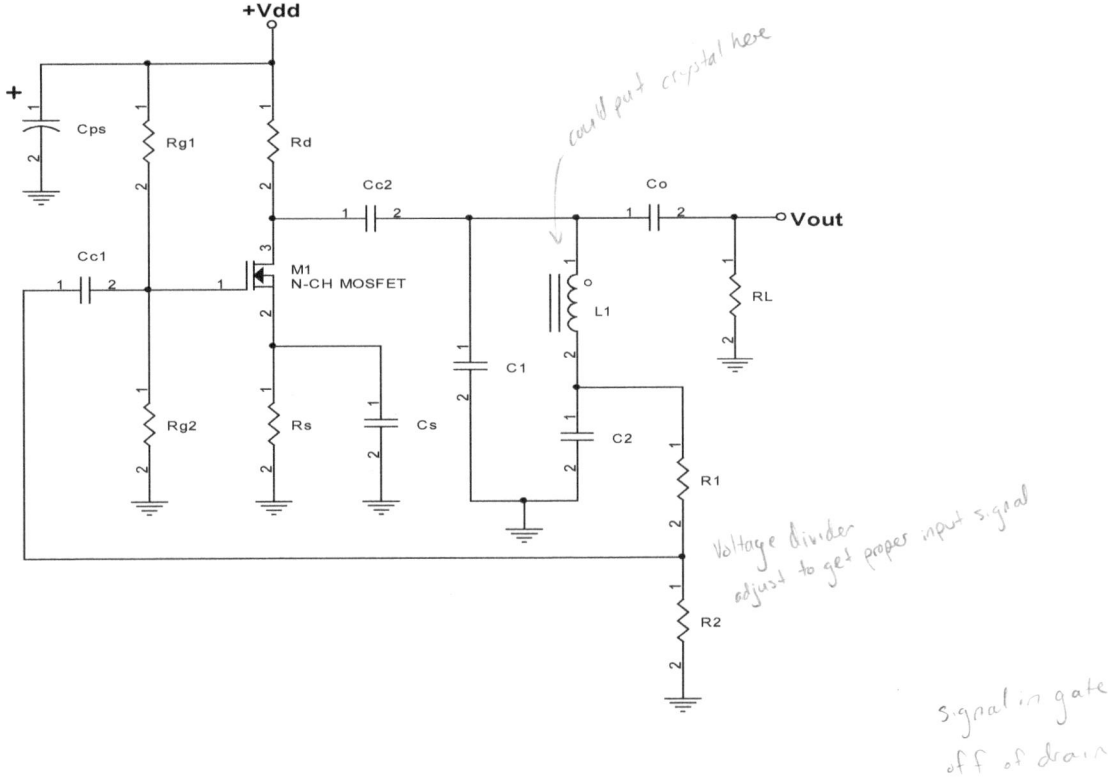

Figure 13.1: MOSFET Colpitts LC Sinusoidal Oscillator (Parallel Resonant)

In general, the approach to the design of this oscillator is as follows;

1.) Capacitors Cps, Cc1, Cc2, Cs and Co are all chosen to be much larger in value that C1 and C2, so that each of these five capacitors looks to be an electrical short circuit (zero ohms) to the frequency of the oscillating AC Signals.

2.) Capacitor C2 is chosen to be much larger than C1 to reduce the amount of impedance loading on the L1, C1 & C2 Resonant Circuit.

3.) The Values of Resistors R1 and R2 are chosen to set the Closed-Loop-Voltage Gain to One, which maintains oscillation and allows for the production of an Undistorted Oscillating Signal. Resistors R1 and R2 can be replaced with a potentiometer for a more precise setting of the Closed-Loop-Voltage Gain.

4.) This Colpitts Oscillator operates under the Condition of High Impedance Parallel Resonant. This is determined by inspection of this circuit. The Resonant circuit of L1, C1 & C2 must produce a high impedance at it's Resonant Frequency, in order provide sufficient voltage gain for the closed-feedback-loop of the oscillator. Series would produce a very low impedance thru L1 & C2, which would yield a near zero voltage gain.

5.) The Common-Source MOSFET amplifier produces 180 degrees of phase shift and capacitors C1 & C2, together produces 180 degrees. Therefore, the total phase shift around the closed-feedback-loop is equal to 360 or Zero degrees. This is shown by the following equations.

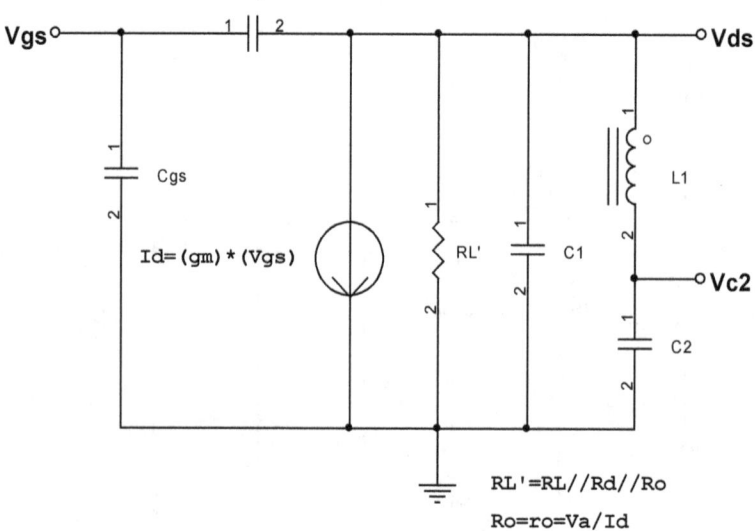

Figure 13.2: MOSFET Colpitts Sinusoidal Oscillator Equivalent Circuit without R1 and R2 for A(s)

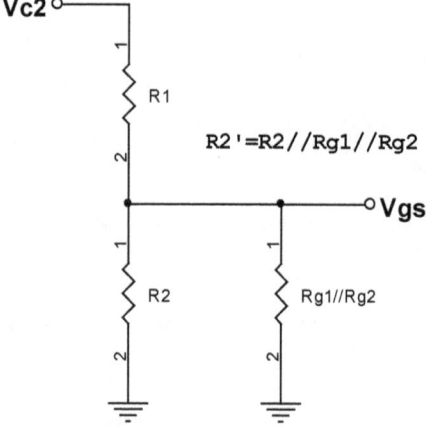

Figure 13.3: MOSFET Colpitts Sinusoidal Oscillator R1 and R2 Equivalent Circuit for B(s)

"DC-ANALYSIS"

$$V_{GG} = \frac{V_{DD} R_{g_2}}{R_{g_1} + R_{g_2}}$$

$$V_{GS} = V_{GG} - I_D R_s$$

$$I_D = K(V_{GS} - V_{Th})^2 \left[1 + \frac{V_{DS}}{V_A}\right]$$

$$gm = 2K(V_{GS} - V_{Th})\left[1 + \frac{V_{DS}}{V_A}\right] = 2K\sqrt{\frac{I_D}{K\left[1 + \frac{V_{DS}}{V_A}\right]}}\left[1 + \frac{V_{DS}}{V_A}\right]$$

$$V_{DS} = V_{DD} - I_D R_d - I_D R_s$$

$$R_0 = r_0 = \frac{V_A}{I_D}$$

"AC-ANALYSIS"

$$Vc_2(s) = \left[\frac{Vds(s)}{sL_1 + \frac{1}{sC_2}}\right] \cdot \left[\frac{1}{sC_2}\right]$$

$$Vds(s) = -gm\, Vgs(s)\left[\frac{1}{\frac{1}{R_L'} + sC_1 + \frac{1}{sL_1 + \frac{1}{sC_2}}}\right]$$

$$Vc_2(s) = -gm\, Vgs(s)\left[\frac{1}{sC_2}\right] \cdot \left[\frac{1}{\left(\frac{1}{R_L'} + sC_1\right)\left(sL_1 + \frac{1}{sC_2}\right) + 1}\right]$$

$$A(s) = \frac{Vc_2(s)}{Vgs(s)} = \frac{-gm/sC_2}{\frac{sL_1}{R_L'} + \frac{1}{sC_2 R_L'} + s^2 L_1 C_1 + \frac{C_1}{C_2} + 1}$$

$$A(s) = \frac{Vc_2(s)}{Vgs(s)} = \frac{-gm\, R_L'}{s^2 L_1 C_2 + 1 + s^3 L_1 C_1 C_2 R_L' + sC_1 R_L' + sC_2 R_L'}$$

LOOP PHASE SHIFT:

Rule #2

@ LOOP PHASE SHIFT $= 0° \rightarrow (C_1+C_2)R_L' = \omega^2 L_1 C_1 C_2 R_L'$

$\therefore \omega = \omega_0 = \dfrac{1}{\sqrt{L_1 \left[\dfrac{C_1 C_2}{C_1+C_2}\right]}}$ ← freq

$L(\omega_0) = A(\omega_0)\beta(\omega_0) = \dfrac{-gm R_L' \left[\dfrac{R_2'}{R_1+R_2'}\right]}{1-\omega^2 L_1 C_2} = 1$

$L(\omega_0) = \dfrac{-gm R_L' \left[\dfrac{R_2'}{R_1+R_2'}\right]}{1 - \left[\dfrac{L_1 C_2}{L_1\left[\dfrac{C_1 C_2}{C_1+C_2}\right]}\right]} = \dfrac{-gm R_L' \left[\dfrac{R_2'}{R_1+R_2'}\right]}{1 - \left[\dfrac{C_1+C_2}{C_1}\right]} = \dfrac{-gm R_L' \left[\dfrac{R_2'}{R_1+R_2'}\right]}{(1-1) - \dfrac{C_2}{C_1}} = 1$

$L(\omega_0) = A(\omega_0)\beta(\omega_0) = gm R_L' \left[\dfrac{R_2'}{R_1+R_2'}\right]\left[\dfrac{C_1}{C_2}\right] = 1$

$A(\omega_0) = gm R_L' \left[\dfrac{C_1}{C_2}\right]$

feedback

$A_f(\omega_0) = \dfrac{A(\omega_0)}{1-A(\omega_0)\beta(\omega_0)} = \dfrac{gm R_L' \left[\dfrac{C_1}{C_2}\right]}{1-gm R_L'\left[\dfrac{C_1}{C_2}\right]\left[\dfrac{R_2'}{R_1+R_2'}\right]}$

$A_f(\omega_0) = \dfrac{gm R_L'\left[\dfrac{C_1}{C_2}\right]}{0} = \infty$ @ $\rightarrow gm R_L'\left[\dfrac{C_1}{C_2}\right]\left[\dfrac{R_2'}{R_1+R_2'}\right] = 1 \rightarrow$ LOOP GAIN

Rule #1

Adjust R_2 until equal to 1

$s = j\omega$ for a sine wave

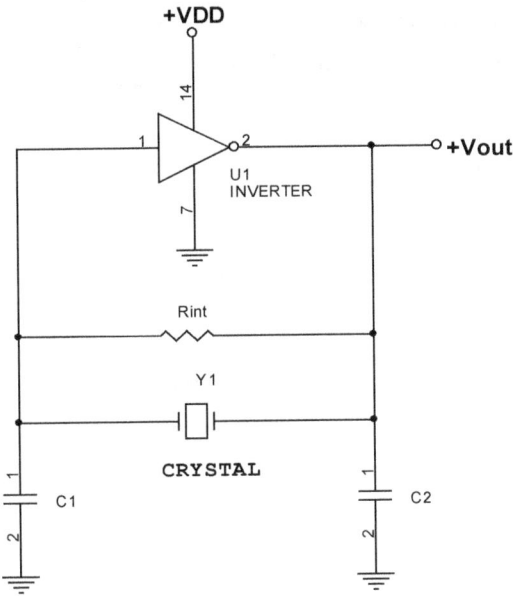

Figure 13.4: Pierce/Colpitts Crystal CMOS Inverter Squarewave Oscillator (Parallel Resonant)

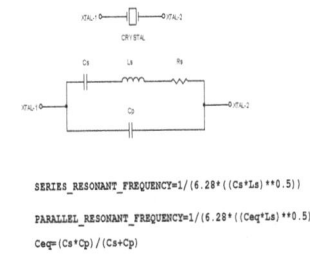

```
SERIES_RESONANT_FREQUENCY=1/(6.28*((Cs*Ls)**0.5))
PARALLEL_RESONANT_FREQUENCY=1/(6.28*((Ceq*Ls)**0.5))
Ceq=(Cs*Cp)/(Cs+Cp)
```

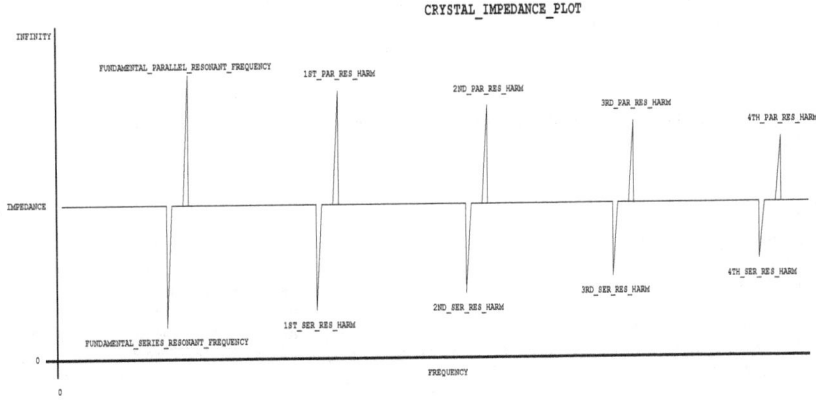

Figure 13.5: Crystal Impedance (Ohms) versus Frequency (Hertz)

Upon inspection of the Crystal Impedance versus Frequency Plot, one can determine that any crystal oscillator circuit can be designed to operate at either it is Fundamental Series (Low Impedance) or Fundamental Parallel (High Impedance) Resonance Frequency. In addition, with the use of Tuned Amplifier Circuits, one can cause a crystal oscillator circuit to operate at any of its Harmonic Series or Parallel Resonance Frequencies. Tuned Amplifier Harmonic Oscillators are practically limited to approximately the crystals 9th Harmonic. Therefore, the circuit configuration of the Pierce Crystal

Oscillator determines whether the crystal operates in Series or Parallel Resonance. In addition, by inspection of the crystal model equation, one can see that the value of the Series Resonant Frequency is always lower than the value of the Parallel Resonant Frequency, because the Series Resonant circuit contains an equivalent capacitance value that is less than that of the Parallel Resonant circuit. The Pierce/Colpitts Crystal Squarewave Sinusoidal Oscillator operates at the Parallel Resonant Frequency of the crystal.

Upon inspection of the Crystal Impedance versus Frequency Plot, one can determine that any crystal oscillator circuit can be designed to operate at either its Fundamental Series (Low Impedance) or Fundamental Parallel (High Impedance) Resonance Frequency. In addition, with the use of Tuned Amplifier Circuits, we can cause a crystal oscillator circuit to operate at any of its Harmonic Series or Parallel Resonance Frequencies. Tuned Amplifier Harmonic Oscillators are practically limited to approximately the crystal's 9th Harmonic. Thus, the circuit configuration of the Pierce Crystal Oscillator determines whether the crystal operates in Series or Parallel Resonance. In addition, by inspection of the crystal model equation, we can see that the value of the Series Resonant Frequency is always lower than the value of the Parallel Resonant Frequency, because the Series Resonant circuit contains an equivalent capacitance value, which is less than that of the Parallel Resonant Circuit. The Pierce/Colpitts Crystal Squarewave Output Oscillator operates at the Parallel Resonant Frequency of the crystal. The signal at the input of the CMOS Inverter is actually a sinewave, due the second order RLC behavior of the crystal. As many additional CMOS Inverters that are necessary to make a square wave with sharper edges can follow the output of the CMOS Inverter. This usually requires the addition of one or two CMOS Inverters.

Applying basic criteria to specifics of the circuit

Using the **Basic Criteria for Oscillation**, let us discuss the **Pierce/Colpitts Crystal CMOS Inverter Squarewave Oscillator** (Parallel Resonant) of **Figure 13.4**.

Pierce/Colpitts LC or **Crystal - CMOS Inverter, BJT** and **MOSFET Sinusoidal Oscillator** analysis is as follows;

Point #1:

In a CMOS Inverter, the Input Voltage and Output Voltage Levels are essentially the same. Therefore, the **"Separated" Open-Feedback-Loop Voltage-Gain (AVSOL) is equal to one**. The BJT and MOSFET Sinusoidal Oscillators must use a potentiometer to adjust the **"Separated" Open-Feedback-Loop Voltage-Gain (AVSOL) to be made** equal to **one**.

Point #2:

From the analysis of the Pierce/Colpitts LC or Crystal BJT or MOSFET or CMOS Inverter Sinusoidal Oscillators, it can be determined that Capacitors (C1) and (C2), together, create a phase shift of 180 degrees. Since, by definition, the CMOS Inverter, BJT Common-Emitter and MOSFET Common-Source circuits also generate 180 degrees of phase shift, the total "Separated" Open-Feedback-Loop Phase Shift is equal to 360 or Zero degrees. In a circuit containing capacitors (C1) & (C2), in the Colpitts configuration, the Crystal can only be operating at its Fundamental Parallel Resonant Frequency. Because, the crystal produces a very High Impedance between its leads, which allows capacitors (C1) & (C2) to produce the necessary 180 degrees of phase difference between Vout & Vfb. The Fundamental Series Resonant Frequency, which produces a very Low Impedance between the leads of the crystal, would prevent capacitors (C1) & (C2) from producing the necessary 180 degrees of phase shift.

Point #3:

A Parallel or Series Resonant Crystal or Ceramic Resonator or a Tuned LC Circuit will create Frequency Selection.

Point #4:

The crystal model shows that a crystal is a **2nd Order device**. Therefore, any circuit using a crystal as a frequency selective device will generate a Sinusoidal waveform. The Pierce-Colpitts Crystal or LC Sinusoidal Oscillator can use a Bipolar-Junction Transistor (BJT), in the Common-Emitter Amplifier configuration, or a MOSFET in the Common-Source Amplifier configuration, as the active device, as opposed to a CMOS inverter. Because the transconductance of the MOSFET or of the Bipolar-Junction Transistor (BJT) changes with current and voltage changes, the gain of the MOSFET or BJT varies. This means that as the output signal increases, the open loop voltage gain of BJT Common-Emitter Amplifier or the MOSFET Common-Source Amplifier decreases and vice versa. This action, in conjunction with the presence a Crystal or an LC Resonant Circuit, produces a sinusoidal output waveform.

Point #5:

This point does not apply because the crystal is a 2nd Order device.

Point #6:

This oscillator starts oscillating with noise that occurs with the application of DC power to the circuit.

HYSTERESIS COMPARATOR VARIABLE DUTYCYCLE CONSTANT FREQUENCY RC OSCILLATOR

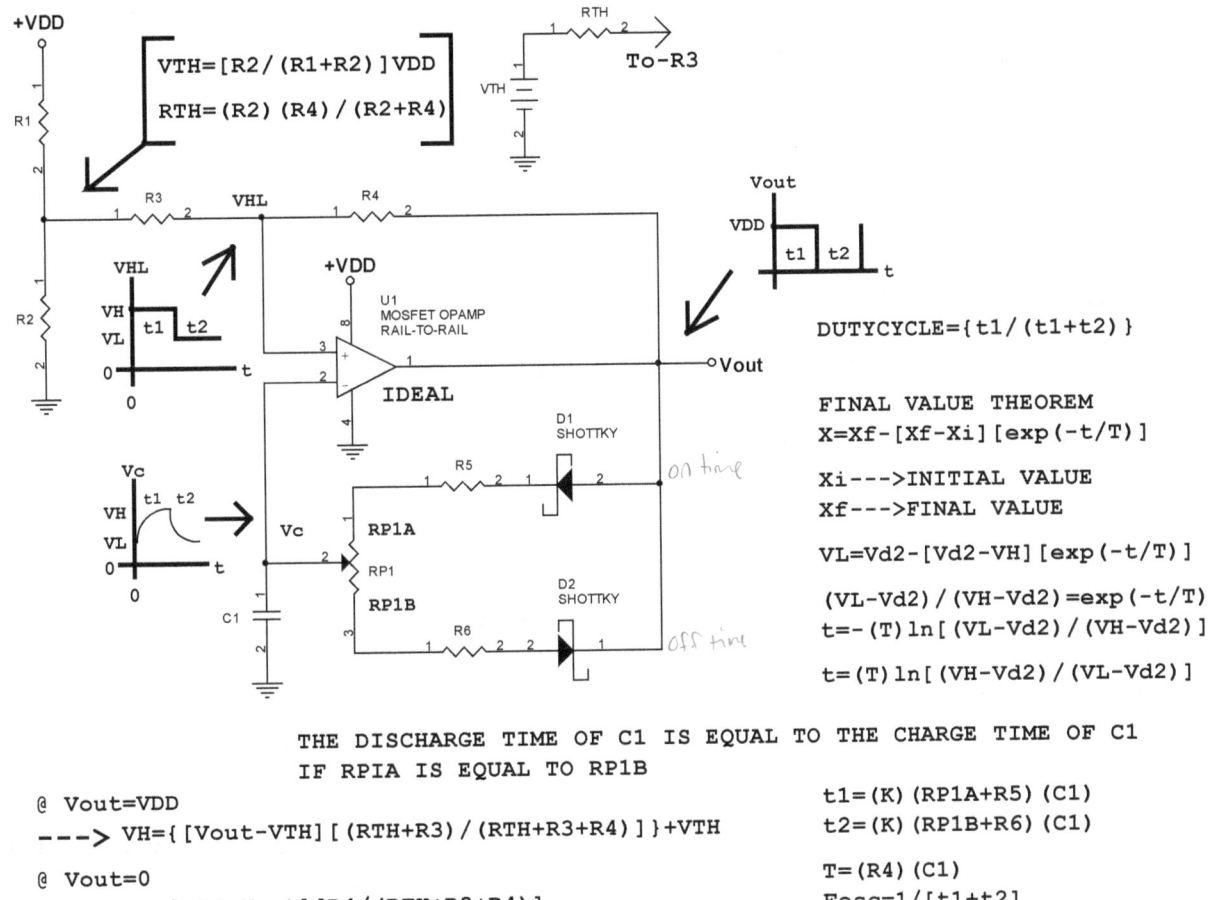

$$VTH = [R2/(R1+R2)]VDD$$
$$RTH = (R2)(R4)/(R2+R4)$$

DUTYCYCLE = {t1/(t1+t2)}

FINAL VALUE THEOREM
$$X = Xf - [Xf - Xi][\exp(-t/T)]$$

Xi ---> INITIAL VALUE
Xf ---> FINAL VALUE

$$VL = Vd2 - [Vd2 - VH][\exp(-t/T)]$$

$$(VL - Vd2)/(VH - Vd2) = \exp(-t/T)$$

$$t = -(T)\ln[(VL - Vd2)/(VH - Vd2)]$$

$$t = (T)\ln[(VH - Vd2)/(VL - Vd2)]$$

THE DISCHARGE TIME OF C1 IS EQUAL TO THE CHARGE TIME OF C1 IF RP1A IS EQUAL TO RP1B

@ Vout=VDD
---> VH = {[Vout-VTH][(RTH+R3)/(RTH+R3+R4)]} + VTH

@ Vout=0
---> VL = [VTH-Vout][R4/(RTH+R3+R4)]

t1 = (K)(RP1A+R5)(C1)
t2 = (K)(RP1B+R6)(C1)

T = (R4)(C1)
Fosc = 1/[t1+t2]

Figure 13.6: The Hysteresis Comparator Variable DutyCycle Constant Frequency RC-Oscillator

The **RC-Oscillator** of **Figure 13.6** is a **1st Order, Single Time Constant Oscillator,** which generates a **Squarewave** Output Waveform from the exponential waveform and Dual Internal Hysteresis Threshold Voltage Levels. The DutyCycle of the oscillator is changed by simultaneous increasing and decreasing the RC Time Constants in the paths of Shottky Diodes D1 and D2 with potentiometer (RP1).

THE IDEAL COMPARATOR

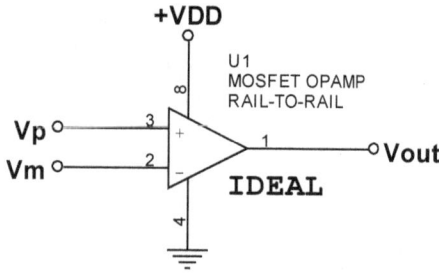

Vout=Av[(Vp)-(Vm)]

Av=Infinity ---> Open-Loop Voltage-Gain

Voutmax=VDD   Rin=Infinity
Voutmin=0   Iin=0

                     Rout=0
                     Iout_source=Infinity
                     Iout_sink=Infinity

Figure 13.7: The Ideal Comparator

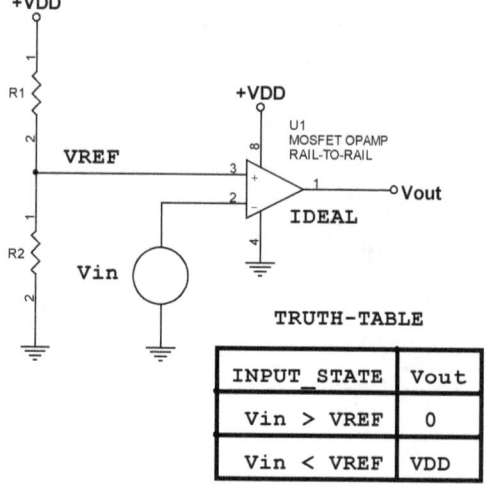

VTH=[R2/(R1+R2)][VDD]   VTH=[R2/(R1+R2)][VDD]

Figure 13.8: The Inverting and Non-Inverting Threshold Comparator Circuits

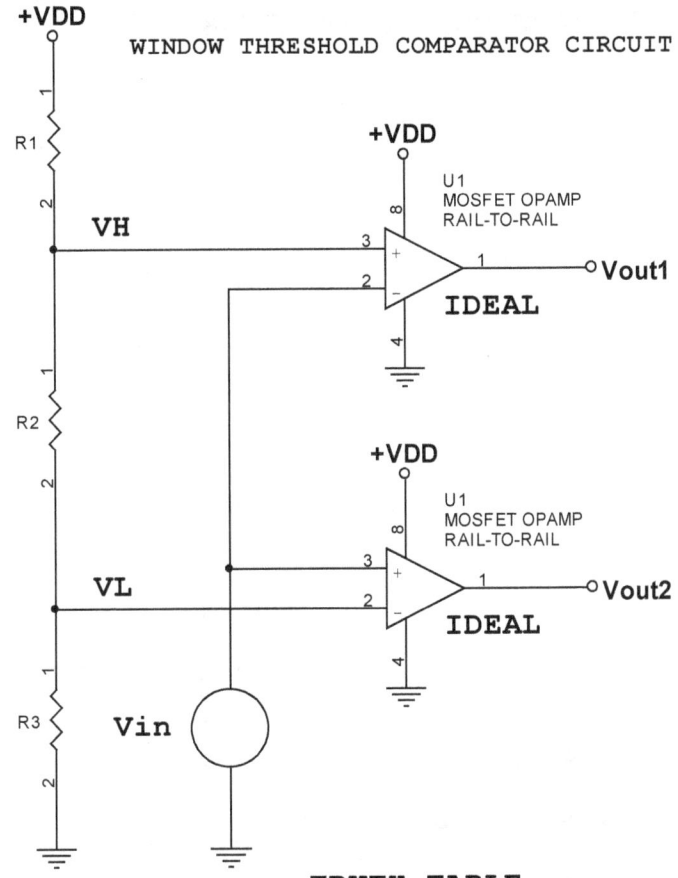

WINDOW THRESHOLD COMPARATOR CIRCUIT

TRUTH-TABLE

INPUT_STATE	Vout1	Vout2
Vin > VH	0	VDD
VL < Vin < VH	VDD	VDD
Vin < VL	VDD	0

$$VH = [(R2+R3)/(R1+R2+R3)][VDD]$$

$$VL = [R3/(R1+R2+R3)][VDD]$$

Figure 13.9: The Window Threshold Comparator Circuit

NON-INVERTING HYSTERESIS COMPARATOR

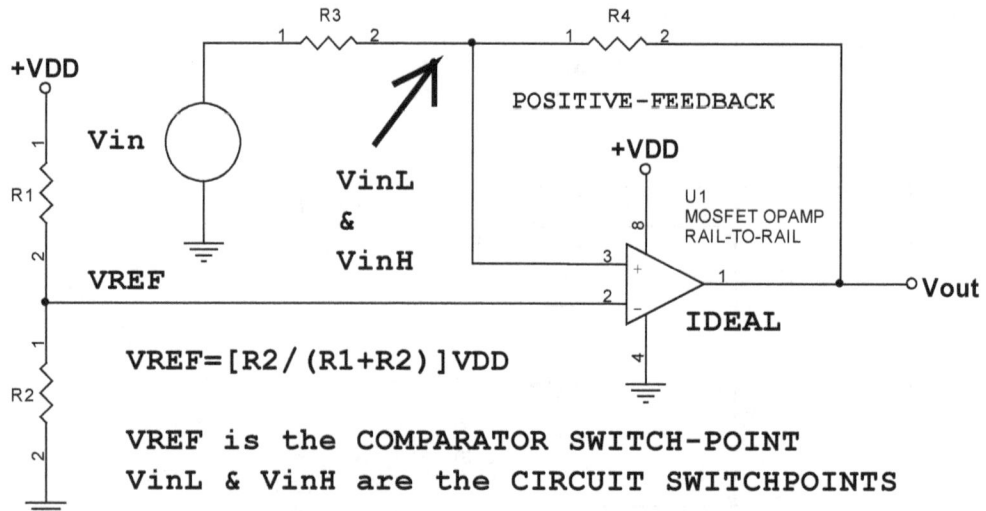

VREF = [R2/(R1+R2)]VDD

VREF is the COMPARATOR SWITCH-POINT
VinL & VinH are the CIRCUIT SWITCHPOINTS

TRUTH-TABLE

INPUT_STATE	Vout
Vin > VinH	VDD
VinL < Vin < VinH	NO-CHANGE
Vin < VinL	0

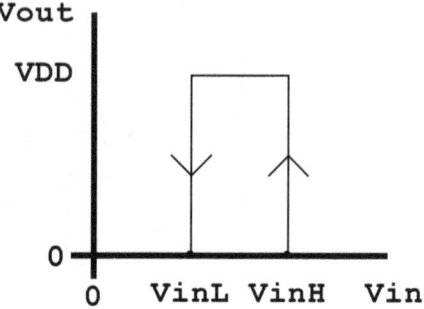

@ Vout=VDD

VREF={[R3/(R3+R4)][Vout-VinL]}+VinL

VREF={[R3/(R3+R4)]Vout}+{[1-[R3/(R3+R4)]]VinL}

VinL={VREF-[((R3/(R3+R4))Vout]}/{1-[R3/(R3+R4)]}

---> VinL={VREF[(R3+R4)/R4]}-{Vout[R3/R4]}

@ Vout=0

VREF=[R4/(R3+R4)][VinH-Vout]

---> VinH=[(R3+R4)/R4][VREF-Vout]

Figure 13.10: The Non-Inverting Noise Rejecting Hysteresis Comparator Circuit

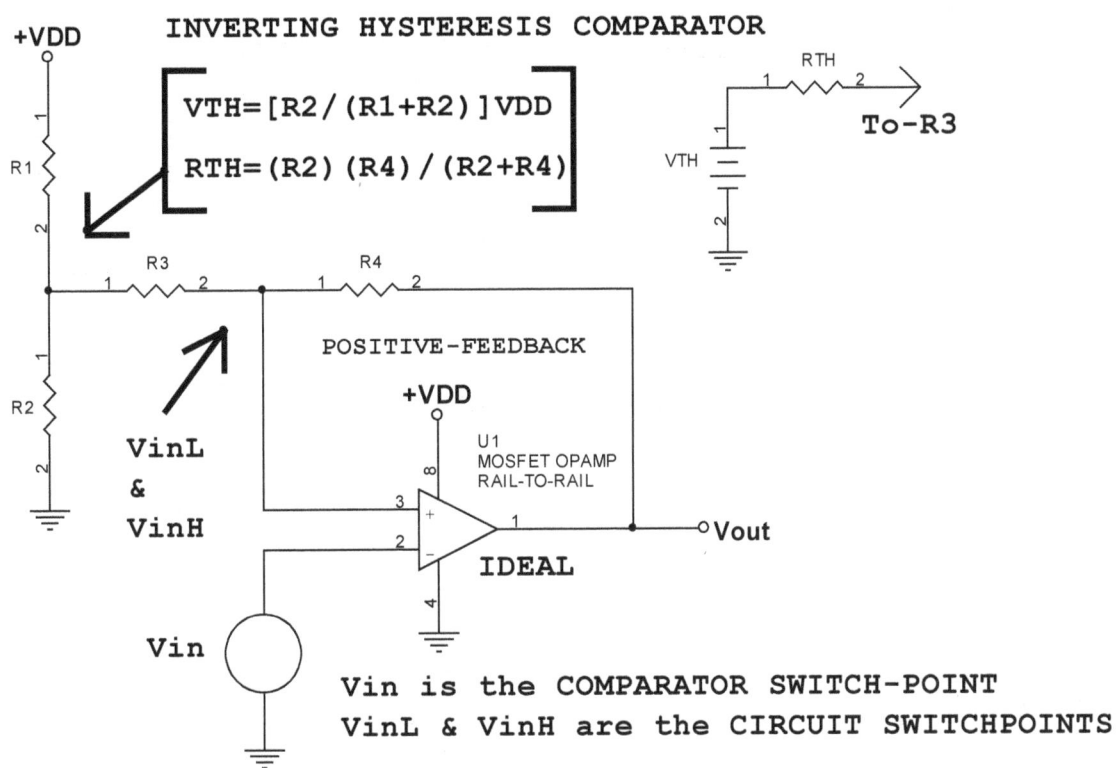

INVERTING HYSTERESIS COMPARATOR

$$VTH = [R2/(R1+R2)]VDD$$
$$RTH = (R2)(R4)/(R2+R4)$$

Vin is the COMPARATOR SWITCH-POINT
VinL & VinH are the CIRCUIT SWITCHPOINTS

TRUTH-TABLE

INPUT_STATE	Vout
Vin > VinH	0
VinL < Vin < VinH	NO-CHANGE
Vin < VinL	VDD

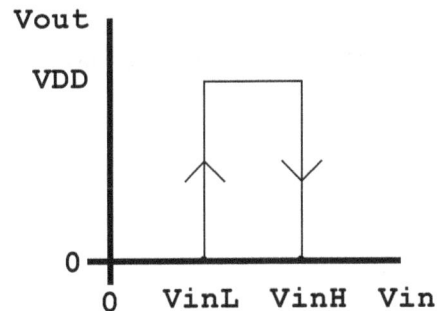

@ Vout=VDD
---> VinH = { [Vout−VTH] [(RTH+R3)/(RTH+R3+R4)] } + VTH

@ Vout=0
---> VinL = [VTH−Vout] [R4/(RTH+R3+R4)]

Figure 13.11: The Inverting Noise rejecting Hysteresis Comparator Circuit

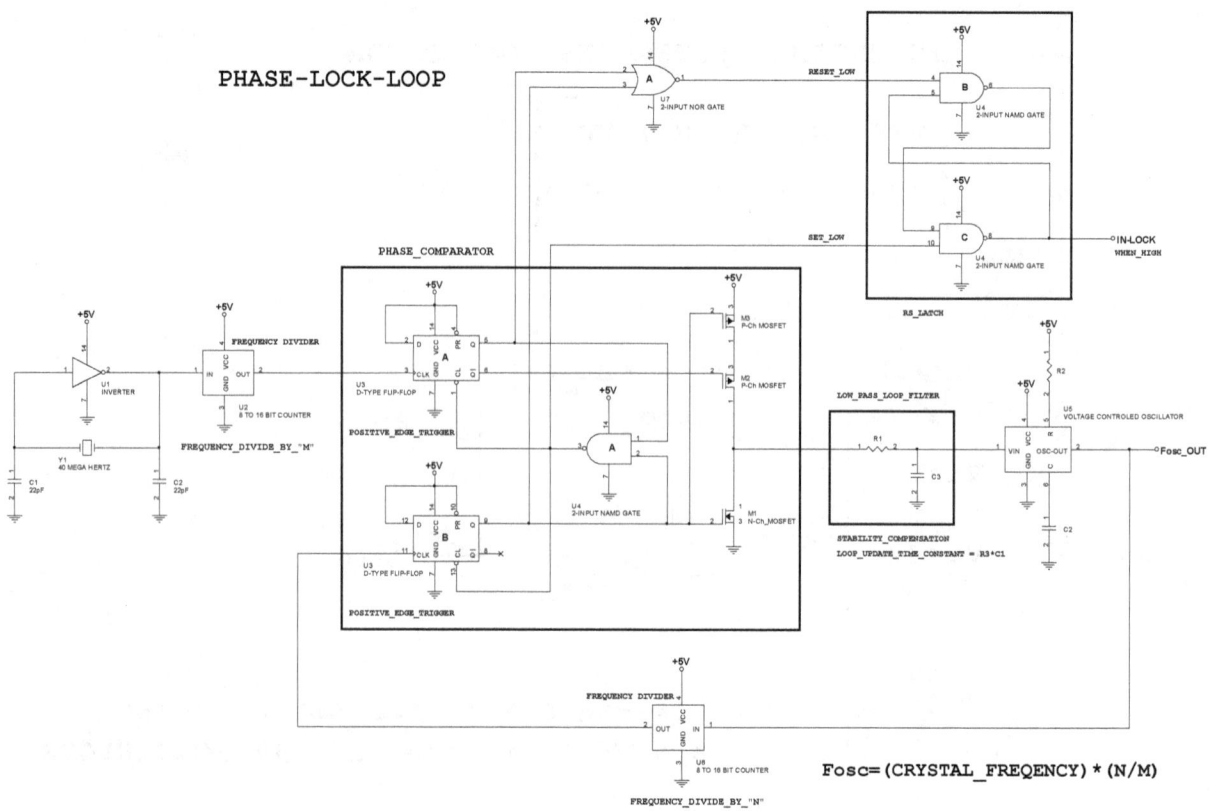

Figure 13.12: The Phase-Lock-Loop with Output Frequency Synthesis and Phase Lock Indicator

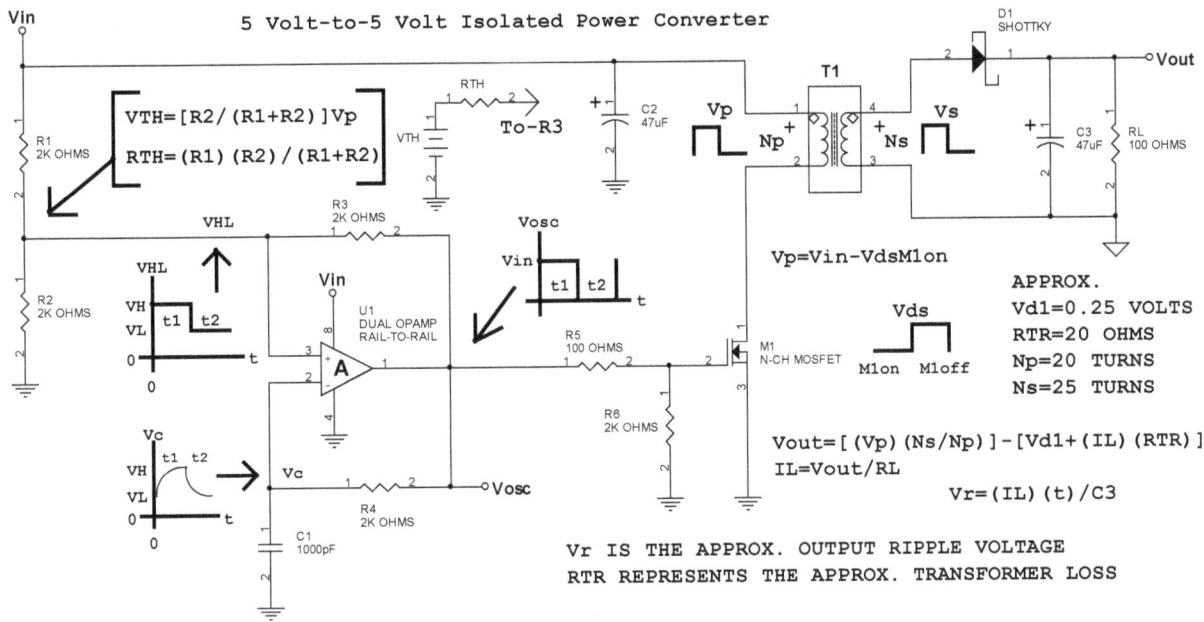

@ Vosc=Vin
---> VH={[RTH/(RTH+R3)][Vosc-VTH]}+VTH

@ Vosc=0
---> VL=[R3/(RTH+R3)][VTH-Vosc]

FINAL VALUE THEOREM
X=Xf-[Xf-Xi][exp(-t/T)]

Xi--->INITIAL VALUE
Xf--->FINAL VALUE

VL=0-[0-VH][exp(-t/T)]
VL/VH=exp(-t/T)
t=-(T)ln[VL/VH]
t=(T)ln[VH/VL]

T=(R4)(C1)

Fosc=1/[t1+t2]

Figure 13.13: The Isolated Low Power Converter

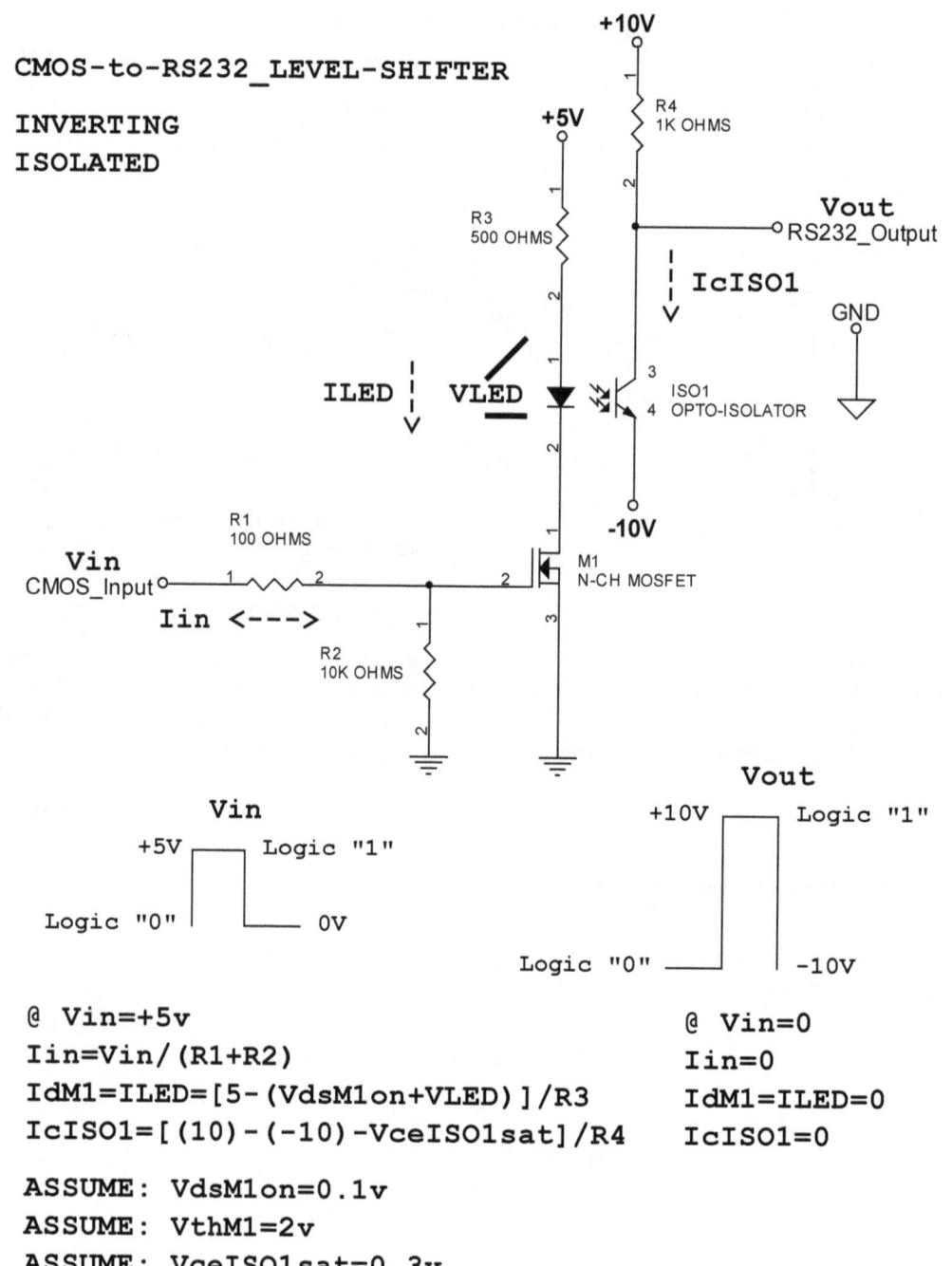

CMOS-to-RS232_LEVEL-SHIFTER
INVERTING
ISOLATED

@ Vin=+5v
Iin=Vin/(R1+R2)
IdM1=ILED=[5-(VdsM1on+VLED)]/R3
IcISO1=[(10)-(-10)-VceISO1sat]/R4

@ Vin=0
Iin=0
IdM1=ILED=0
IcISO1=0

ASSUME: VdsM1on=0.1v
ASSUME: VthM1=2v
ASSUME: VceISO1sat=0.3v

Figure 13.14: The Inverting Isolated CMOS-to-RS232 Level-Shifting Circuit

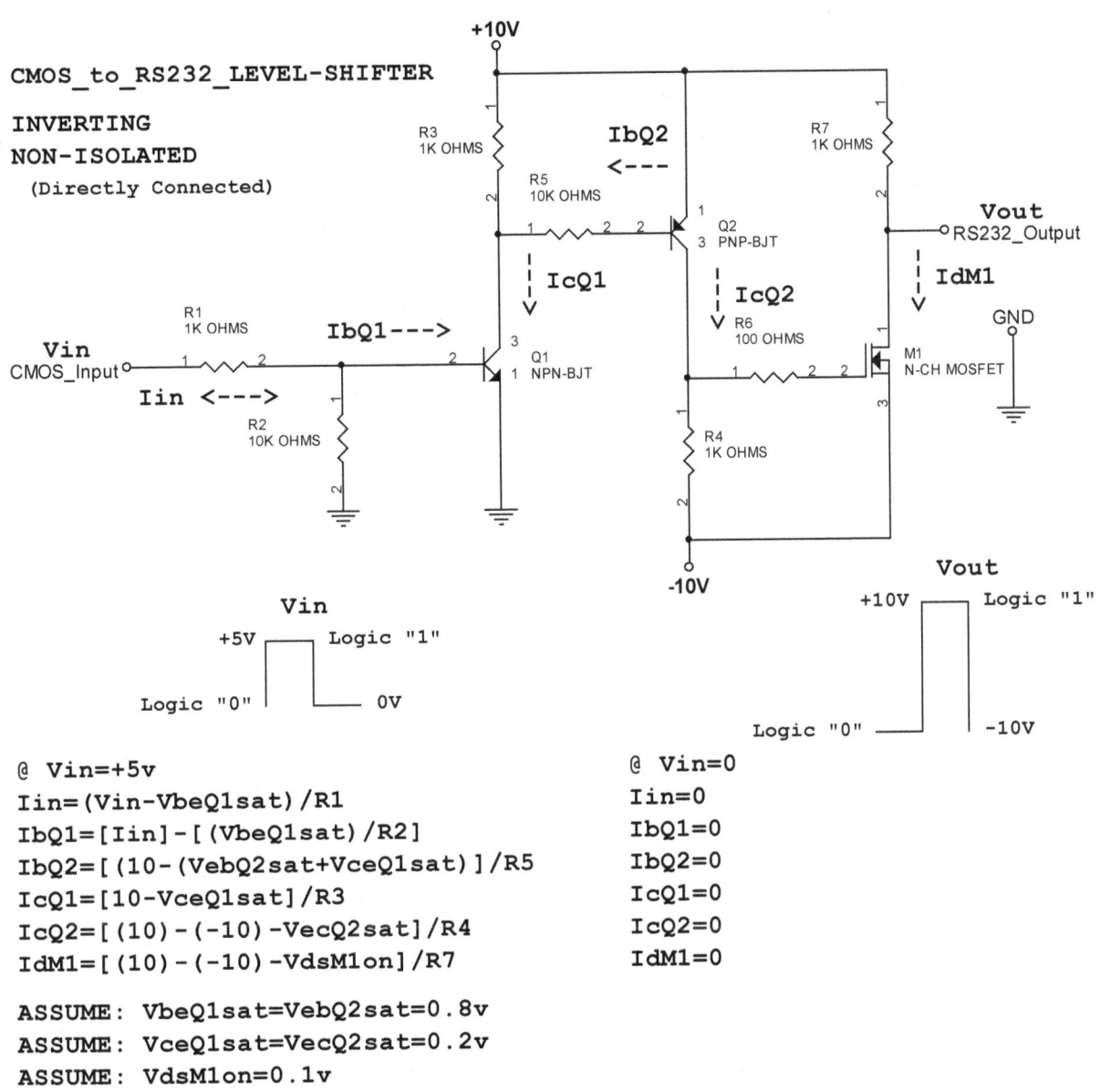

```
CMOS_to_RS232_LEVEL-SHIFTER
INVERTING
NON-ISOLATED
   (Directly Connected)
```

@ Vin=+5v

Iin=(Vin−VbeQ1sat)/R1
IbQ1=[Iin]−[(VbeQ1sat)/R2]
IbQ2=[(10−(VebQ2sat+VceQ1sat)]/R5
IcQ1=[10−VceQ1sat]/R3
IcQ2=[(10)−(−10)−VecQ2sat]/R4
IdM1=[(10)−(−10)−VdsM1on]/R7

@ Vin=0

Iin=0
IbQ1=0
IbQ2=0
IcQ1=0
IcQ2=0
IdM1=0

ASSUME: VbeQ1sat=VebQ2sat=0.8v
ASSUME: VceQ1sat=VecQ2sat=0.2v
ASSUME: VdsM1on=0.1v
ASSUME: VthM1=2v

Figure 13.15: The Inverting Non-Isolated CMOS-to-RS232 Level-Shifting Circuit

CMOS-to-RS232_LEVEL-SHIFTER

NON-INVERTING
ISOLATED

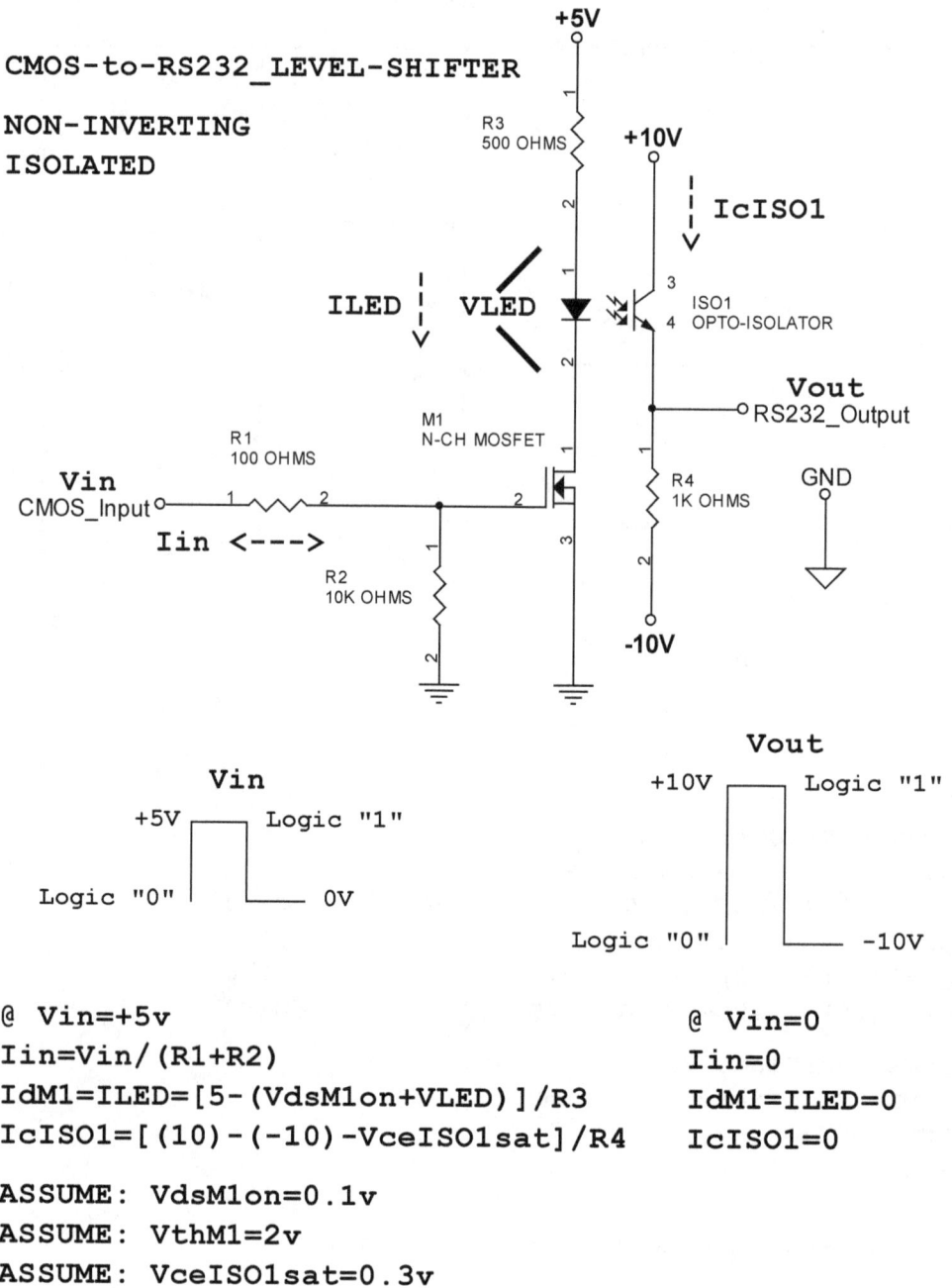

@ Vin=+5v
Iin=Vin/(R1+R2)
IdM1=ILED=[5-(VdsM1on+VLED)]/R3
IcISO1=[(10)-(-10)-VceISO1sat]/R4

@ Vin=0
Iin=0
IdM1=ILED=0
IcISO1=0

ASSUME: VdsM1on=0.1v
ASSUME: VthM1=2v
ASSUME: VceISO1sat=0.3v

Figure 13.16: The Non-Inverting Isolated CMOS-to-RS232 Level-Shifting Circuit

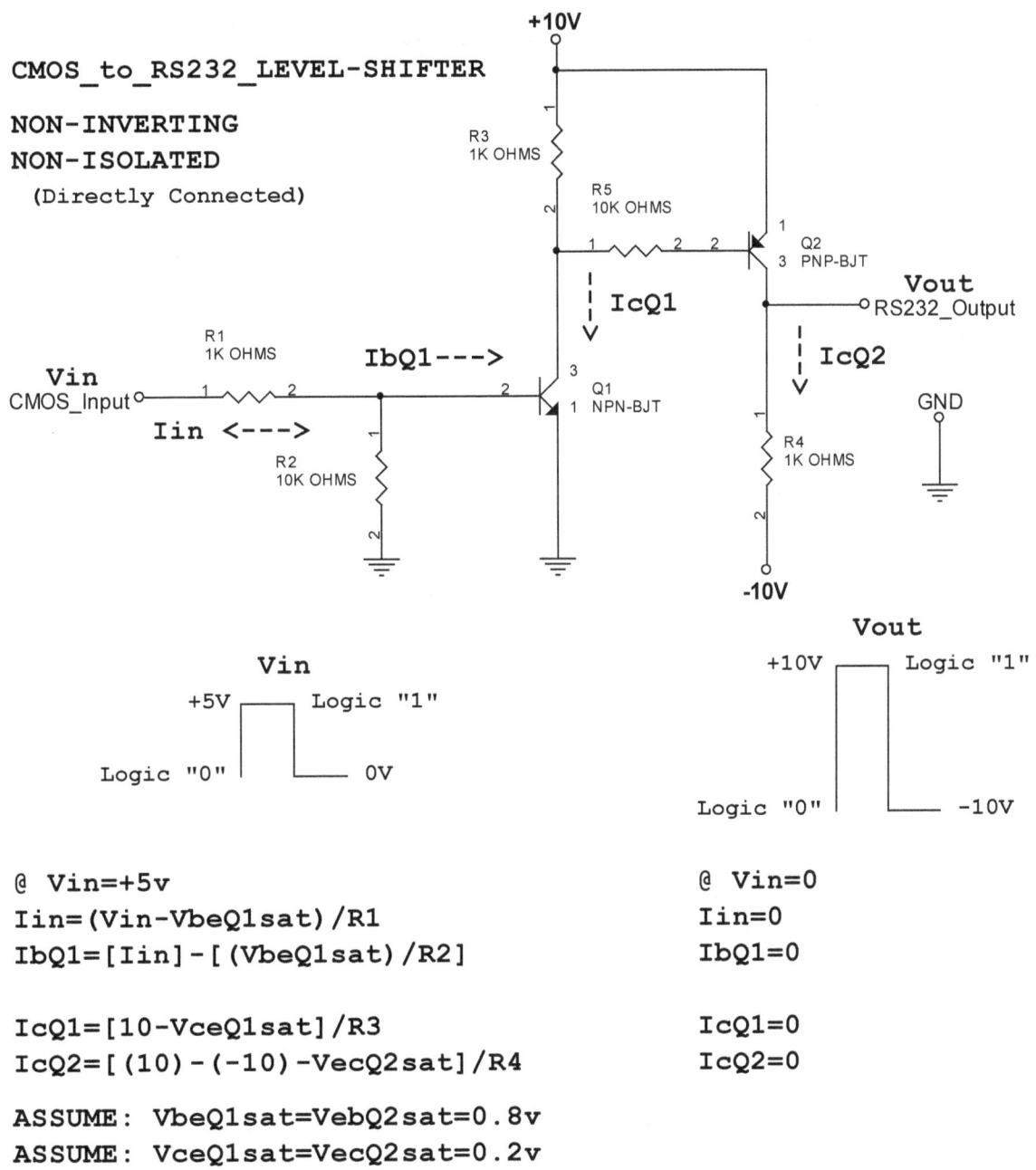

Figure 13.17: The Non-Inverting Non-Isolated CMOS-to-RS232 Level-Shifting Circuit

RS232-to-CMOS_LEVEL-SHIFTER

INVERTING
ISOLATED

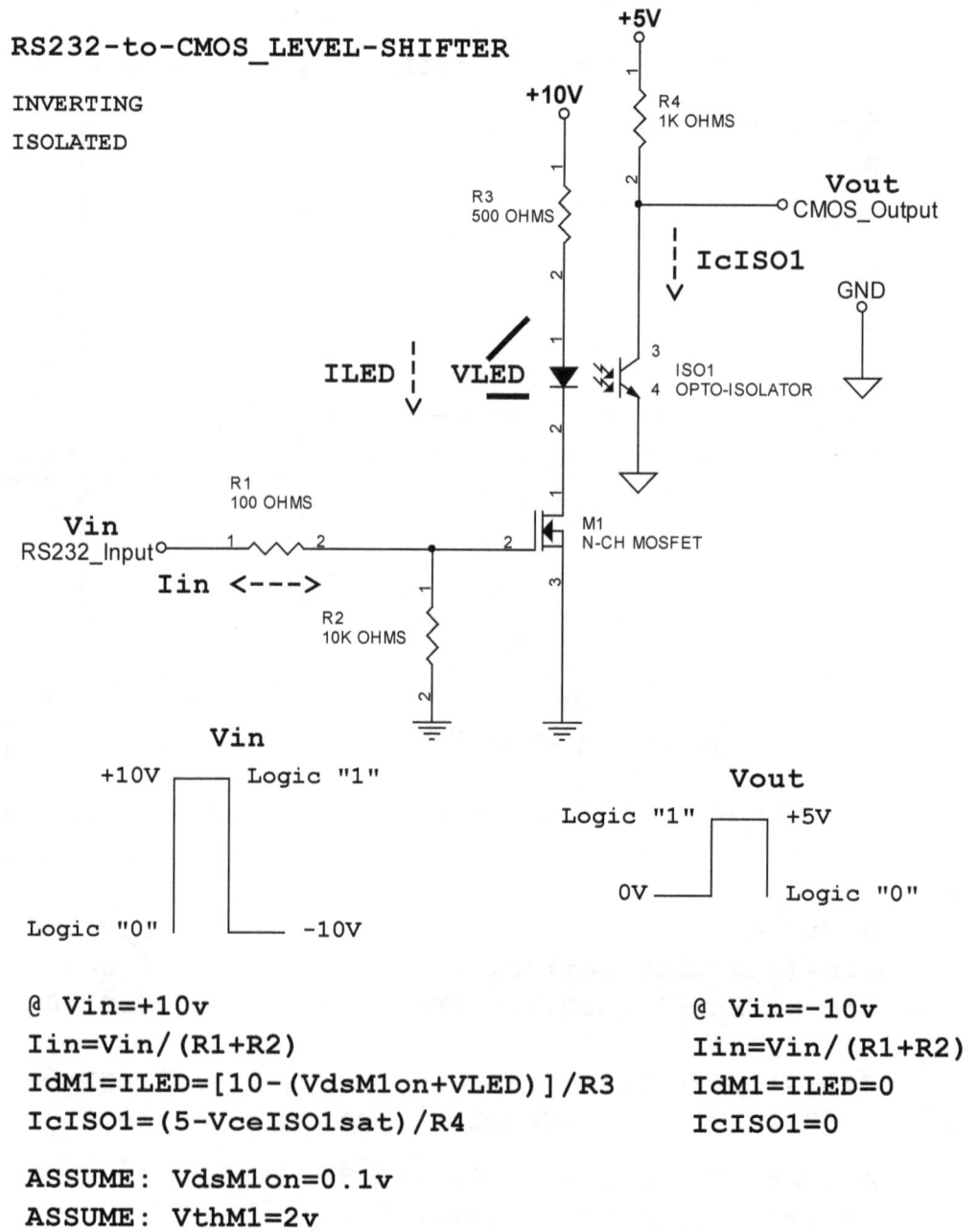

@ Vin=+10v
Iin=Vin/(R1+R2)
IdM1=ILED=[10-(VdsM1on+VLED)]/R3
IcISO1=(5-VceISO1sat)/R4

ASSUME: VdsM1on=0.1v
ASSUME: VthM1=2v
ASSUME: VceISO1sat=0.3v

@ Vin=-10v
Iin=Vin/(R1+R2)
IdM1=ILED=0
IcISO1=0

Figure 13.18: The Inverting Isolated RS232-to-CMOS Level-Shifting Circuit

RS232-to-CMOS_LEVEL-SHIFTER

INVERTING
NON-ISOLATED
(Directly Connected)

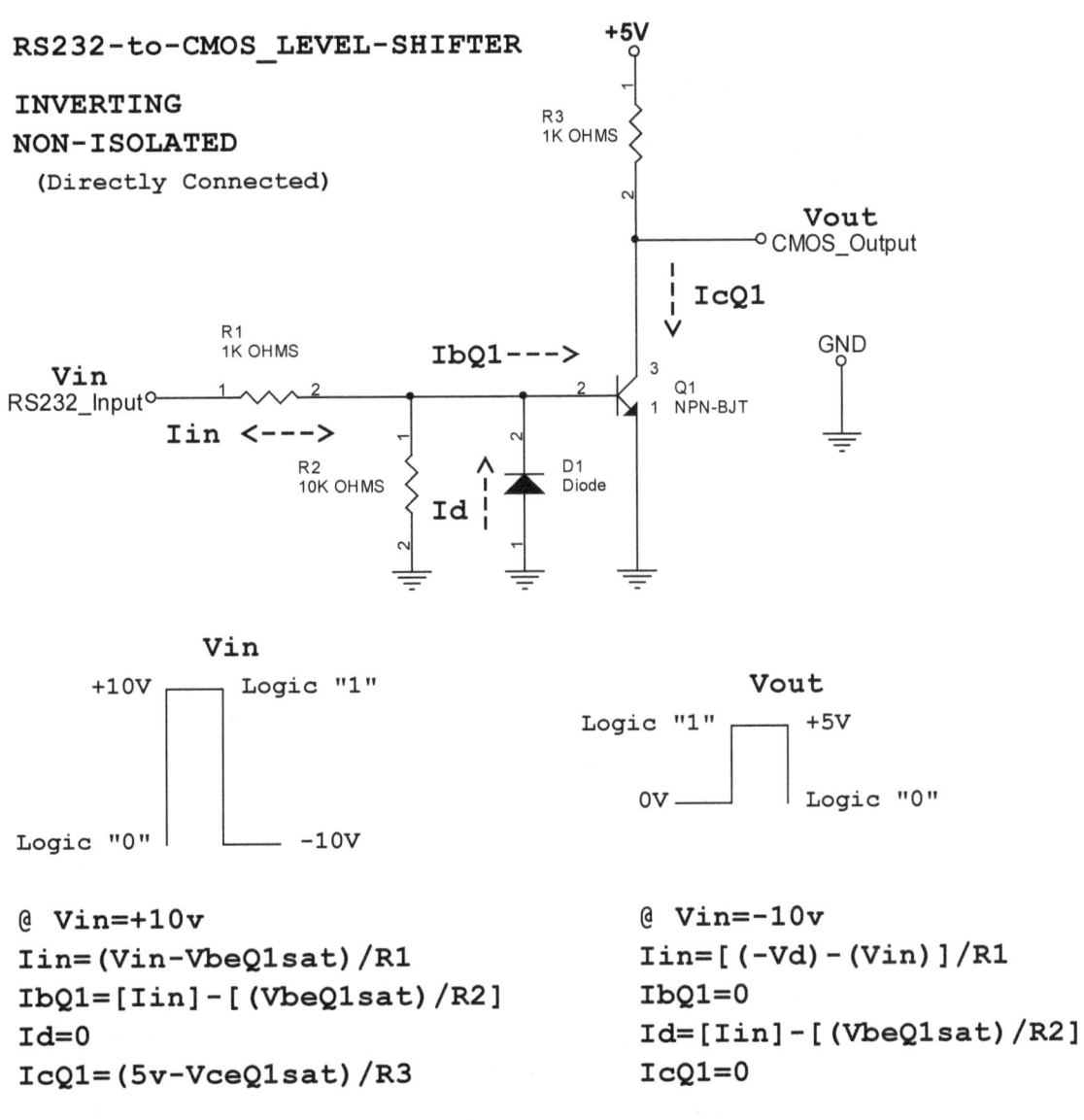

@ Vin=+10v
Iin=(Vin-VbeQ1sat)/R1
IbQ1=[Iin]-[(VbeQ1sat)/R2]
Id=0
IcQ1=(5v-VceQ1sat)/R3

@ Vin=-10v
Iin=[(-Vd)-(Vin)]/R1
IbQ1=0
Id=[Iin]-[(VbeQ1sat)/R2]
IcQ1=0

ASSUME: VbeQ1sat=VebQ2sat=0.8v
ASSUME: VceQ1sat=VecQ2sat=0.2v

Figure 13.19: The Inverting Non-Isolated RS232-to-CMOS Level-Shifting Circuit

RS232-to-CMOS_LEVEL-SHIFTER

NON-INVERTING ISOLATED

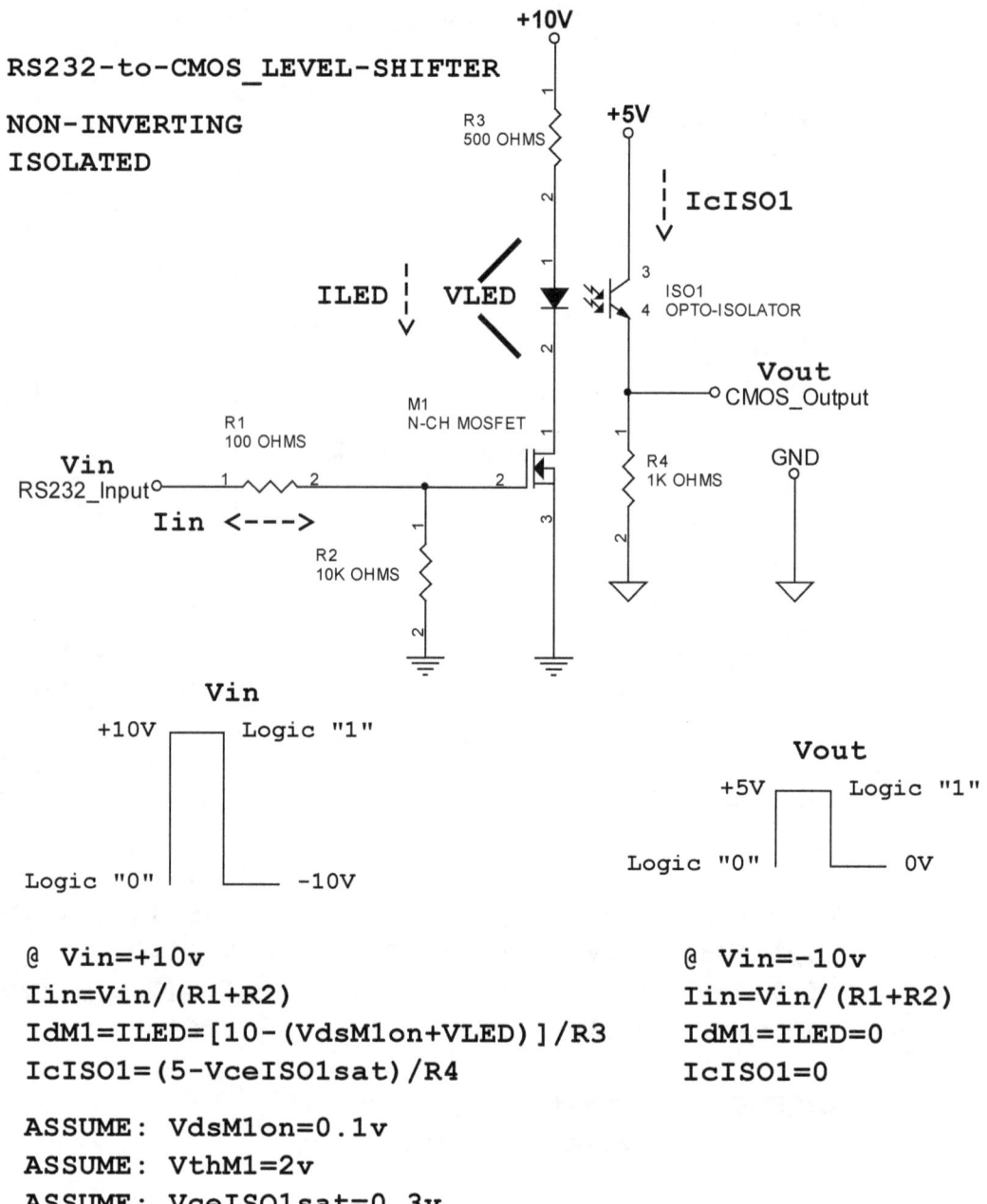

@ Vin=+10v
Iin=Vin/(R1+R2)
IdM1=ILED=[10-(VdsM1on+VLED)]/R3
IcISO1=(5-VceISO1sat)/R4

ASSUME: VdsM1on=0.1v
ASSUME: VthM1=2v
ASSUME: VceISO1sat=0.3v

@ Vin=-10v
Iin=Vin/(R1+R2)
IdM1=ILED=0
IcISO1=0

Figure 13.20: The Non-Inverting Isolated RS232-to-CMOS Level-Shifting Circuit

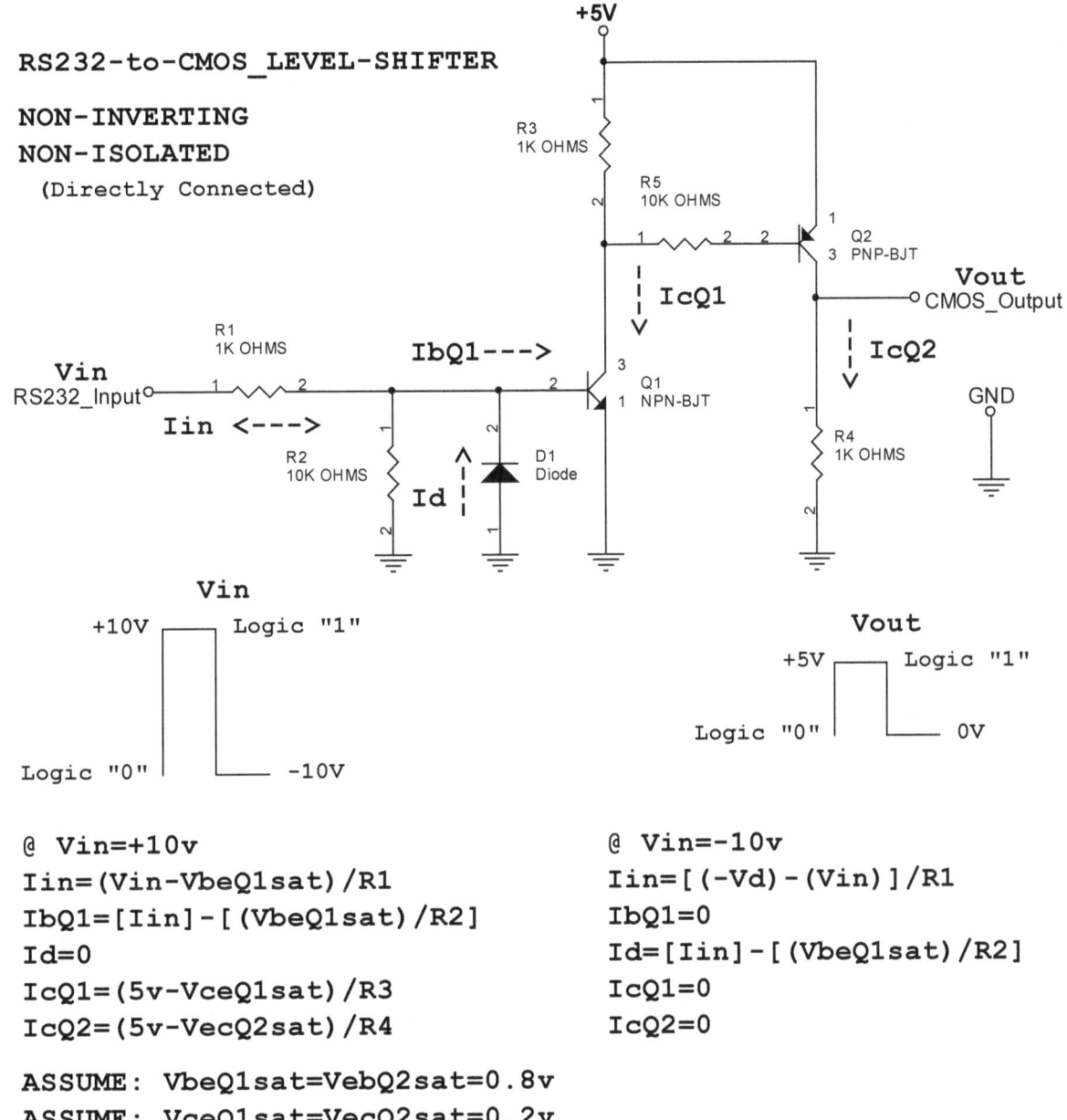

RS232-to-CMOS_LEVEL-SHIFTER

NON-INVERTING
NON-ISOLATED
(Directly Connected)

@ Vin=+10v
Iin=(Vin-VbeQ1sat)/R1
IbQ1=[Iin]-[(VbeQ1sat)/R2]
Id=0
IcQ1=(5v-VceQ1sat)/R3
IcQ2=(5v-VecQ2sat)/R4

@ Vin=-10v
Iin=[(-Vd)-(Vin)]/R1
IbQ1=0
Id=[Iin]-[(VbeQ1sat)/R2]
IcQ1=0
IcQ2=0

ASSUME: VbeQ1sat=VebQ2sat=0.8v
ASSUME: VceQ1sat=VecQ2sat=0.2v

Figure 13.21: The Non-Inverting Non-Isolated RS232-to-CMOS Level-Shifting Circuit

Appendix - A

The Experiment Board Schematics,

Bill-of-Materials

and

The PC-Card Layout Drawings

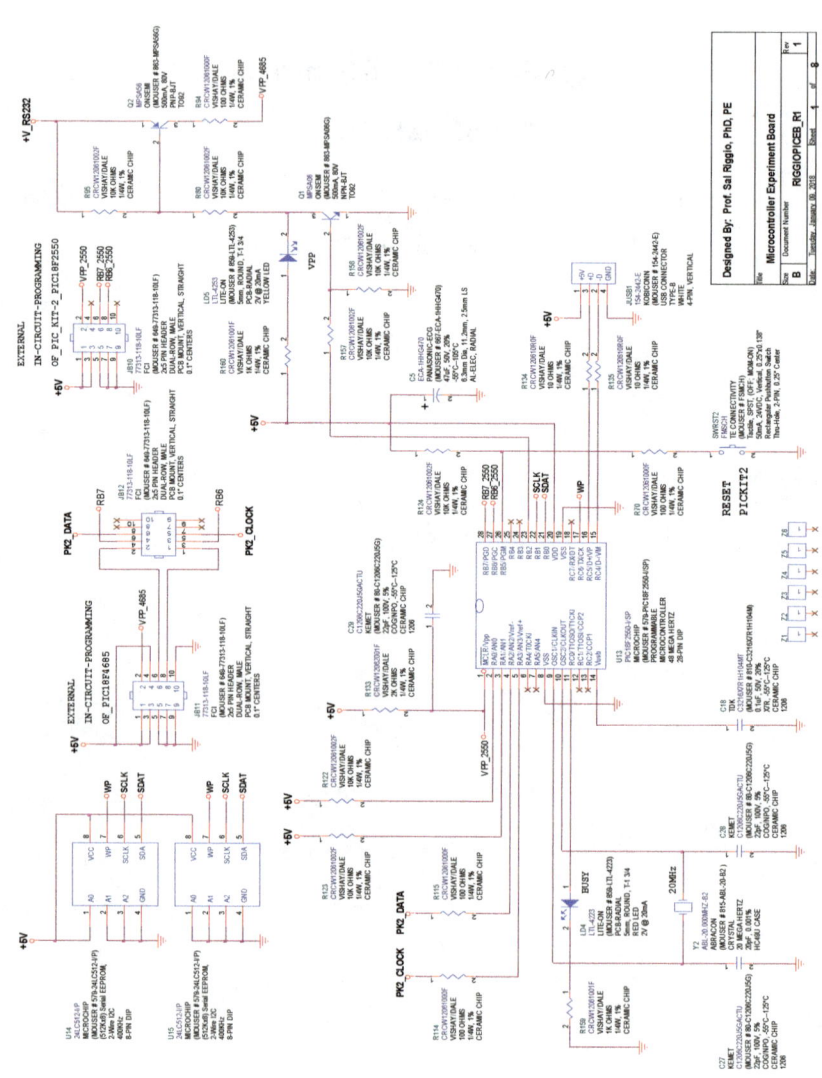

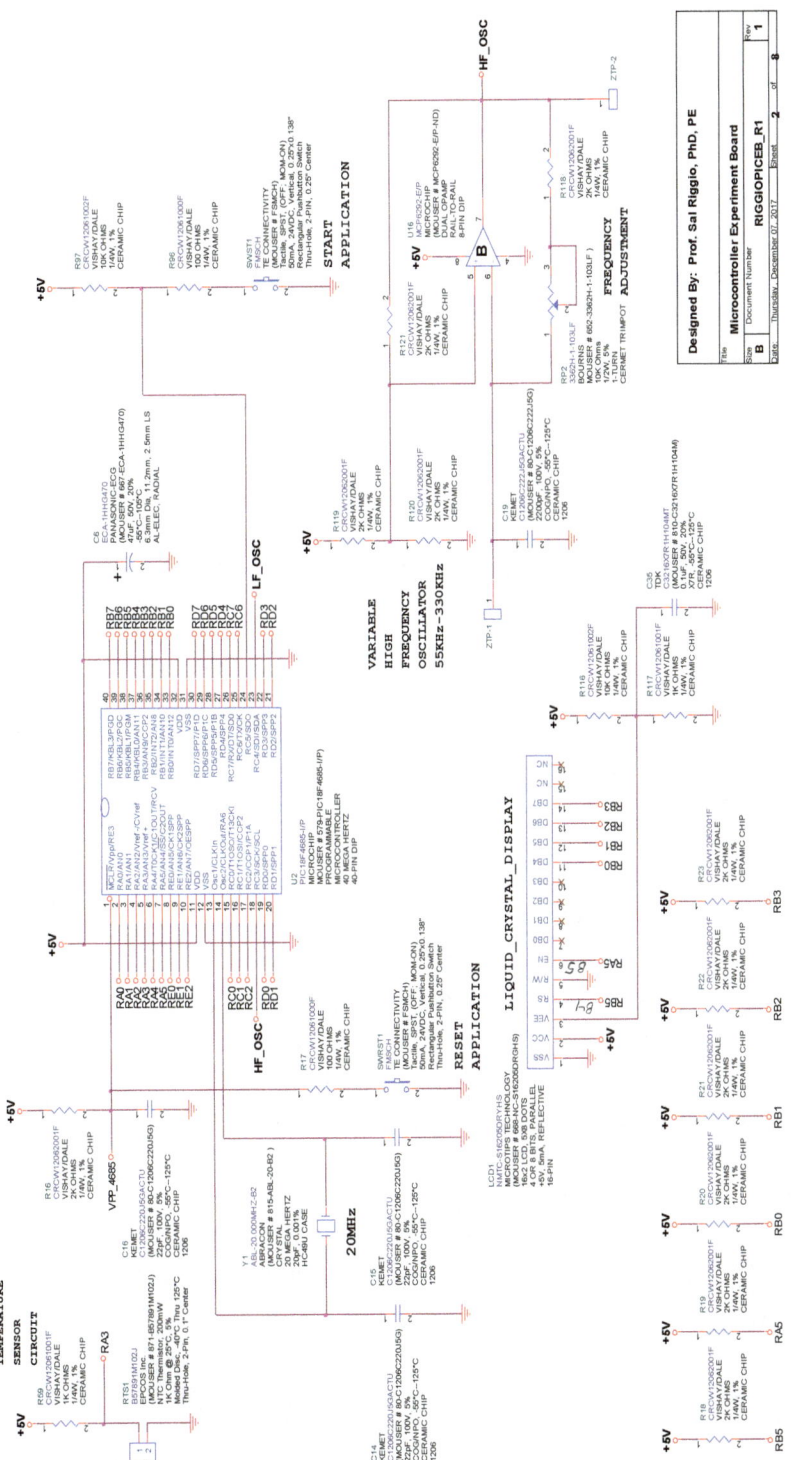

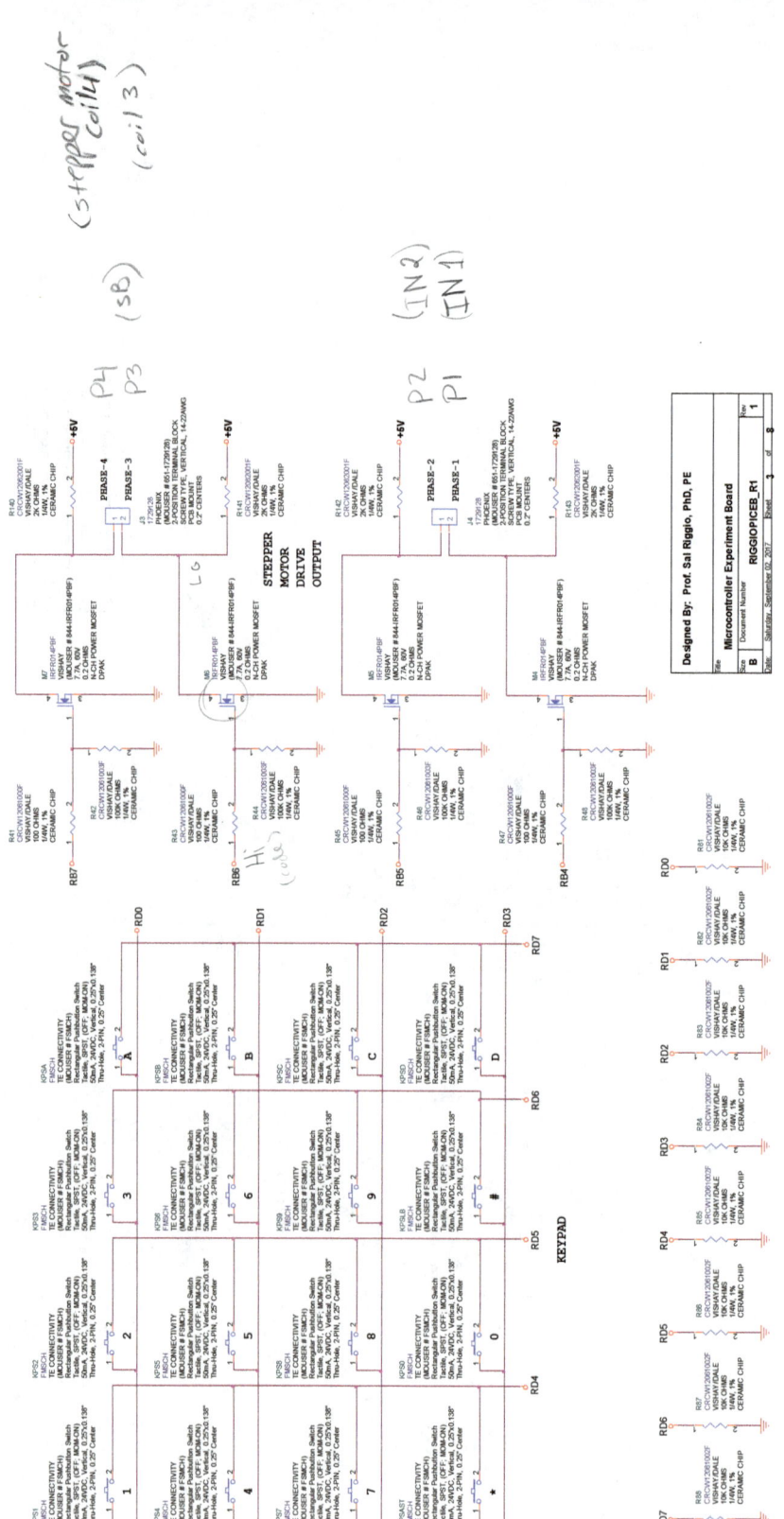

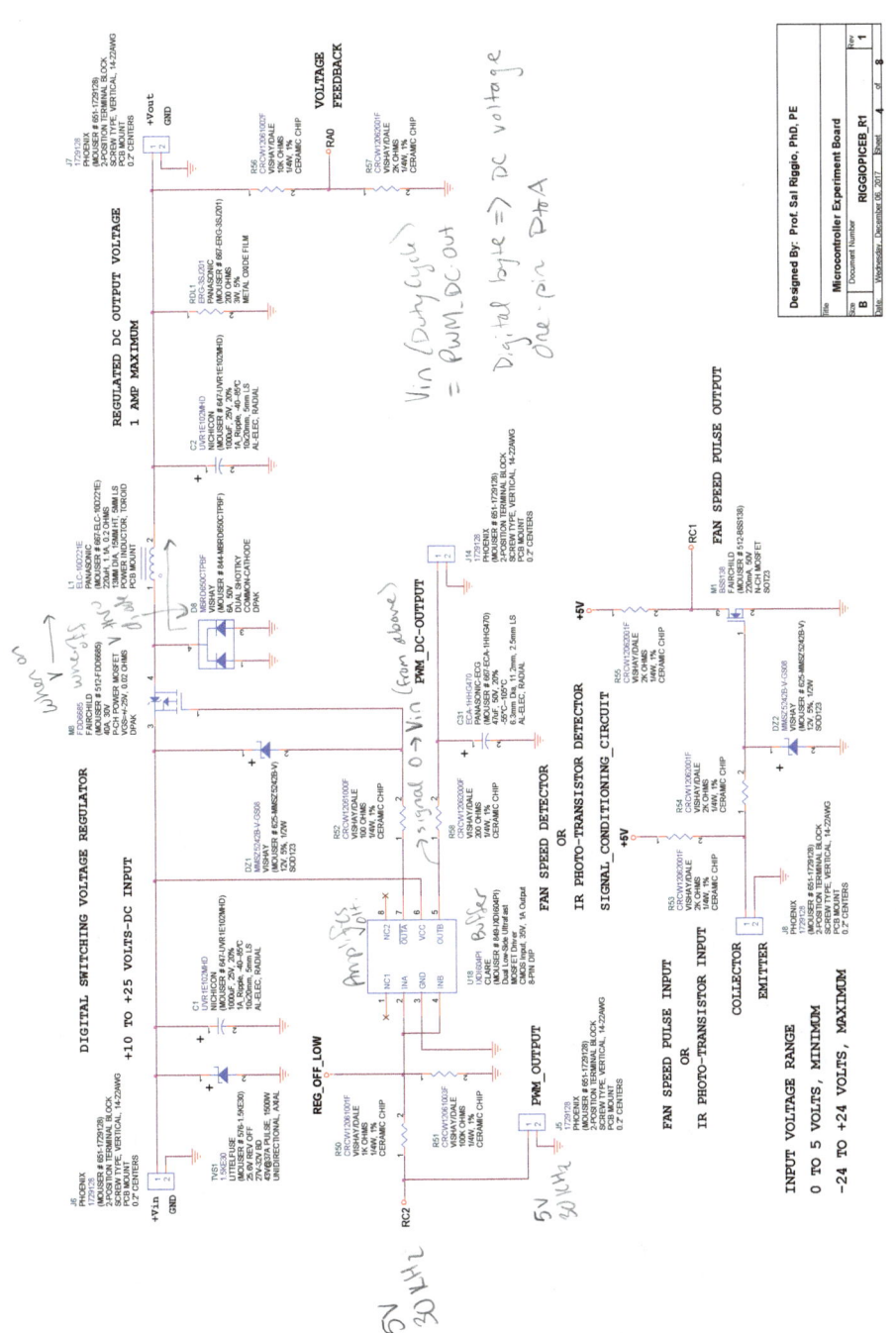

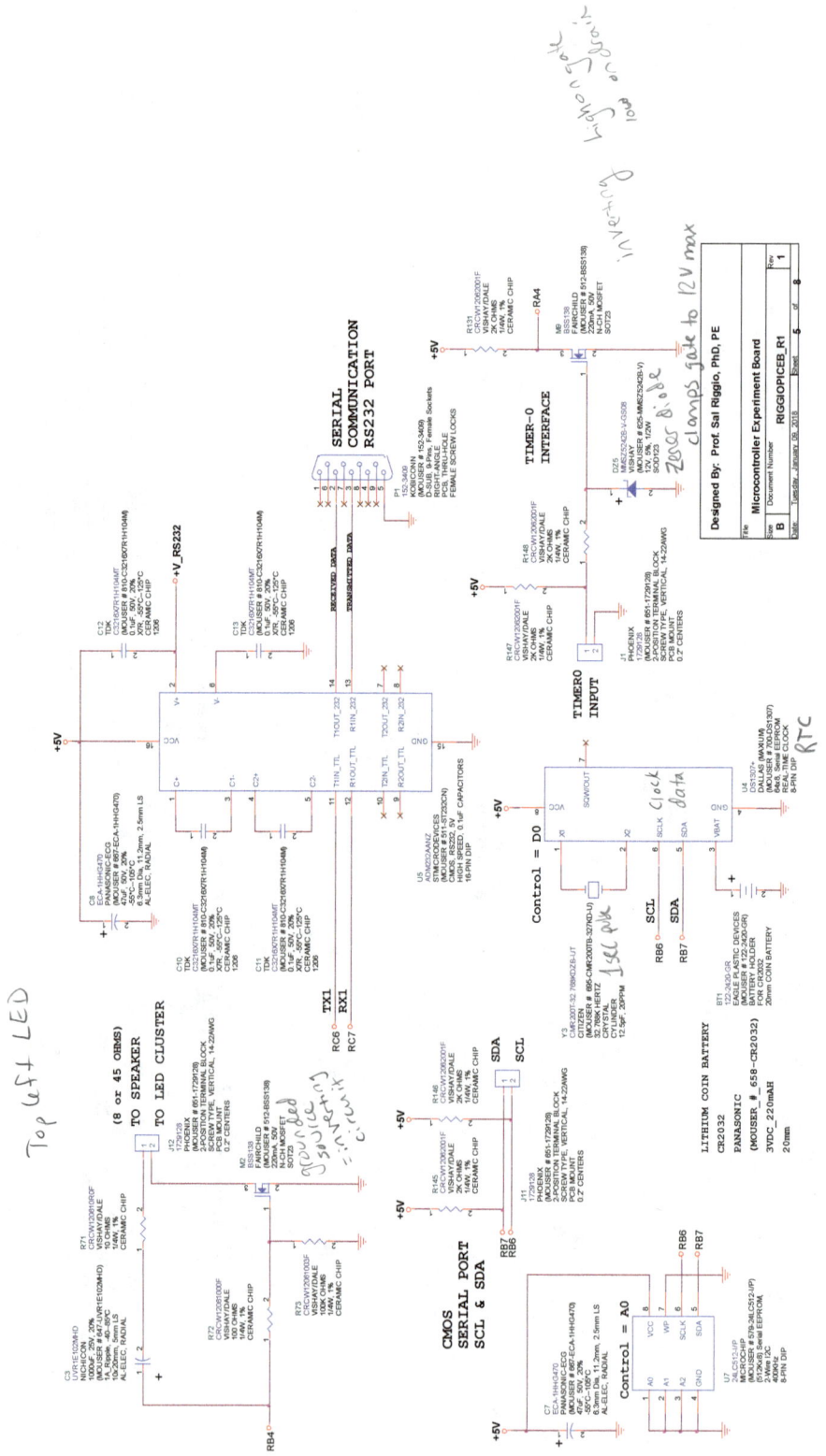

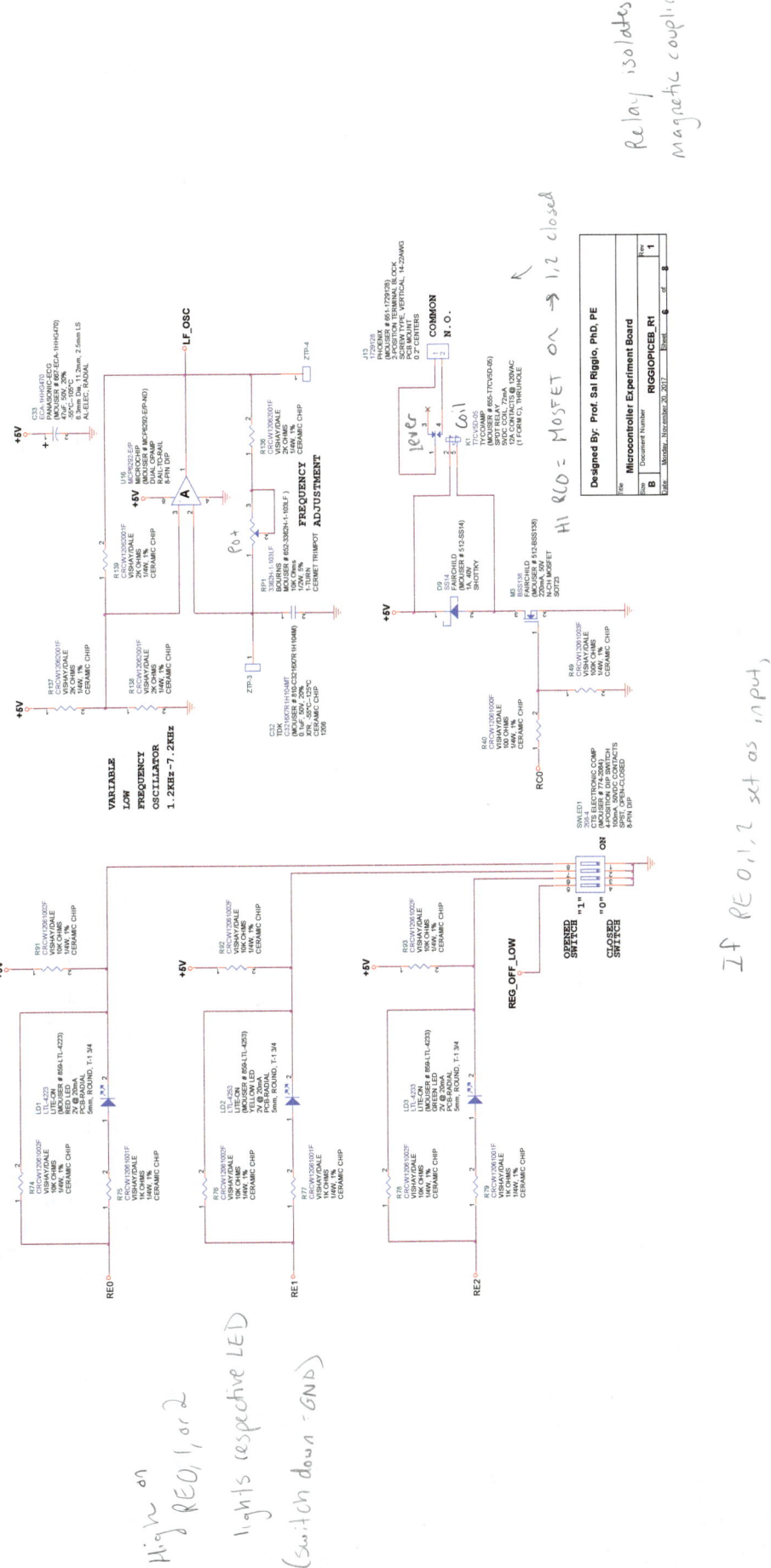

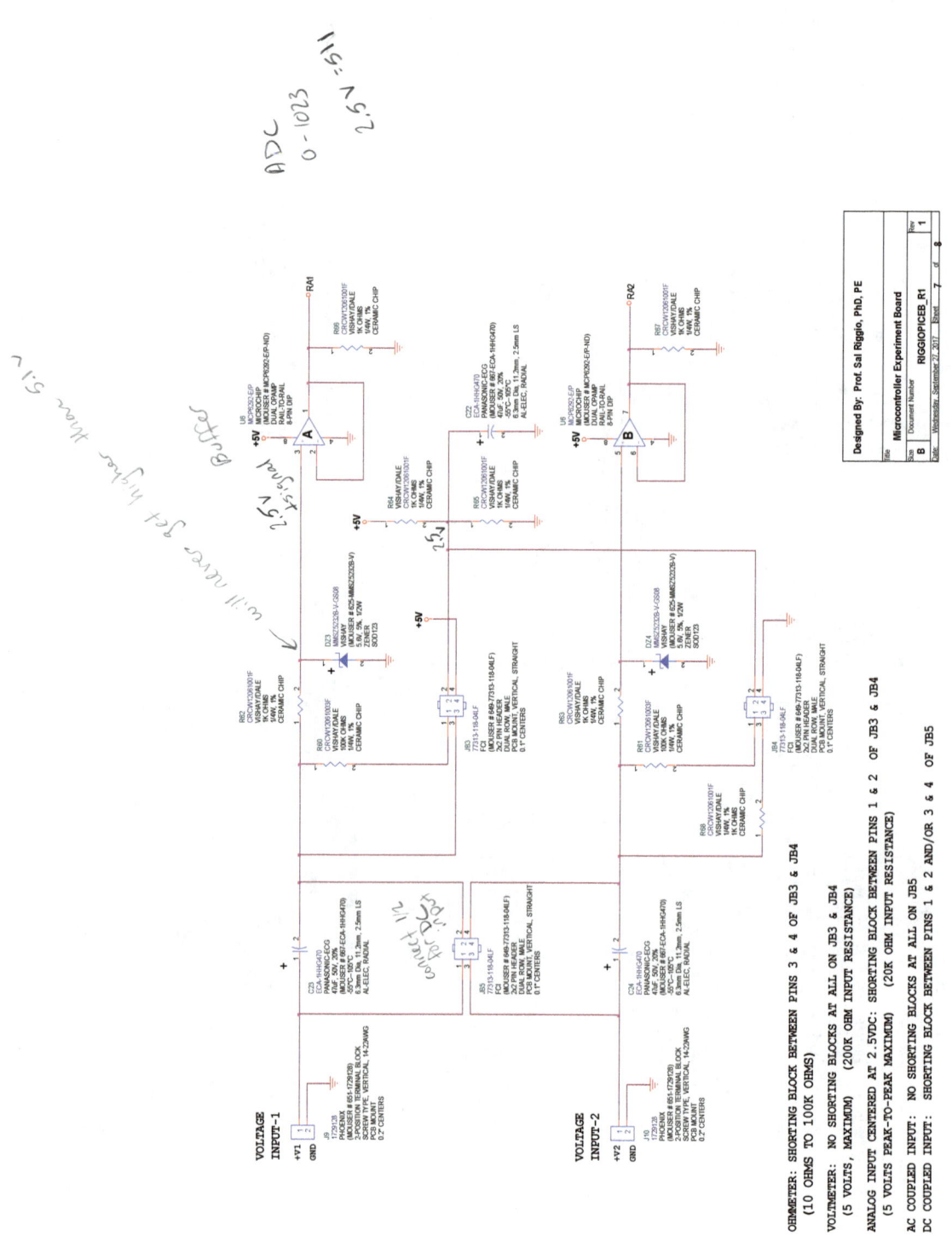

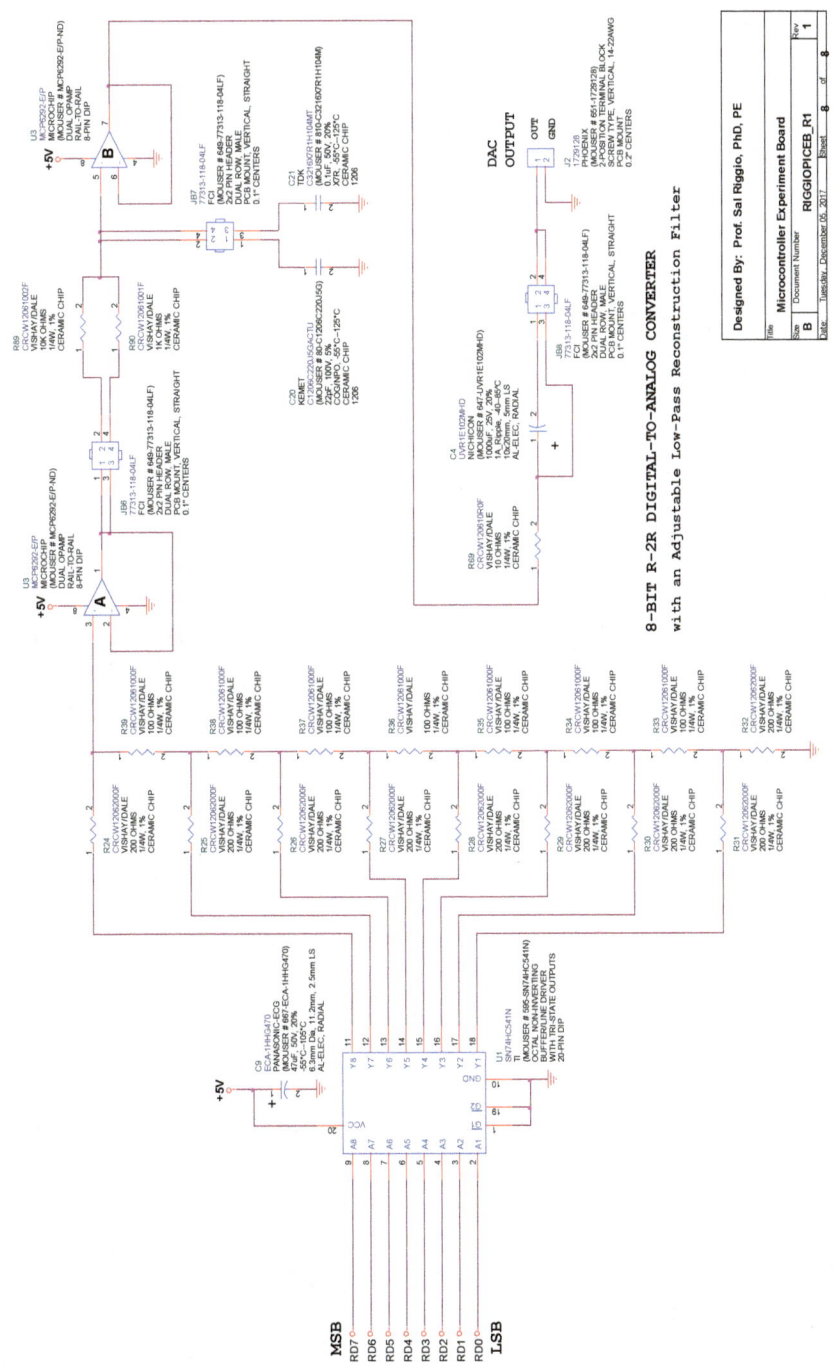

Bill-of-Materials (One PC-Card) — Page 1 of 2

RiggioPICEB_R1 — Microcontroller / Experiment Board

Item	Quan	Reference Designator	Description	Mfr	Description	Mouser Part	Description	Description
1	1	BT1	122-2420-GR	EAGLE PLASTIC DEVICES	BATTERY HOLDER	(MOUSER # 122-2420-GR)	FOR CR2032	BATTERY
2	3	C1,C2,C3,C4	UVR1E102MHD	NICHICON	1000uF, 25V, 20%	(MOUSER # 647-UVR1E102MHD)	10x20mm, 5mm LS	AL-ELEC, RADIAL
3	10	C5,C6,C7,C8,C9,C22,C23,C24,C31,C33	ECA-1HHG470	PANASONIC-ECG	47uF, 50V, 20%	(MOUSER # 667-ECA-1HHG470)	6.3mm Dia, 11.2mm, 2.5mm LS	AL-ELEC, RADIAL
4	8	C10,C11,C12,C13,C18,C21,C32,C35	C3216X7R1H104MT	TDK	0.1uF, 50V, 20%	(MOUSER # 810-C3216X7R1H104M)	X7R, -55°C~125°C	CERAMIC CHIP
5	7	C14,C15,C16,C20,C27,C28,C29	C1206C220J5GACTU	KEMET	22pF, 100V, 5%	(MOUSER # 80-C1206C220J5G)	COG/NPO, -55°C~125°C	CERAMIC CHIP	1206
6	1	C19	C1206C222J5GACTU	KEMET	2200pF, 100V, 5%	(MOUSER # 80-C1206C222J5G)	COG/NPO, -55°C~125°C	CERAMIC CHIP	1206
7	3	D21,D22,D25	MMS2524ZB-V-GS08	VISHAY	12V, 5%, 1/2W	(MOUSER # 625-MMS2524ZB-V)	ZENER	SOD123
8	2	D23,D24	MMS25232B-V-GS08	VISHAY	5.6V, 5%, 1/2W	(MOUSER # 625-MMS25232B-V)	ZENER	SOD123
9	1	D8	MBRD650CTPBF	VISHAY	6A, 50V	(MOUSER # 844-MBRD650CTPBF)	DUAL SHOTTKY	COMMON-CATHODE
10	1	D9	SS14	FAIRCHILD	1A, 40V	(MOUSER # 512-SS14)	SHOTTKY	DPAK
11	6	JB3,JB4,JB5,JB6,JB7,JB8	77313-118-04LF	FCI	2x2 PIN HEADER	(MOUSER # 649-77313-118-04LF)	PCB MOUNT, VERTICAL, STRAIGHT	0.1" CENTERS
12	3	JB10,JB11,JB12	77313-118-10LF	FCI	2x6 PIN HEADER	(MOUSER # 649-77313-118-10LF)	PCB MOUNT, VERTICAL, STRAIGHT	0.1" CENTERS
13	1	JUSB1	154-2442-E	KOBICONN	USB CONNECTOR	(MOUSER # 154-2442-E)	TYPE-B	4-PIN, VERTICAL
14	14	J1,J2,J3,J4,J5,J6,J7,J8,J9,J10,J11,J12,J13,J14	1729128	PHOENIX	2-POSITION TERMINAL BLOCK	(MOUSER # 651-1729128)	SCREW TYPE, VERTICAL, 14-22AWG	PCB MOUNT	0.2" CENTERS
15	19	SWRST1,SWRST2,SWST1,KPS0,KPS1,KPS2,KPS3,KPS4,KPS5,KPS6,KPS7,KPS8,KPS9,KPSA,KPSB,KPSC,KPSD	FMSCH	TE CONNECTIVITY	Rectangular Pushbutton Switch	(MOUSER # FSMCH)	Tactile, SPST, (OFF, MOM-ON)	Thru-Hole, 2-PIN, 0.25" Center
16	1	K1	T7CV5D-05	TYCO/AMP	SPDT RELAY	(MOUSER # 655-T7CV5D-05)	5VDC COIL, 72mA	(1 FORM C), THRUHOLE
17	1	LCD1	NMTC-S16205DRYHS	MICROTIPS TECHNOLOGY	16x2 LCD, 5X8 DOTS	(MOUSER # 668-NC-S16205DRGHS)	+5V, 5mA, REFLECTIVE	16-PIN
18	2	LD1,LD4	LTL-4223	LITE-ON	RED LED	(MOUSER # 859-LTL-4223)	2V @ 20mA	PCB-RADIAL	5mm, ROUND, T-1 3/4
19	1	LD2,LD5	LTL-4253	LITE-ON	YELLOW LED	(MOUSER # 859-LTL-4253)	2V @ 20mA	PCB-RADIAL	5mm, ROUND, T-1 3/4
20	1	LD3	LTL-4233	LITE-ON	GREEN LED	(MOUSER # 859-LTL-4233)	2V @ 20mA	PCB-RADIAL	5mm, ROUND, T-1 3/4
21	1	L1	ELC-10D221E	PANASONIC	220uF, 1.1A, 0.2 OHMS	(MOUSER # 667-ELC-10D221E)	POWER INDUCTOR, TOROID	PCB MOUNT
22	4	M1,M2,M3,M9	BSS138	FAIRCHILD	220mA, 50V	(MOUSER # 512-BSS138)	N-CH MOSFET	SOT23
23	4	M4,M5,M6,M7	IRFR014PBF	VISHAY	7.7A, 60V	(MOUSER # 844-IRFR014PBF)	N-CH POWER MOSFET	DPAK
24	1	M8	FDD6685	FAIRCHILD	40A, 30V	(MOUSER # 512-FDD6685)	P-CH POWER MOSFET	DPAK
25	1	P1	152-3409	KOBICONN	D-SUB, 9-Pins, Female Sockets	(MOUSER # 152-3409)	PCB, THRU-HOLE	FEMALE SCREW LOCKS
26	1	Q1	MPSA06	ONSEMI	500mA, 80V	(MOUSER # 863-MPSA06G)	NPN-BJT	TO92
27	1	Q2	MPSA56	ONSEMI	500mA, 80V	(MOUSER # 863-MPSA56G)	PNP-BJT	TO92
28	1	RDL1	ERG-3SJ201	PANASONIC	200 OHMS	(MOUSER # 667-ERG-3SJ201)	3W, 5%	METAL OXIDE FILM
29	2	RP1,RP2	3362H-1-104LF	BOURNS	100K Ohms	(MOUSER # 652-3362H-1-104LF)	1/2W, 5%	CERMET TRIMPOT
30	1	RTS1	B57891M102J	EPCOS Inc.	NTC Thermistor, 200mW	(MOUSER # 871-B57891M102J)	1-TURN	Molded Disc., -40°C Thru 125°C	Thru-Hole, 2-Pin, 0.1" Center
31	29	R16,R18,R19,R20,R21,R22,R23,R53,R54,R55,R57,R118,R119,R120,R121,R131,R133,R136,R137,R138,R139,R140,R141,R142,R143,R145,R146,R147,R148	CRCW12062001F	VISHAY/DALE	2K OHMS	(MOUSER # 71-CRCW12062K00FKEB)	1/4W, 1%	CERAMIC CHIP
32	20	R17,R33,R34,R35,R36,R37,R38,R39,R40,R41,R44,R45,R47,R52,R70,R72,R94,R96,R114,R115	CRCW12061000F	VISHAY/DALE	100 OHMS	(MOUSER # 71-CRCW1206100RFKEB)	1/4W, 1%	CERAMIC CHIP
33	10	R24,R25,R26,R27,R28,R29,R30,R31,R32,R58	CRCW12062000F	VISHAY/DALE	200 OHMS	(MOUSER # 71-CRCW1206200RFKEB)	1/4W, 1%	CERAMIC CHIP
34	9	R42,R44,R46,R48,R49,R51,R60,R61,R73	CRCW12061003F	VISHAY/DALE	100K OHMS	(MOUSER # 71-CRCW1206100KFKEB)	1/4W, 1%	CERAMIC CHIP
35	16	R50,R59,R62,R63,R64,R65,R66,R67,R68,R75,R77,R79,R90,R117,R159,R160	CRCW12061001F	VISHAY/DALE	1K OHMS	(MOUSER # 71-CRCW12061K00FKEB)	1/4W, 1%	CERAMIC CHIP
36	25	R56,R74,R76,R78,R80,R81,R82,R83,R84,R85,R86,R87,R88,R89,R91,R92,R93,R95,R97,R116,R122,R123,R124,R157,R158	CRCW12061002F	VISHAY/DALE	10K OHMS	(MOUSER # 71-CRCW120610K0FKEB)	1/4W, 1%	CERAMIC CHIP
37	4	R69,R71,R134,R135	CRCW12061R0F	VISHAY/DALE	10 OHMS	(MOUSER # 71-CRCW120610R0FKEB)	1/4W, 1%	CERAMIC CHIP

Item	Quan	Reference Designator	RiggioPICEB_R1 Description	Microcontroller Description	Experiment Board Description	Bill-of-Materials (One PC-Card) Description	Page 2 of 2 Description	Description	Description
38	1	SWLED1	208-4	CTS ELECTRONIC COMP	CTS ELECTRONIC COMP	(MOUSER # 774-2084)	4-POSITION DIP SWITCH	100mA, 50VDC CONTACTS	8-PIN DIP
39	1	TVS1	1.5KE30	LITTELFUSE	LITTELFUSE	(MOUSER # 576-1.5KE30)	25.6V REV OFF	27V-32V BD	UNIDIRECTIONAL, AXIAL
40	1	U1	SN74HC541N	TI	TI	(MOUSER # 595-SN74HC541N)	OCTAL NON-INVERTING	BUFFER/LINE DRIVER WITH TRI-STATE OUTPUTS	20-PIN DIP
41	1	U2	PIC18F4685-I/P	MICROCHIP	MICROCHIP	(MOUSER # 579-PIC18F4685-I/P)	PROGRAMMABLE	MICROCONTROLLER 40 MEGA HERTZ	40-PIN DIP
42	3	U3,U6,U16	MCP6292-E/P	MICROCHIP	MICROCHIP	(MOUSER # MCP6292-E/P-ND)	DUAL OPAMP	RAIL-TO-RAIL HIGH SPEED, 0.1uF CAPACITORS	8-PIN DIP
43	1	U4	DS1307+	DALLAS (MAXIUM)	DALLAS (MAXIUM)	(MOUSER # 700-DS1307)	512, 64x8, Serial EEPROM	REAL-TIME CLOCK	8-PIN DIP
44	1	U5	ADM232AANZ	STMICRODEVICES	STMICRODEVICES	(MOUSER # 511-ST232CN)	CMOS, RS232, 5V		16-PIN DIP
45	3	U7,U14,U15	24LC512-I/P	MICROCHIP	MICROCHIP	(MOUSER # 579-24LC512-I/P)	512K Serial EEPROM, (64Kx8)	2-Wire I2C 400KHz	8-PIN DIP
46	1	U13	PIC18F2550-I/SP	MICROCHIP	MICROCHIP	(MOUSER # 579-PIC18F2550-I/SP)	PROGRAMMABLE	MICROCONTROLLER 48 MEGA HERTZ	28-PIN DIP
47	1	U18	IXD1604PI	CLARE	CLARE	(MOUSER # 849-IXD1604PI)	Dual Low-Side Ultrafast	MOSFET Driver CMOS Input, 35V, 1A Output	8-PIN DIP
48	2	Y1,Y2	ABL-20.000MHZ-B2	ABRACON	ABRACON	(MOUSER # 815-ABL-20-B2)	CRYSTAL	20 MEGA HERTZ 20pF, 0.001%	HC49U CASE
49	1	Y3	CMR200T-32.768KDZB-UT	CITIZEN	CITIZEN	(MOUSER # 695-CMR200TB-327KD-U)	CRYSTAL	32.768K HERTZ 12.5pF, 20PPM	CYLINDER
50	4	ZTP-1, ZTP-2, ZTP-3, ZTP-4	5005K	KEYSTONE	KEYSTONE	(DIGIKEY # 5005K-ND)	RED, .063"		
51	6		2-881545-2	TE CONNECTIVITY	TE CONNECTIVITY	(MOUSER # 571-28815452)	2-Pin Jumper, 0.1" Centers	Shunt	
52	2		68000-416HLF	FCI	FCI	(MOUSER # 649-68000-416HLF)	1x16 Pin Strip Header, 0.1" Centers	For Liquid-Crystal-Display	
53	1 Pack		1421T2	HAMMOND	HAMMOND	(MOUSER # 546-1421T2)	Rubber Feet, Self-Adhesive	24 Rubber Feet in a Pack	
54	8		4808-3004-CP	3M	3M	(MOUSER # 517-4808-3004-CP)	IC-Socket, 8-Pin Dip		
55	1		4816-3004-CP	3M	3M	(MOUSER # 517-4816-3004-CP)	IC-Socket, 16-Pin Dip		
56	1		4820-3004-CP	3M	3M	(MOUSER # 517-4820-3004-CP)	IC-Socket, 20-Pin Dip		
57	1		4828-3004-CP	3M	3M	(MOUSER # 517-4828-3004-CP)	IC-Socket, 28-Pin Dip		
58	1		4840-6000-CP	3M	3M	(MOUSER # 517-4840-6000-CP)	IC-Socket, 40-Pin Dip		
59	1		CR2032	PANASONIC	PANASONIC	(MOUSER # 658-CR2032)	Coin Battery, Lithium, 3V, 200mA	20mm x 3.2mm	

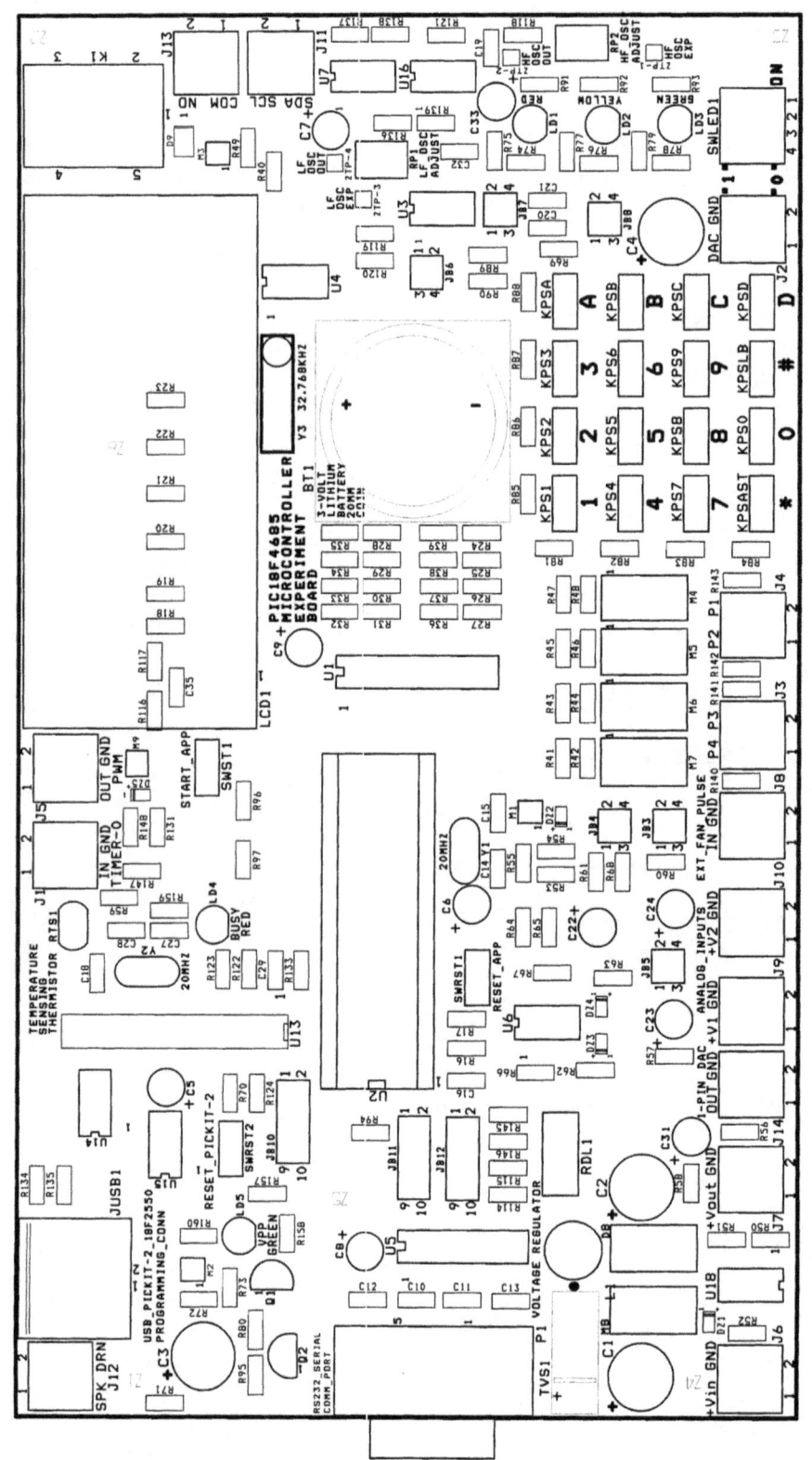

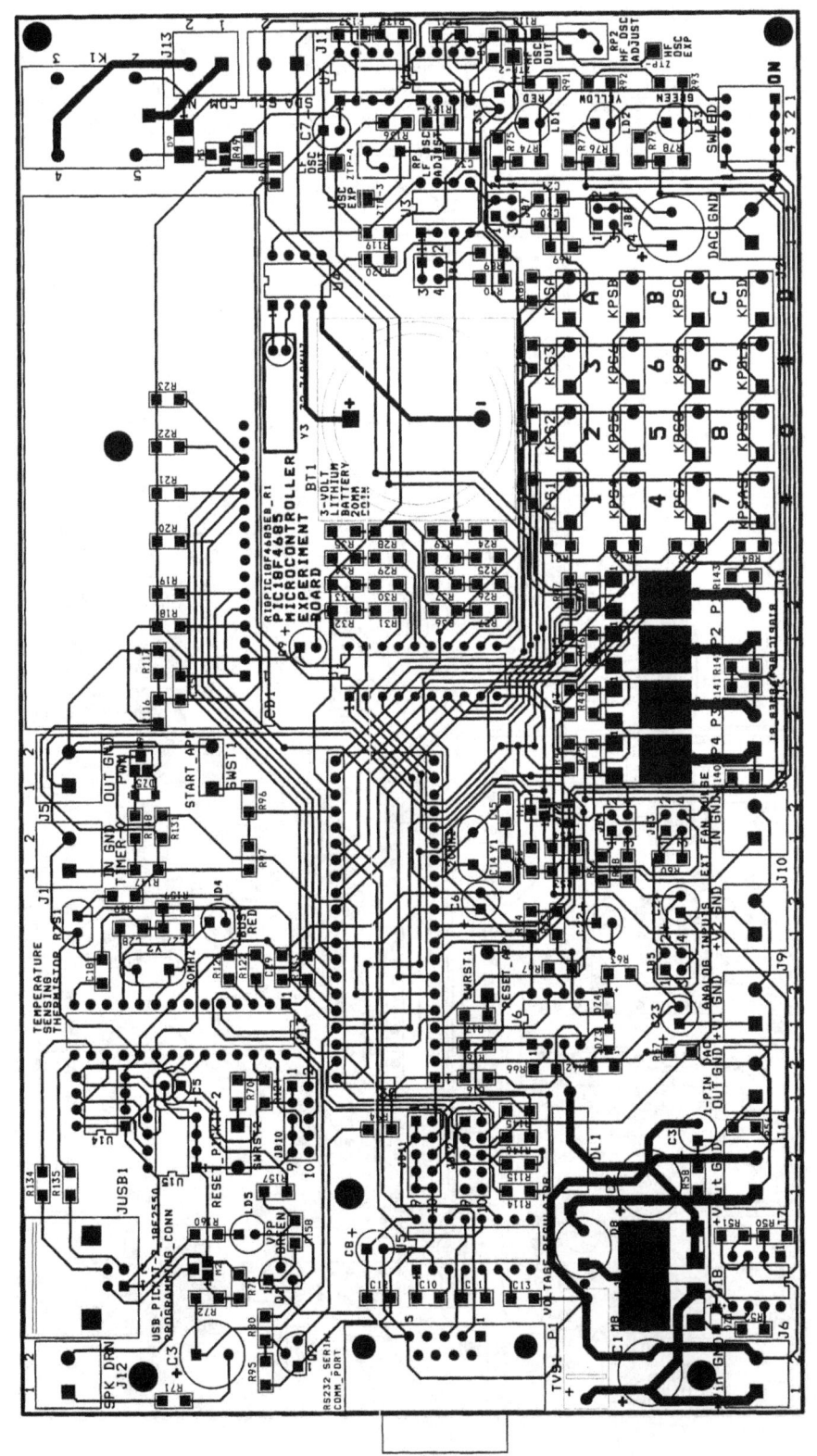

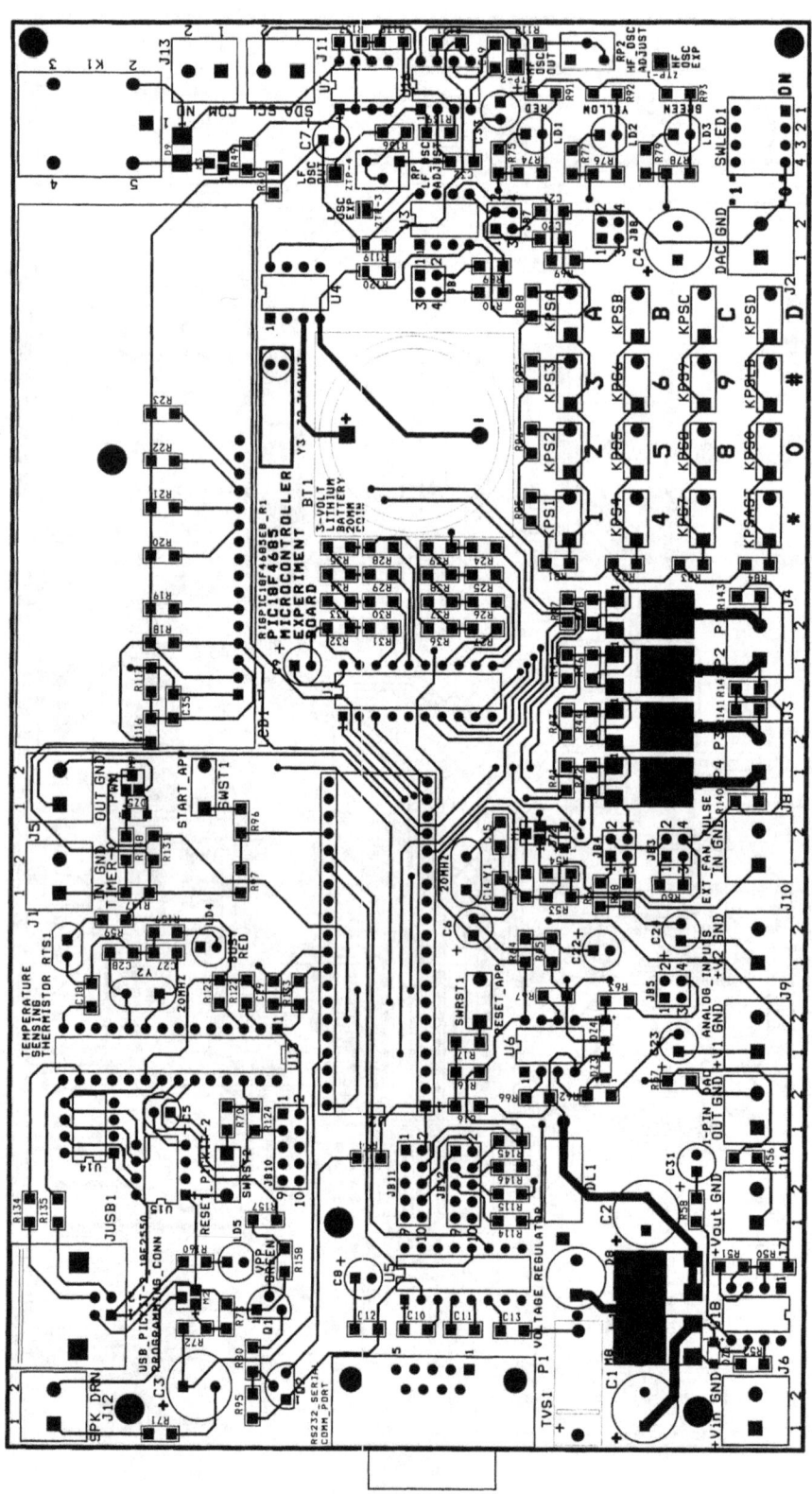

TOP-SIDE WIRING

BOTTOM-SIDE WIRING

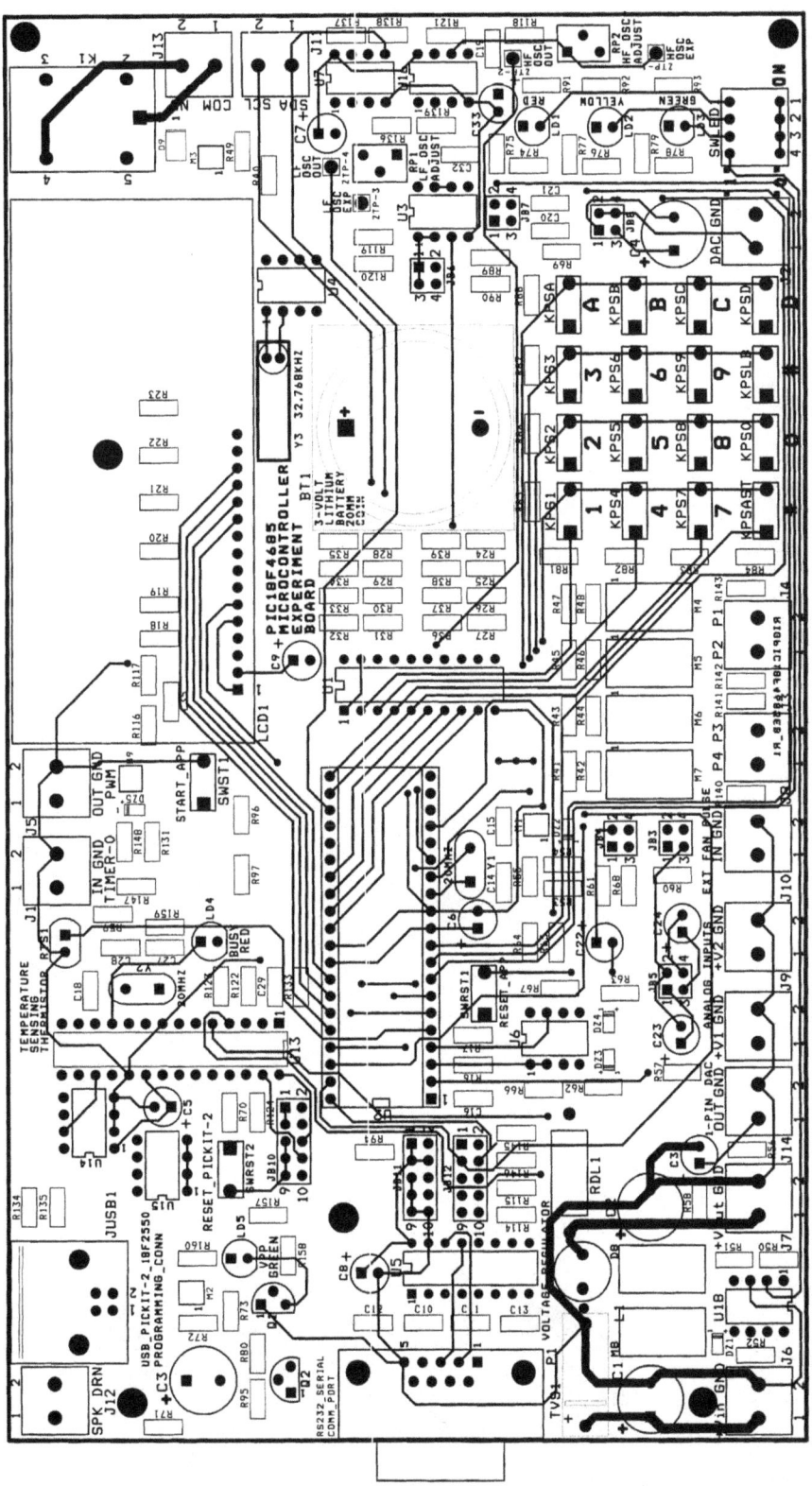

211

Appendix - B

The Lab Board Schematics,
Bill-of-Materials
and
The PC-Card Layout Drawings

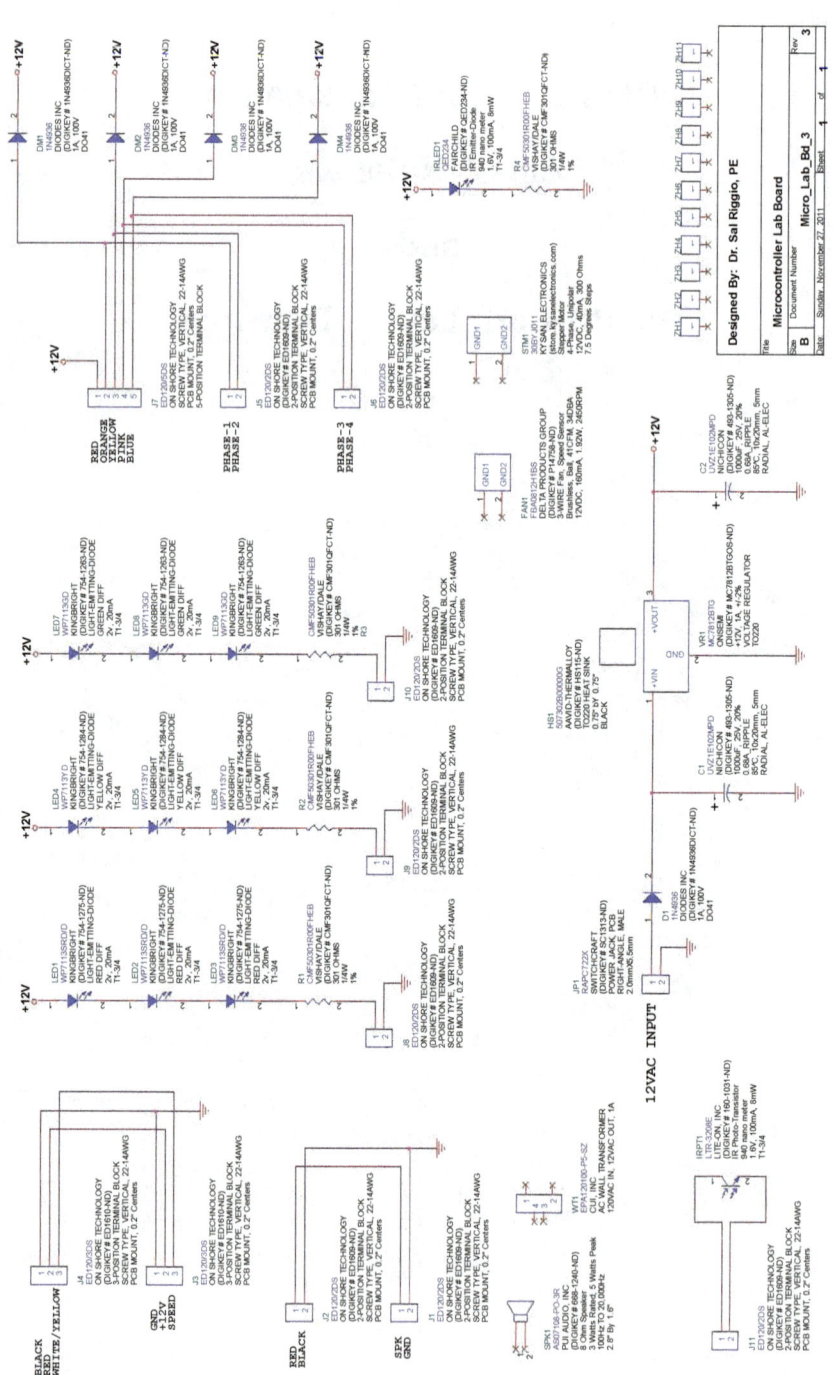

EMCONTLABBD_R3   Emebbed Controller   Lab Board   Bill-of-Materials for One PC-Card   Page 1 of 1

Item	Quan	Reference Designator	Description		Description		Description		Description
1	2	C1,C2	UVZ1E102MPD	NICHICON	(DIGIKEY# 493-1305-ND)	1000uF, 25V, 20%	0.68A_RIPPLE	85°C, 10x20mm, 5mm	RADIAL, AL-ELEC
2	5	DM1,D1,DM2,DM3,DM4	1N4936	DIODES INC	(DIGIKEY# 1N4936DICT-ND)	1A, 100V	DO41		
3	1	FAN1	FBA0812H1BS	DELTA PRODUCTS GROUP	(DIGIKEY# P14758-ND)	3-WIRE Fan, Speed Sensor	Brushless, Ball, 41CFM, 34DBA	12VDC, 160mA, 1.92W, 2450RPM	3.15"x3.15"x1"
4	1	HS1	507302B00000G	AAVID-THERMALLOY	(DIGIKEY# HS115-ND)	TO220 HEAT SINK	0.75" bY 0.75"	BLACK	
5	1	IRLED1	QED234	FAIRCHILD	(DIGIKEY# QED234-ND)	IR Emitter-Diode	940 nano meter	1.6V, 100mA, 8mW	T1-3/4
6	1	IRPT1	LTR-3208E	LITE-ON, INC	(DIGIKEY# 160-1031-ND)	IR Photo-Transistor	940 nano meter	1.6V, 100mA, 8mW	T1-3/4
7	1	JP1	RAPC722X	SWITCHCRAFT	(DIGIKEY# SC1313-ND)	POWER JACK, PCB	2.0mmX5.5mm	RIGHT-ANGLE, MALE	
8	8	J1,J2,J5,J6,J8,J9,J10,J11	ED1202DS	ON SHORE TECHNOLOGY	(DIGIKEY# ED1609-ND)	2-POSITION TERMINAL BLOCK	SCREW TYPE, VERTICAL, 22-14AWG	PCB MOUNT, 0.2" Centers	
9	2	J3,J4	ED1203DS	ON SHORE TECHNOLOGY	(DIGIKEY# ED1610-ND)	3-POSITION TERMINAL BLOCK	SCREW TYPE, VERTICAL, 22-14AWG	PCB MOUNT, 0.2" Centers	
10	1	J7	ED1205DS	ON SHORE TECHNOLOGY	(DIGIKEY# ED2228-ND)	5-POSITION TERMINAL BLOCK	SCREW TYPE, VERTICAL, 22-14AWG	PCB MOUNT, 0.2" Centers	
11	3	LED1,LED2,LED3	WP7113SRD/D	KINGBRIGHT	(DIGIKEY# 754-1275-ND)	LIGHT-EMITTING-DIODE	RED DIFF	2v, 20mA	T1-3/4
12	3	LED4,LED5,LED6	WP7113YD	KINGBRIGHT	(DIGIKEY# 754-1284-ND)	LIGHT-EMITTING-DIODE	YELLOW DIFF	2v, 20mA	T1-3/4
13	3	LED7,LED8,LED9	WP7113GD	KINGBRIGHT	(DIGIKEY# 754-1263-ND)	LIGHT-EMITTING-DIODE	GREEN DIFF	2v, 20mA	T1-3/4
14	4	R1,R2,R3,R4	CMF50301R00FHEB	VISHAY/DALE	(DIGIKEY# CMF301QFCT-ND)	301 OHMS	1/4W	1%	
15	1	SPK1	AS07108-PO-3R	PUI AUDIO, INC	(DIGIKEY# 668-1240-ND)	8 Ohm Speaker	100Hz TO 20,000Hz	3 Watts Rated, 5 Watts Peak	2.8" By 1.6"
16	1	STM1	30BYJ011	KYSAN ELECTRONICS	(store.kysanelectronics.com)	Stepper Motor	7.5 Degrees Steps	4-Phase, Unipolar	12VDC, 40mA, 300 Ohms
17	1	VR1	MC7812BTG	ONSEMI	(DIGIKEY# MC7812BTGOS-ND)	+12V, 1A, +/-2%	VOLTAGE REGULATOR	TO220	
18	1	WT1	EPA120100-P5-SZ	CUI, INC	(DIGIKEY# T1128-P5-ND)	AC WALL TRANSFORMER	120VAC IN, 12VAC OUT, 1A		

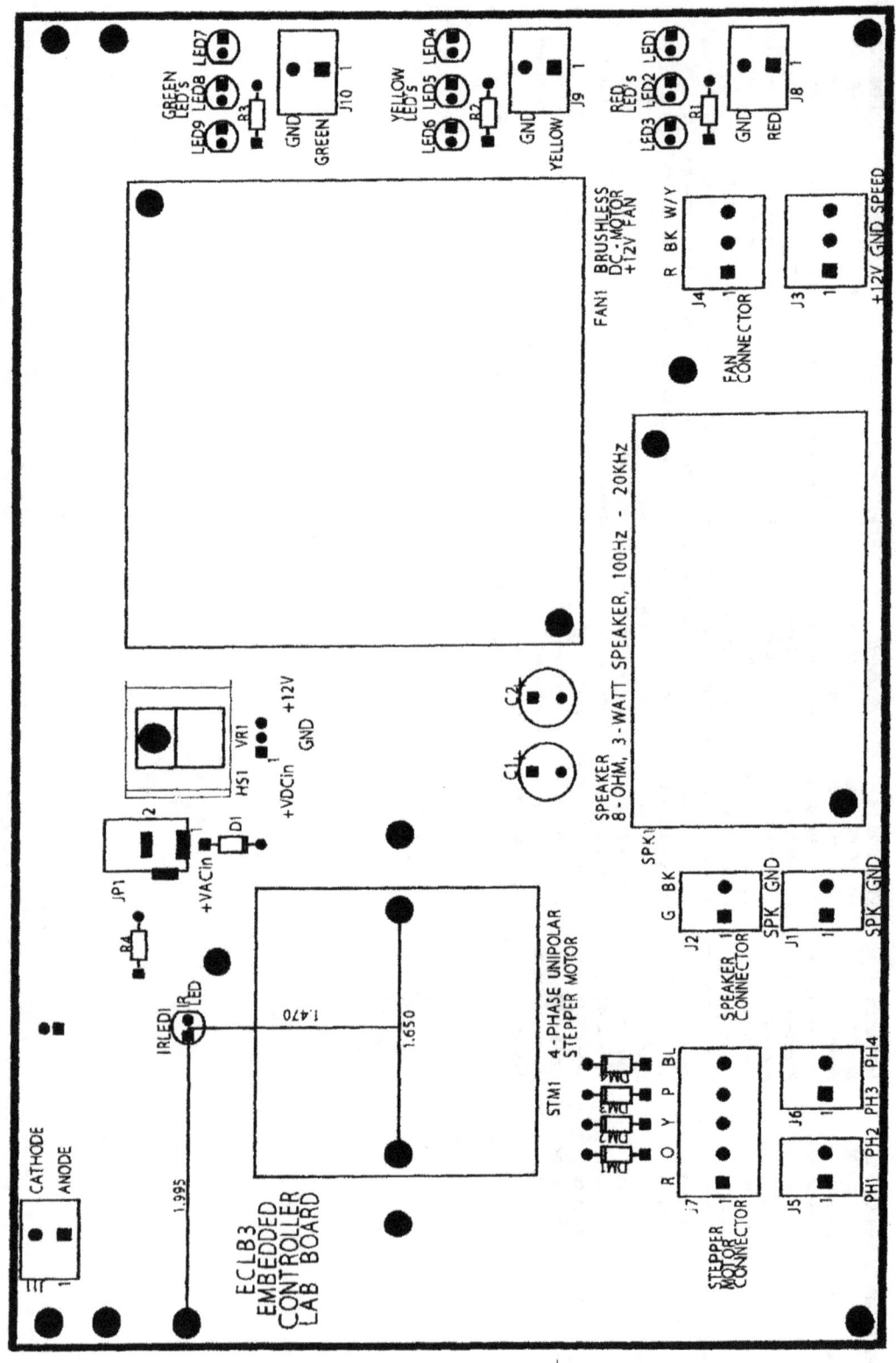

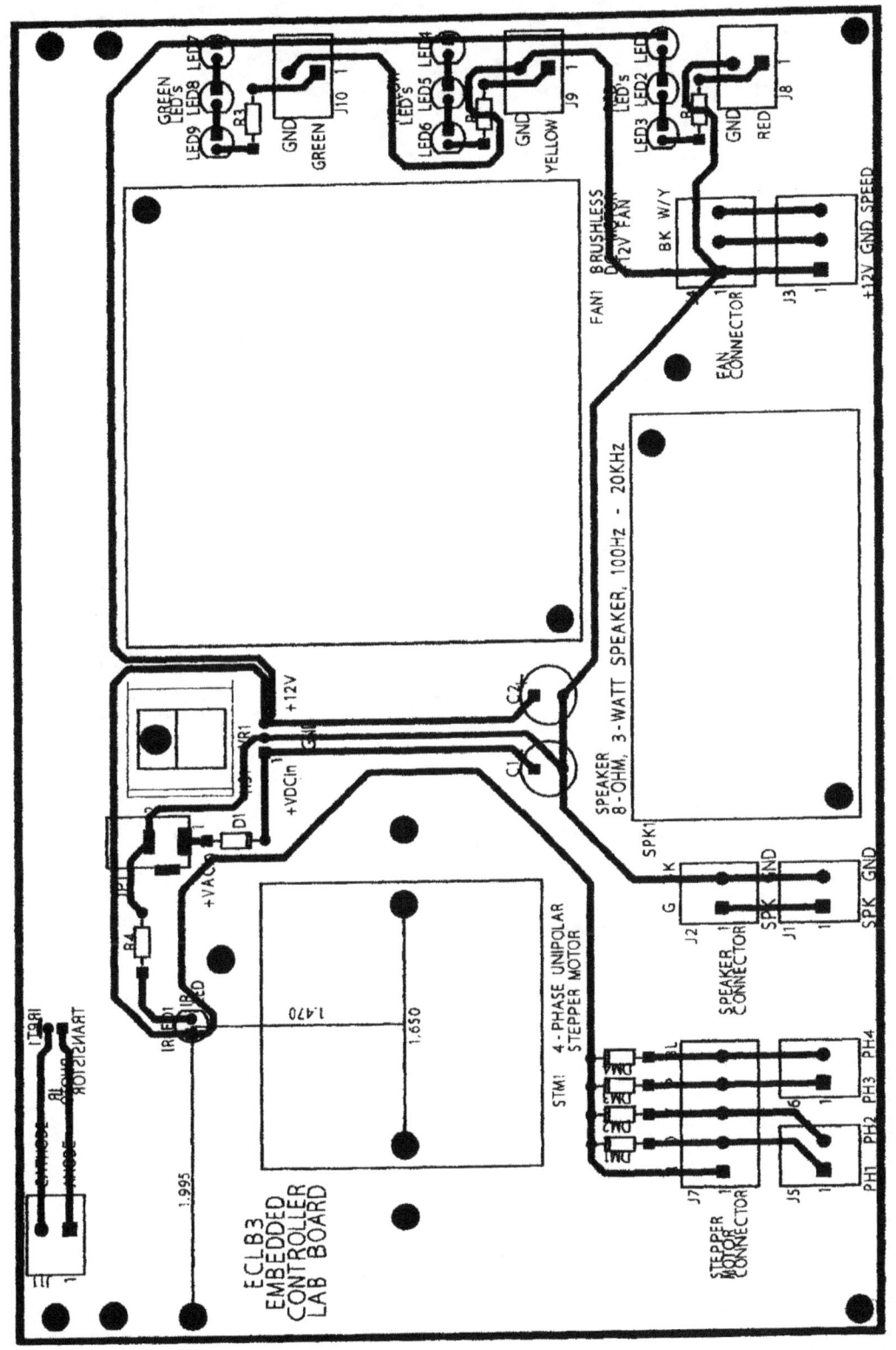

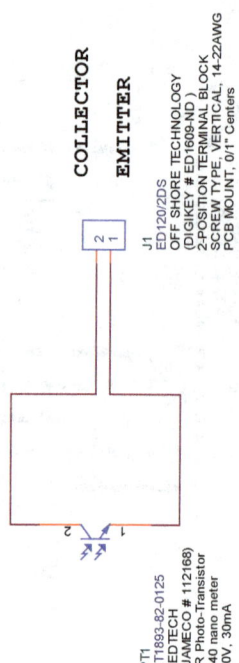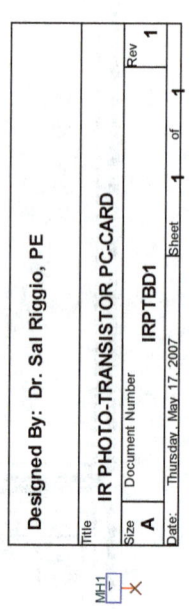

Appendix - C

The Z-Transform as it Relates to Digital Filters

The mathematics used to explain the operation of two different types of Digital Filters from the Discrete Time Difference Equation representation to the Frequency Domain Magnitude & Phase representation, is shown in this appendix.

The First Type of Digital Filter is called the Infinite-Impulse-Response (IIR) Filter.

The Second Type of Digital Filter is called the Finite-Impulse-Response (FIR) Filter.

The **Infinite-Impulse-Response** Filter is a **Causal Filter,** because it calculates its instantaneous output value y(n) from previous output values {y(n-1), y(n-2 etc.} along with current and previous input values {x(n), x(n-1), x(n-2) etc.} and its instantaneous output value y(n) does not and cannot depend upon Predicted Future output values {y(n+1), y(n+2) etc.).

The **Finite-Impulse Response** Filter is a **Causal Filter,** because it calculates its instantaneous output value y(n) from current and previous input values, only, {x(n), x(n-1), x(n-2) etc.} and its instantaneous output value y(n) does not and cannot depend upon Predicted Future output values {y(n+1), y(n+2) etc.).

A Causal Filter cannot have a Non-Zero Output Value until it is given a Non-Zero Input Value.

Discrete Time Sequences X(n) are converted into the Frequency Domain H(f) thru the Z-Transform. The Definition of the Z-Transform is as follows;

$$X(z) = Z[x(n)] = \sum_{n=-\infty}^{\infty} x(n) z^{-n} \qquad (C.1)$$

The following is the derivation of a general rule from the Z-Transform, which allows one to mathematically define the operational performance and response of most Digital Filters.

$$\begin{aligned}
Z[x(n-m)] &= \sum_{n=-\infty}^{\infty} z^{-n} x(n-m) & &(C.2) \\
&= \sum_{p+m=-\infty}^{p+m=+\infty} z^{-p-m} x(p) & &p = n-m,\ n = p+m \\
&= \sum_{p=-\infty}^{p=+\infty} z^{-p-m} x(p) & &m\ \text{is a constant} \\
&= z^{-m} \sum_{p=-\infty}^{p=+\infty} z^{-p} x(p) & & \\
&= z^{-m} Z[x(n)] & &(C.3)
\end{aligned}$$

Therefore, it follows that if X(z) is the Z-Transform of X(n), then {Zexp-3}{X(z)} is the Z-Transform of x(n-3).

If all of the **Signs** between the **Terms** within a **Difference Equation** are the **Same** (All Positive, or All Negative), the difference equation represents a **Low-Pass-Filter**.

If all of the **Signs** between the **Terms** within a **Difference Equation** **Alternate** from Negative to Positive or Positive to Negative, the difference equation represents a **High-Pass-Filter**.

Two examples of **Averaging Digital Filters** with **M=5 Terms** are shown below.

M is the **Number of Terms** to be Averaged with **No Filter Response** existing at M=1.

The following is an **Approximation** of the **Filter Response**;

1.) **First** Null Frequency = (Sampling Frequency)/(M)
2.) The **Subsequent** Null Frequencies are **Integer** Harmonics of the **First Null** Frequency.
3.) The **Cutoff** Frequency of the Filter = (First Null Frequency)/2

Digital Low-Pass-Filter

Difference Equation:
Y(n)={X(n)-X(n-1)-X(n-2)-X(n-3)-X(n-4)}/5   (C4)

Z-Transform:
Y(z)={X(z)-[(zexp-1)(X(z)])-[(zexp-2)(X(z)])-[(zexp-3)(X(z)])-[(zexp-4)X(z)]}/5   (C5)

Z-Domain Transfer Function:
H(z)=Y(z)/X(z)= {1-(zexp-1)-(zexp-2)-(zexp-3)-(zexp-4)}/5   (C6)

Frequency-Domain Transfer Function:
H(f)={1-Cos[2(Pi)f/fs]-Cos[4(Pi)f/fs]-Cos[6(Pi)f/fs]-Cos[8(Pi)f/fs]}/5
+ j{Sin[2(Pi)f/fs]+Sin[4(Pi)f/fs]+Sin[6(Pi)f/fs]+Sin[8(Pi)f/fs]}/5   (C7)

Magnitude of H(F):
|H(F)|={[{1-Cos[2(Pi)f/fs]-Cos[4(Pi)f/fs]-Cos[6(Pi)f/fs]-Cos[8(Pi)f/fs]}/5][exp2]
+ [{Sin[2(Pi)f/fs]+Sin[4(Pi)f/fs]+Sin[6(Pi)f/fs]+Sin[8(Pi)f/fs]}/5][exp2]}{exp1/2}   (C8)

Phase-Angle of H(F):
∡(F)=ArcTan[{Sin[2(Pi)f/fs]+Sin[4(Pi)f/fs]+Sin[6(Pi)f/fs]+Sin[8(Pi)f/fs]} /
{1-Cos[2(Pi)f/fs]-Cos[4(Pi)f/fs]-Cos[6(Pi)f/fs]-Cos[8(Pi)f/fs]}]   (C9)

Digital High-Pass-Filter

Difference Equation:
Y(n)={X(n)-X(n-1)+X(n-2)-X(n-3)+X(n-4)}/5   (C10)

Z-Transform:
Y(z)={X(z)-[(zexp-1)(X(z)])+[(zexp-2)(X(z)])-[(zexp-3)(X(z)])+[(zexp-4)X(z)]}/5   (C11)

Z-Domain Transfer Function:
H(z)=Y(z)/X(z)= {1-(zexp-1)+(zexp-2)-(zexp-3)+(zexp-4)}/5   (C12)

Frequency-Domain Transfer Function:

$H(f) = \{1 - \cos[2(\pi)f/fs] + \cos[4(\pi)f/fs] - \cos[6(\pi)f/fs] + \cos[8(\pi)f/fs]\}/5$ (C13)
$+ j\{-\sin[2(\pi)f/fs] + \sin[4(\pi)f/fs] - \sin[6(\pi)f/fs] + \sin[8(\pi)f/fs]\}/5$

Magnitude of H(F):

$|H(F)| = \{[\{1 - \cos[2(\pi)f/fs] + \cos[4(\pi)f/fs] - \cos[6(\pi)f/fs] + \cos[8(\pi)f/fs]\}/5]^2$ (C14)
$+ [\{-\sin[2(\pi)f/fs] + \sin[4(\pi)f/fs] - \sin[6(\pi)f/fs] + \sin[8(\pi)f/fs]\}/5]^2\}^{1/2}$

Phase-Angle of H(F):

$\angle H(F) = \arctan[\{-\sin[2(\pi)f/fs] + \sin[4(\pi)f/fs] - \sin[6(\pi)f/fs] + \sin[8(\pi)f/fs]\}/$ (C15)
$\{1 - \cos[2(\pi)f/fs] + \cos[4(\pi)f/fs] - \cos[6(\pi)f/fs] + \cos[8(\pi)f/fs]\}]$

The following **MATLAB Code** will **Plot** the above **Magnitude** and **Phase** Equations of either Filter.

The Nyquist Theorem states that one must Sample a **Band-limited** Input Signal at a Rate which is, at least, Slightly Greater than **Two** times the **Highest** Frequency within the **Spectrum** of the **Input Signal** to **Prevent Aliasing.** Aliasing is the generation of unwanted **Harmonics** and **Subharmonics** of the Spectrum of the Input Signal. Additionally, an Analog Low-Pass Reconstruction Filter must be placed at the output of the Digital Filter for the same reason.

```
Clear all;
fs = 20000;
f = logspace(0,4,100000);
```
For **Low**-Pass-Filter use the following equation in **Matlab**
```
Hf = (1-exp(-j*2*pi.*f/fs)-exp(-j*4*pi.*f/fs)-exp(-j*6*pi.*f/fs)-exp(-j*8*pi.*f/fs))/5
```
For **High**-Pass-Filter use the following equation in **Matlab**
```
Hf = (1-exp(-j*2*pi.*f/fs)+exp(-j*4*pi.*f/fs)-exp(-j*6*pi.*f/fs)+exp(-j*8*pi.*f/fs))/5
magHf = abs(Hf);
phaseHf = angle(Hf);
subplot(2,1,1);
semilogx(f,magHf);
grid on;
xlabel('f (Hz)');
ylable('|H(f)|');
title('gain');
subplot(2,1,2);
semilogx(f,PhaseHf*180/Pi);
grid on;
xlable('f (Hz)');
ylable('\angleH(f)(degrees)');
title('Phase in degrees');
```

DIGITAL LOW-PASS-FILTER RESPONSE

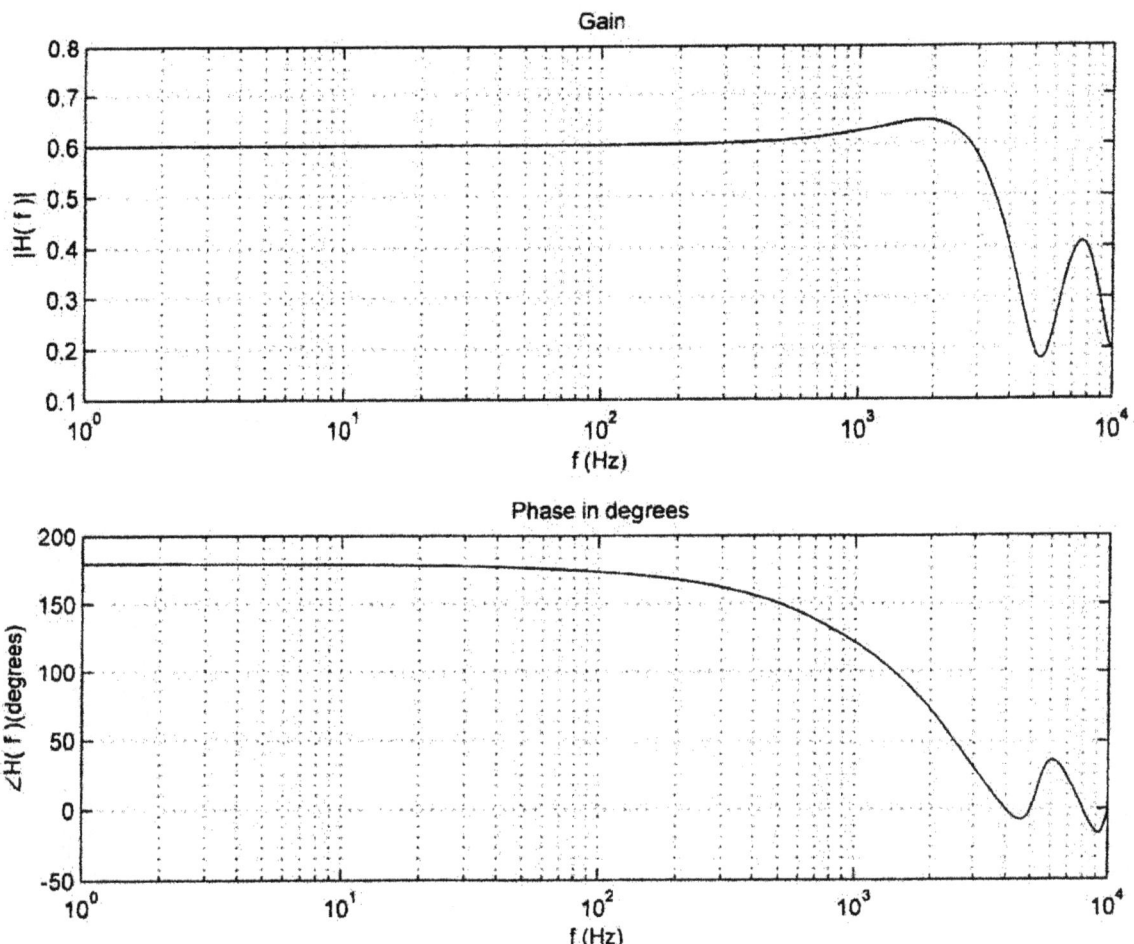

DIGITAL HIGH-PASS-FILTER RESPONSE

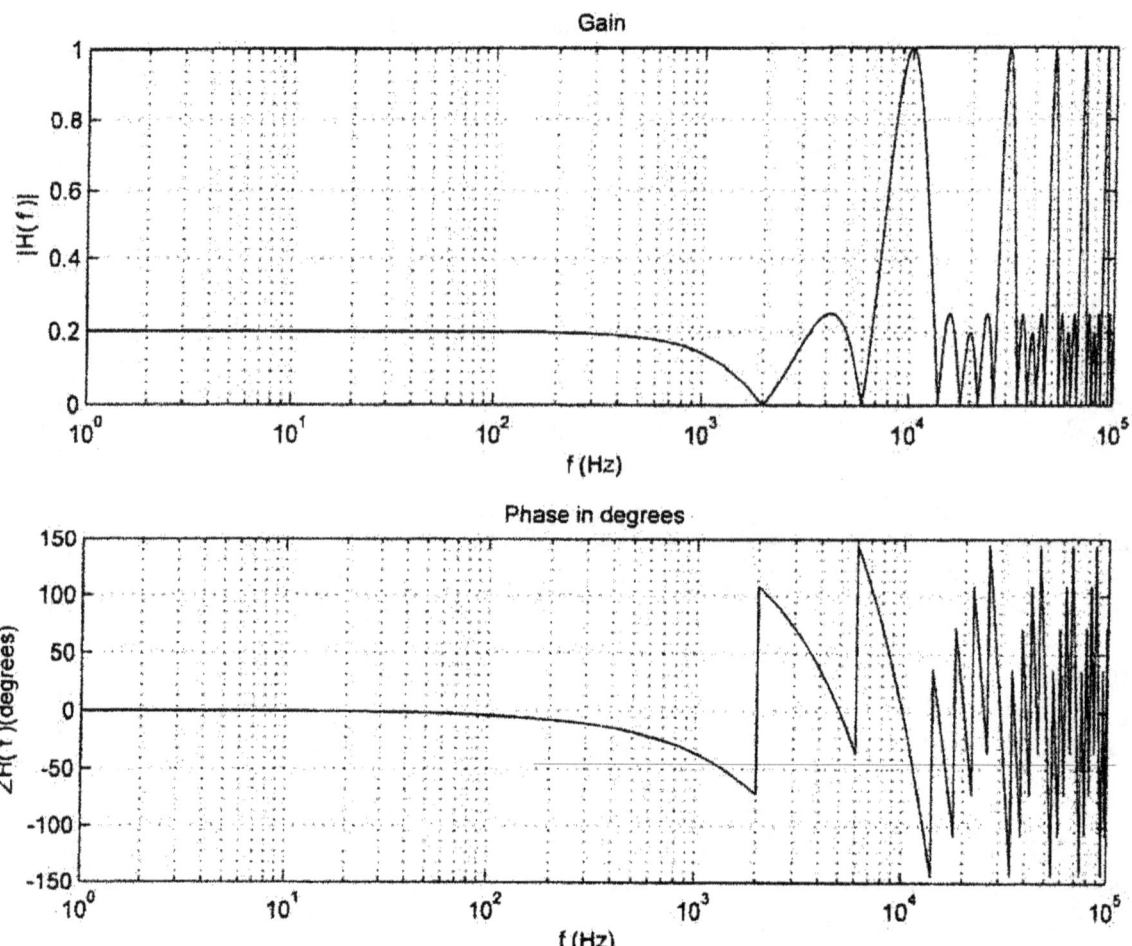

Bibliography

"Embedded Design with the PIC18F452 Microcontroller", John B. Peatman,
Prentice-Hall, 2003, ISBN: 0-13-046213-6

"Programming & Customizing PICmicro Microcontrollers", Myke Predko,
McGraw-Hill, 2000, ISBN: 0-07-136172-3

"PIC Microcontrollers and Embedded Systems", M. Mazidi, R. McKinlay, D. Causey,
Prentice-Hall, 2008, ISBN: 10-0-13-119404-6

"Applying PIC18 Microcontrollers", Barry B. Brey,
Prentice-Hall, 2008, ISBN: 10-0-13-088546-0

"PIC Microcontroller: An Introduction to Software and Hardware Interfacing",
Han-Way Huang, Delmar Cengage Learning, 2007, ISBN: 10-1-4018-3967-3

"Designing Embedded Systems with PIC Microcontrollers, Principles and Applications",
Tim Wilmshurst, Newnes, ISBN: 978-1-85617-750-4

"Embedded Microcontroller Systems Real Time Interfacing", Jonathan W. Valvano,
Cengage Learning, 2007, ISBN: 10-1-111-42625-2

"The PIC Microcontroller, Your Personal Introductory Course", John Morton,
Newnes, 2005, ISBN: 978-0-7506-6664-0

"PIC Projects, A Practical Approach", Hassan Parchizadeh, Branislav Vuksanovic, Wiley,
2009, ISBN: 978-0-470-69461-9

18238

CPSIA information can be obtained
at www.ICGtesting.com
Printed in the USA
LVHW02s0016310818
588678LV00005B/21/P